ATOMIC AND MOLECULAR ORBITAL THEORY

ATOMIC
AND
MOLECULAR
ORBITAL
THEORY

PETER O'D. OFFENHARTZ
Assistant Professor of Chemistry
Amherst College

McGRAW-HILL
BOOK COMPANY
New York
St. Louis
San Francisco
Düsseldorf
London
Mexico
Panama
Sydney
Toronto

ATOMIC AND MOLECULAR ORBITAL THEORY

Library of Congress Catalog Card Number 74-89790

47608

1 2 3 4 5 6 7 8 9 0 M A M M 7 9 8 7 6 5 4 3 2 1 0

This book was set in Modern by The Maple Press Company, and printed on
permanent paper and bound by The Maple Press Company. The designer was
Jack Ellis; the drawings were done by B. Handelman Associates, Inc. The editors were
James L. Smith and Maureen McMahon. Matt Martino supervised the production.

TO MY PARENTS

PREFACE

This book is written in an attempt to provide first-year graduate students in physical chemistry with a reasonably rigorous, unified, and self-contained introduction to molecular orbital theory. A prior undergraduate course in quantum chemistry has been assumed, but the text is entirely suitable for use with advanced undergraduates since no prerequisites in quantum mechanics are *formally* necessary for understanding the material. Indeed, I have used the text in the undergraduate physical chemistry course at Amherst College. The book is intended for use as a text in a one-semester course; enough extra material has been provided to allow a partial selection of topics, as well as further independent study by the student.

The selection of topics for the first graduate course in quantum chemistry is a difficult task. On one hand, one is tempted to teach the useful concepts of valence without going deeply into the theory, and on the other hand, to be rigorous at the expense of chemical concepts. In this text I have tried to develop chemically useful ideas by discussing at length some of the latest advances in semiempirical molecular orbital theory, while providing the necessary theoretical background in some detail. In order to carry out this approach in a one-semester text, I have omitted many topics conventionally treated at this level, for instance, time-dependent theory, the interaction of radiation and matter, and elementary spectroscopy. To preserve an orderly development, valence concepts are slighted, and the valence-bond theory is mentioned only in passing. This is in keeping with modern research trends and is hopefully a useful didactic approach. It is my personal prejudice that valence concepts are better appreciated by a student with a good background in theory and are easily learned by the student working on his own. I also feel that the other topics omitted are better treated in separate courses in spectroscopy and advanced quantum chemistry.

The order of presentation of material is perhaps somewhat unusual. In the first place, a chapter on the history of quantum mechanics is included, primarily for reasons of enjoyment—the thirty-odd years after 1900 were undoubtedly the most turbulent and exciting in the history of physical science. The chapter leads rather naturally to the Dirac formulation, and here I have tried to present some of the mathematical and philosophical concepts behind the postulatory approach to the Schrödinger equation. The treatment is not always rigorous, the intention being to develop physical concepts rather than to justify all aspects of the Dirac approach.

Angular momentum is discussed next, partially because of the usefulness of angular-momentum theory in the discussion of the rigid rotor and the hydrogen atom and partially because the intrinsic elegance of the theory provides a fine illustration of the Dirac approach. Thus, the elementary applications of the Schrödinger equation can be discussed, presumably for the second time in the student's career, from a more sophisticated point of view. Sections on the harmonic oscillator and rigid rotor are included primarily for traditional reasons; there are few enough problems in quantum chemistry which can be solved exactly.

The theory of molecular orbitals is prefaced with a rather extensive introduction to atoms. Not only are the formal theories of atoms and molecules nearly identical, but a great many types of problems can be solved exactly for atoms, although the corresponding molecular problems are at best difficult. Practice with these atomic problems can help build a student's working knowledge of the subject, as well as his formal knowledge of the theory. Furthermore, a clear understanding of atoms is necessary to grasp many of the concepts used in semiempirical molecular orbital theories; the idea of a valence-state ionization potential is typical. Finally, the theory of atomic structure leads naturally, through a discussion of group theory, to ligand field theory, again an area of great interest to chemists.

Molecular orbital theory per se is covered at three different levels. The theory is introduced, as is conventional, with the first-row diatomic molecules. By including a discussion of the Hartree-Fock treatment of diatomics, the theory is extended to include many modern developments. Then, in Chapter 10, the theory is developed from a less specific point of view, and many useful and general results are derived. These are then applied to several semiempirical theories in Chapter 11, and at this level it is feasible to discuss the approximate treatments critically, with a view both at the requirements of the theory and at the need for chemically useful concepts.

I have not found it possible to preserve a completely even level of difficulty throughout the text. Later chapters are necessarily more difficult than the first few, if only because they depend upon the first few.

Nevertheless, I have tried to keep the chapters somewhat independent of one another, so at least it is possible to proceed to later chapters without having mastered the *details* of a given section. There are two exceptions of note: Chapters 3 and 8, on angular-momentum theory and group theory, respectively, are quite essential to the material which follows. The student must have mastered shift-operator techniques to understand the bulk of Chapter 6, on atomic electronic structure, and group theory is equally crucial to any critical discussion of molecular electronic structure. This, incidentally, makes Chapter 9 the most difficult of all since ligand field theory depends intimately on both the theory of angular momentum and the theory of groups. The reader should probably omit this chapter in a first reading. Nevertheless, even in Chapter 9 I have tried to separate *techniques* from *applications*, and one should be able to understand the uses of ligand field concepts without necessarily being able to follow the details of how they are derived.

Since most sections of the text are at least partially independent, advanced undergraduates (and others learning quantum mechanics for the first time) should not have too much difficulty following the material. I hope, however, that the reader will typically find rather more information than he can digest, but that a student who finds his curiosity stimulated will find enough detail in the text to get a good start in the literature. I do think it is genuinely possible to obtain a true working knowledge of atomic and molecular orbital theory from this text, such that the reader can obtain more than a theoretical knowledge of quantum mechanics and will be able to carry out actual molecular calculations.

This text can be followed by several different approaches. Chapter 9 should provide a useful take-off point for a text such as J. S. Griffith's "Theory of Transition Metal Ions"; Chapters 10 and 11 could be followed by Parr's "Quantum Theory of Molecular Electronic Structure," the series of texts by J. C. Slater, or the recent book by F. L. Pilar, "Elementary Quantum Chemistry." Students not specifically interested in theory may still wish to go on to a study of valence, using the excellent text by J. N. Murrell, S. F. A. Kettle, and J. M. Tedder. It is my hope that after a semester's study of the major portion of the present text, the reader will be in a position to proceed on his own to more advanced treatments in areas of his special interest.

It is a great pleasure to have the opportunity to acknowledge my many debts to teachers, family, and friends who have contributed, in whatever direct or indirect way, to the genesis of this book. First, I thank Gil Haight and Philip George, two truly excellent teachers who got me started as a chemist. Secondly, I acknowledge the influence of four fine theoreticians with whom I had the pleasure to be associated in my post-doctoral work: Christopher Longuet-Higgins, Leslie Orgel, Satoru Sugano, and Martin Gouterman. Next, I owe thanks to my colleagues at the

University of Colorado, where the major portion of the text was written. In particular, I want to thank Joop de Heer, from whose lectures I learned much, and Denis Williams, who was kind enough to read and criticize the entire text. I am also indebted to Peter Yankwich of the University of Illinois, who prepared a very detailed and excellent critique of the first four chapters, and to Miss Rita Mohr, who with astonishing cheeriness typed nearly the entire manuscript. I also thank the editors at McGraw-Hill, including Peter Karsten, who diligently improved so many of my more tortured sentences. Finally, I thank my wife, not only for the usual wifely assistance without which few books would ever be written, but for many useful discussions on the text itself. I count myself among the rare few men who can claim a loving wife, an excellent cook, a sensitive mother, a sharp-eyed proofreader, and a physical chemist—all in one woman!

Peter O'D. Offenhartz

CONTENTS

1

INTRODUCTION:
THE COPENHAGEN INTERPRETATION
OF QUANTUM MECHANICS

There was once a young man who
was sent by his own village to
another town to hear a great
rabbi. When he returned he
reported: "The rabbi spoke three
times; the first talk was brilliant,
clear and simple. I understood
every word. The second was even
better, deep and subtle. I didn't
understand much, but the rabbi
understood all of it. The third talk
was a great and unforgettable
experience. I understood nothing
and the rabbi didn't understand
much either."
A story which Niels Bohr liked

to tell

1.1 THE PHYSICS OF 1900

The honor of founding quantum mechanics belongs to Max Planck and
Albert Einstein, but in its present form the theory is almost entirely a
creation of Niels Bohr and the group which he founded in Copenhagen.
Of course, his original ideas—in particular the famous "Bohr atom"—
have long since been discredited, primarily by Bohr himself, but in his
constant search for more refined and logically satisfying laws he has left
us with a magnificent superstructure which we now call the *postulates of
quantum mechanics*. In this book we shall examine the consequences
of these postulates, but before we do so it is perhaps valuable to examine
their origins, both in a historical context as a response to experimental
observation and in an intellectual context as a refinement of seemingly
contradictory concepts.

Today students have a difficult time understanding why quantum mechanics is such a new branch of science. After all, the subject is absolutely necessary to an understanding of such diverse phenomena as chemical bonding, spectra, the energy of the sun and stars, and the physical properties of objects. It seems impossible that the theoretical problems inherent in the existence of these phenomena could have gone so long unsolved without someone suspecting a flaw in classical Newtonian mechanics; in this light the complacency of nineteenth-century physicists (their positive belief that the science of physics was "closed") seems all the more incredible.

To think this is to misjudge history. A great deal was indeed known at the end of the nineteenth century, but a great deal had not yet even been guessed at. For example, it was suspected that matter consisted of atoms; certainly Dalton's law of combining weights and Avogadro's law of combining gas volumes form nearly overwhelming evidence for atoms and molecules. Yet, in terms of composition or size, no one had any idea what an atom was. The value of Avogadro's number was not known with any certainty, and the types of objects it counted were equally vague. Quantum mechanics was not necessary to explain atoms since the nature of the atom was undefined.

Of course, at this time a good deal of knowledge had been established with great precision. The laws of Newton, including the law of gravitation, were known, leading on the one hand to an extraordinary growth of technology and on the other hand to a wealth of precise astronomical data and a well-developed science of mechanics. In addition, a magnificent theory of electricity and magnetism had recently been systematized and developed by James Clerk Maxwell. The equations of Maxwell had successfully *predicted* the velocity of light and had begun to lead to the development of electromagnetic devices, such as the wireless telegraph and later the radio. The natures of both the gravitational force and the electromagnetic force were thought to be completely understood, and it was natural for scientists to believe that since the basic forces of nature had been elucidated, there was nothing *fundamental* left to be discovered.

Max Planck, born in 1858, was raised in this atmosphere. By the time he had finished his studies at the University of Munich, Maxwell's theories had been fully developed, and Planck was specifically advised against entering research in physics since there seemed so little left to learn. Despite this advice, he decided to devote his life to the study of thermodynamics, not in the hope of discovering something fundamentally new, but in a desire to develop new relationships among the thermodynamic variables. In a most basic way, Planck was a conservative in the German style of the time of Prussian dominance. By all accounts, he regulated his life with a confident logical precision which is rarely observed today, if indeed this sort of ideal of life survived either of the World Wars. Most emphatically, Planck was far from being a revolutionary, and it is a special irony that the constant h, the symbol of the most revolutionary idea in physics, is named for him.

Shortly before 1900, Planck began work on the problem of black-body radiation. As was well known, when a body is heated, it emits radiation, and, to a certain extent, the distribution of the intensity of the radiation (as a function of wavelength) is independent of the nature of the body. (Sharp lines, peaks in the intensity distribution, may also be observed in certain materials, but that is *another* problem!) The intensity distribution of black-body radiation depends only on the temperature to which the body is heated. Heating an object, of course, adds energy to it. If one assumes that matter is composed of atoms and molecules, then heat acts to speed up the motions of the particles. The distribution of velocities can be obtained from Boltzmann's law, which was derived on the basis of Newtonian mechanics in 1886. It states that if n_i is the number of particles with an energy ϵ_i, then the number of particles n_j with an energy ϵ_j is

$$n_j = n_i e^{-(\epsilon_j - \epsilon_i)/kT} \tag{1.1}$$

On the basis of this law one can derive the *distribution* of energies for any given total energy and temperature.

As was well known at the time, radiation is caused by the motion of charged particles. If one naturally assumes that black-body radiation is caused by such motion, one can calculate, using Maxwell's equations, the distribution of intensity of emitted light on the basis of the distribution of energies (i.e., velocities) of the charged particles. In an alternative, more elegant derivation, one can consider the black body as an empty cavity in equilibrium with its (heated) surroundings and containing a certain amount of radiation in the form of standing waves. It is possible to show that the number of standing waves per unit volume which can be placed in the cavity is

$$dn = \left(\frac{8\pi}{c^3}\right) \nu^2 \, d\nu \tag{1.2}$$

where $\nu = c/\lambda$ is the frequency of the wave. On the basis of the Boltzmann equipartition principle, each standing wave should have energy kT ($\frac{1}{2}kT$ for both left and right circularly polarized radiation), so the energy density in the cavity is

$$\rho \, d\nu = \epsilon \, dn = \frac{8\pi kT}{c^3} \nu^2 \, d\nu \tag{1.3}$$

This is the Rayleigh-Jeans law, derived in 1900. Aside from its inability to predict the observed black-body intensity distribution (see Fig. 1.1), it implies that the total energy in the cavity is infinite since the integral of the distribution function is infinite. This is the ultraviolet catastrophe;

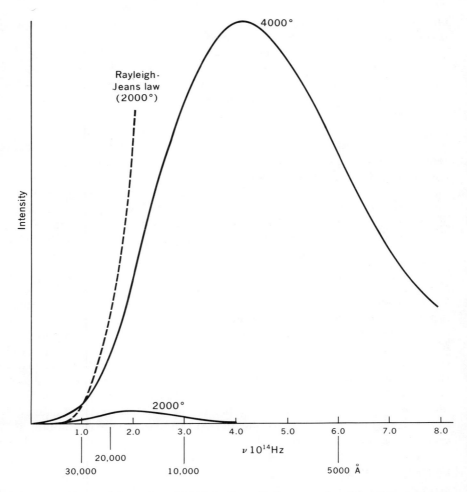

Fig. 1.1 Observed intensity of black-body radiation as a function of temperature and frequency. The intensity distribution predicted by the Rayleigh-Jeans law (the ultraviolet catastrophe) is also shown at 2000°K.

that is, the energy density becomes monotonically large as the frequency increases.

In early 1900, Planck announced his discovery of a formula which completely fitted the experimental observations. He was not able to offer any theoretical justification for the expression at the time, and it was not until late in the year that he was able to claim a derivation. His formula

$$\rho \, d\nu = \frac{8\pi h\nu^3}{c^3} \frac{1}{e^{h\nu/kT} - 1} \, d\nu \qquad (1.4)$$

avoided the ultraviolet catastrophe since at very high frequencies the energy density goes to zero. The integral of the energy from zero to infinity gives

$$\rho_t = \int_0^\infty \rho\, d\nu = \frac{8\pi^5 k^4}{15c^3 h^3}\, T^4 \tag{1.5}$$

which agrees with Stefan's law that the total black-body energy density is proportional to the fourth power of the temperature. At low frequencies, or, equivalently, in regions where h can be considered small, Planck's formula goes over into the Rayleigh-Jeans equation.

Planck claimed a derivation of the formula using Boltzmann statistics and the additional, rather curious assumption $E = h\nu$. He attached no particular significance to the assumption; it was regarded instead as a puzzling device which enabled one to get the right answer. Actually, as demonstrated by Einstein, the Planck black-body formula should not be derived using Boltzmann statistics; Bose-Einstein statistics are required. Furthermore, the assumption $E = h\nu$ is no mere device, but a fundamental law of nature now known as the Planck relation. Einstein showed that the Planck derivation implied that energy is discrete and quantized (Planck had assumed that all the frequencies in the black-body cavity were integral overtones of a fundamental basic frequency), but Planck personally disagreed and thought the idea of discrete energy laughable. No one, perhaps not even Einstein, suspected at first that a basic flaw‡ in Newtonian mechanics had been discovered.

Einstein was able to take the Planck relation a step further and use it to explain an apparently unrelated phenomenon, the photoelectric effect. When a beam of monochromatic light is used to illuminate a metal plate under a vacuum, it is observed that electrons are emitted by the plate. The energy of the emitted electrons can be measured by an arrangement similar to the vacuum tube used in electronic devices; a potential can be applied to the plate to either hinder or facilitate the release of electrons.§ Three simple yet curious rules governing the results of this experiment were discovered by Philipp Lenard and R. A. Millikan. First of all, at a given potential no electrons are emitted until the frequency of light exceeds a certain critical value. In addition, the energy of the electron current is proportional to the difference of the frequency of the light and the above-mentioned critical frequency. Finally, for

‡ Einstein published his first paper on special relativity in 1905 in *Annalen der Physik*. In the same volume he also published a paper on Brownian motion and a third paper, on the photoelectric effect, which employed the Planck relation.

§ Electrons had been discovered by J. J. Thomson, who in 1898 measured their charge-to-mass ratio; it was not until Millikan's oil-drop experiment of 1909 that their absolute charge was determined.

light of a given frequency but differing intensity, the kinetic energy of a given electron remains constant, while the number of electrons increases in direct proportion to the intensity of the incident light. The first two rules can be summarized by the equation

$$T = h(\nu - \nu_0) + V \tag{1.6}$$

where T is the kinetic energy of an electron, V is the voltage applied to the plate, and ν_0 is the frequency of light necessary to detach electrons in the absence of an applied voltage. If, at a given frequency, the voltage is adjusted so that the electron current is just reduced to zero, a plot of this voltage against frequency will be linear, as in Fig. 1.2.

Einstein was able to explain the photoelectric effect entirely on the basis of the Planck relation. If *one* packet or quantum of light energy (a photon) is required to detach *one* electron through a collision, then clearly light with low frequency will not cause any photoelectric current.

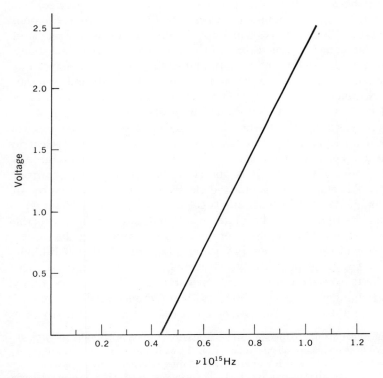

Fig. 1.2 The photoelectric effect for sodium metal. The voltage required to reduce the photoelectric current to zero is shown as a function of frequency.

Furthermore, for light above the critical frequency, the current (or flux of electrons) will be proportional to the intensity, and the kinetic energy of an electron will depend only on the frequency. Finally, the plot mentioned above will have Planck's constant as slope, and it is possible to extract a value of h from Lenard's experiments. Thus, in a clear and direct way, the Planck relation leads to the photoelectric effect, and the evidence for the quantization of light appears in an especially simple way.

1.2 THE EARLY DEVELOPMENT OF QUANTUM MECHANICS: NIELS BOHR

Despite the clarity with which Einstein demonstrated the quantization of radiation, physicists were not quick to investigate the meaning of what Einstein chose to call the Planck theory of radiation. Perhaps the photoelectric effect was too unfamiliar to fire the imagination of physicists of the time; perhaps the wave behavior of light was too firmly established for them to take Einstein seriously. In any event, it was not until the work of Bohr on atomic structure that theoreticians began to turn their attention to the quantum. We may therefore follow the early development of quantum mechanics by studying the way in which Bohr developed his famous model, the so-called Bohr atom.

Niels Bohr personified the idea that people are not what they seem at first glance. By all accounts and from his photographs, he had an unmistakably unintelligent appearance, with a long, dull, drooping face. The way he often spoke did nothing to dispel this impression; if anything, he seemed to speak deliberately in a complicated and obfuscating way. Yet Bohr probably possessed one of the most open, questioning minds that this or any other century has seen.

Even as a student, Bohr had already recognized the implications of the Planck relationship and the ideas of Einstein. This was no mean feat; Planck himself thought the idea of discontinuous energy ridiculous. Bohr reasoned that there were areas to which classical physics did not apply, but he was especially impressed that in certain limits, all the new laws went over into the classical equations. Planck's black-body radiation-density function was identical to the Rayleigh-Jeans function in the limit of low frequency, and the equations of relativity theory gave the same results as those of Newton in the limit of small velocities. From these observations Bohr developed the following principle of correspondence, which he used to test and develop new theories: All new laws had to go over into classical physics in a suitable limit.

Bohr had the enormous good fortune to arrive in England on a research fellowship at just about the time Rutherford and Marsden in Manchester were discovering the nature of the atom. Rutherford had assigned his student Marsden to a rather dull and painstaking experiment, that is, to bombard a thin leaf of gold with the newly discovered α particles (these are nuclei of He, but their specific nature was then unknown) and measure the scattering angles. Radioactivity had been discovered as recently as 1896 by Bequerel and 1898 by the Curies, and Rutherford's group was hard at work investigating every facet of the α and β rays. However routine Marsden's experiment, the result was "as surprising as if bullets were deflected by paper"—a few of the α particles were scattered through astonishingly large angles.

Prior to this experiment, J. J. Thomson at the Cavendish Laboratory in Cambridge had proposed that matter was composed of a fluid of positive charge in which electrons were embedded. But such a fluid could not possibly deflect the positively charged α particles; Rutherford was quick to argue that this experiment proved atoms consisted of heavy, dense, positive nuclei surrounded by far lighter electrons. According to legend, Rutherford was heard to exclaim, "Now I know what the atom looks like."‡

Rutherford's picture of the atom both pleased and puzzled Bohr. If the atom were indeed planetary, then it violated classical physics; an orbiting charge ought to act as an oscillating current, continuously emitting electromagnetic radiation until the electrons collapsed into the nucleus. On the other hand, if there were other forces holding the electrons in place, such forces had never been observed in nature before. In any case, to complicate the matter, Einstein had argued that radiation was *not* continuous.

Bohr's method of attack was through his principle of correspondence. Starting with the simplest system, the hydrogen atom, he argued first that as the electron moved far away from the nucleus, classical physics ought to apply. Secondly, he assumed that there were no special forces involved and that the classical law of attraction of unlike charges applied. Thirdly, for simplicity he decided to allow the electrons only (planar) circular orbits. Finally, he restrained the possible values of the angular momenta of the electron, integrated around a circular path, to integral multiples of Planck's constant. Writing the velocity of an electron as§

$$\mathbf{v} = \dot{r}\hat{\mathbf{r}} + r\dot{\theta}\hat{\boldsymbol{\theta}} \tag{1.7}$$

where $\hat{\mathbf{r}}$ and $\hat{\boldsymbol{\theta}}$ are unit vectors and $\dot{r} = dr/dt$, we have

$$T = \tfrac{1}{2}mv^2 = \tfrac{1}{2}m\mathbf{v} \cdot \mathbf{v} = \tfrac{1}{2}m(\dot{r}^2 + r^2\dot{\theta}^2) \tag{1.8}$$

‡ Jeans was later to say of the atom, as if in reply, "Today we not only have no perfect model, but we know that it is of no use to search for one."
§ The symbols used in this text are largely conventional, but the confused reader may find the Index of Symbols at the end of this book to be of value.

The classical Lagrangian L is given by $T - V$ (Becker, p. 325),‡ so

$$L = \tfrac{1}{2}m(\dot{r}^2 + r^2\dot{\theta}^2) + \frac{e^2}{r} \tag{1.9}$$

For a circular orbit, $\dot{r} = 0$ and $\dot{\theta}$ is constant. By definition, the angular momentum of the electron is given by

$$p_\theta = \frac{\partial L}{\partial \dot{\theta}} = mr^2\dot{\theta} \tag{1.10}$$

Integrating over the path of the electron, we have

$$\oint p_\theta \, d\theta = nh = 2\pi p_\theta \tag{1.11}$$

so $p_\theta = nh/2\pi \equiv n\hbar$. (The constant $\hbar$ is also often referred to as Planck's constant.) In Lagrangian form the equations of motion are

$$\begin{aligned}
\frac{d}{dt}\left(\frac{\partial L}{\partial \dot{\theta}}\right) &= \frac{d}{dt}(mr^2\dot{\theta}) = \frac{\partial L}{\partial \theta} = 0 \\
\frac{d}{dt}\left(\frac{\partial L}{\partial \dot{r}}\right) &= 0 = \frac{\partial L}{\partial r} = -\frac{e^2}{r^2} + mr\dot{\theta}^2
\end{aligned} \tag{1.12}$$

so we can solve at once for the energies and radii of the orbits; thus

$$\begin{aligned}
\frac{e^2}{r} &= mr^2\dot{\theta}^2 = \frac{n^2\hbar^2}{mr^2} \\
r &= \frac{n^2\hbar^2}{me^2} \\
E &= T + V = -\frac{n^2\hbar^2}{2mr^2} = -\frac{me^4}{2\hbar^2}\frac{1}{n^2}
\end{aligned} \tag{1.13}$$

This formula triumphantly predicted the energy pattern observed for the hydrogen atom. Bohr assumed that radiation was absorbed or emitted by the system only when the electron changed orbits. He set the energy change involved equal to $h\nu$ and obtained the equation

$$\Delta E = h\nu = \frac{me^4}{2\hbar^2}\left(\frac{1}{n^2} - \frac{1}{m^2}\right) \tag{1.14}$$

or, since $\nu = c/\lambda$,

$$\frac{1}{\lambda} = R\left(\frac{1}{n^2} - \frac{1}{m^2}\right) \tag{1.15}$$

‡ This refers to the work by Becker as listed in the Bibliography at the end of the book. Additional references are listed at the end of each chapter as a guide to further reading. This latter group is not cross-referenced in the text.

where

$$R = \frac{2\pi^2 m e^4}{h^3 c} = 109{,}737 \text{ cm}^{-1} \tag{1.16}$$

R is known experimentally as the Rydberg constant. The predicted value is in agreement with the observed. Setting $n = 1$ and allowing m to take on all positive integral values, one obtains the experimentally known Lyman series; with $n = 2$, the Balmer series; and so forth. A sketch of the energy levels and the possible transitions among them is given in Fig. 1.3; the fact that the energies of a relatively large number of transitions can be expressed as differences between a relatively small number of energy levels is known as the Ritz combination principle, and was known to spectroscopists before Bohr's work.

Despite the wonderful agreement with experiment, Bohr was dissatisfied. Although others refined his theory to include elliptical orbits and to try to extend the theory to many-electron systems, Bohr concerned

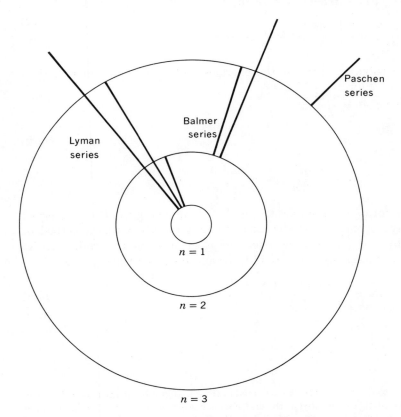

Fig. 1.3 Bohr orbits in the hydrogen atom. Absorption and emission processes are indicated schematically by straight lines.

himself with what the model did *not* predict. A noticeable defect of the theory was its failure to even hint at how intensities of absorption or emission should be calculated. But to Bohr, a more fundamental defect was the completely *ad hoc* nature of the theory. It offered no intrinsic reasonableness, no picture of physics, no fundamental laws of mechanics which encompassed the general classical equations of motion.

Bohr first advanced his model of the atom in 1912, just a year after the discovery of the nucleus. The model lasted over 10 years, which in retrospect seems surprising when we consider the number of physicists who took it up. Bohr himself used the model to develop an empirical shell structure for many-electron systems in order to explain the chemical regularities of the periodic table; in 1916 G. N. Lewis announced his octet rule, which may be considered the earliest expression of quantum chemistry. But the problem of the intensities of the hydrogen lines resisted solution, and the extension of Bohr's model to a calculation of the spectrum of helium was never accomplished.

1.3 THE COPENHAGEN INTERPRETATION AND THE HEISENBERG UNCERTAINTY PRINCIPLE

The problem began to budge in the early 1920s. Aided in large part by a grant from the Carlsberg Brewery (which, every theoretician is obliged to remark, produces some of the finest beer in the world), Bohr began to collect around him most of the world's theoretical physicists—or so it seems! The list included Wolfgang Pauli, Werner Heisenberg, P. A. M. Dirac, J. Robert Oppenheimer, J. C. Slater, N. F. Mott, Edward Teller, L. Landau, H. A. Kramers, Otto Klein, Felix Bloch, George Gamow, Paul Ehrenfest, H. Casimir, R. C. Tolman, Y. Nishina, Carl von Weizsäcker, Erwin Schrödinger, and many others. All these men at one time or another visited Bohr's institute in Copenhagen, and all were to make some contribution to Bohr's understanding of the conceptual basis of quantum mechanics. The successful attack came from two different approaches, the wave mechanics of Schrödinger and the matrix mechanics of Heisenberg.

Schrödinger's mechanics had its origin in a suggestion by Louis Victor, Duc de Broglie, that particles were "guided" in their motions by certain "pilot waves." The wavelengths of these pilots, said de Broglie, were given by

$$\lambda = \frac{h}{mv} \tag{1.17}$$

and he was able to show that Bohr's quantum conditions could be obtained from this alternative assumption. Thus, mathematically the de Broglie hypothesis was not essentially new, but conceptually it represented a different approach. Experimentalists, on the one hand, were stimulated to find evidence for the existence of the pilot waves,‡ whereas Schrödinger, on the other hand, was motivated to write a Maxwell-type wave equation for them. The derivation of the Schrödinger equation in this manner is straightforward and is sometimes given in the older texts as a derivation of quantum mechanics. But although such a treatment is perhaps defensible, it leaves many questions unanswered: What is the nature of the pilot wave? What is the physical origin of the guidance? These were the questions Bohr put to Schrödinger, and Schrödinger was unable to find satisfactory solutions.

The de Broglie hypothesis was suggested in 1924, and Schrödinger's equation first appeared in 1926. In the same year, Heisenberg published his first paper on matrix mechanics and, working with Pauli, Max Born, and Pascual Jordan, was able to make the same predictions as Schrödinger. But Heisenberg's model, if the Heisenberg treatment can be said to contain *any* model, was completely different from that of wave mechanics. No waves appear in matrix mechanics; indeed, no "explanation" seems to appear at all. Thus the pilot waves of de Broglie soon lost their fundamental significance since it was now possible to formulate a quantum theory without them.

The resolution between these two conflicting forms of quantum mechanics was worked out by Pauli, Bohr, and Heisenberg. Each had a major influence on what is now known as the Copenhagen interpretation. The major contribution of Pauli was to find logical and philosophical flaws in every *other* interpretation. He had a biting and sarcastic wit, and there was little he would put up with. He was equally rude to everyone, which made him somewhat more tolerable; it was easier to be criticized by Pauli if you had heard him in argument with Bohr or Einstein.§

Bohr was best in argument. Although slow and stubborn, he persisted in believing that quantum mechanics has to consist of more than equations and rules, and by forcing Heisenberg and Pauli to think and think again, Bohr was able to elicit from them an entirely new way of looking at nature.

Heisenberg deserves the major share of credit for the new theory. All along he had insisted that atoms could not be visualized with con-

‡ In 1927, C. Davisson and L. H. Germer verified that a beam of electrons is diffracted by a crystal much like x-ray radiation. The modern interpretation of this experiment is not that it confirms the existence of pilot waves, but rather that under certain conditions particles can behave as waves. This is one of the most striking examples of the *wave-particle duality*. The "wavelength" of a particle behaving as a wave is correctly given by the de Broglie relation.

§ A famous story relates the first meeting of Pauli and Ehrenfest. Pauli was, as usual, rude. Ehrenfest remarked, "You know, I like your publications better than I like you." Pauli was not to be outdone. "Strange, my feeling about you is just the opposite." The two soon became fast friends.

ventional models and pictures. This kind of reasoning led him to formulate the uncertainty principle, a development which was to revolutionize quantum mechanics. At last physicists had begun to ask the right questions.

Suppose, Heisenberg asked, one tried to measure, at the same time, the position and momentum (or velocity) of a particle. In case after case (via *gedanken* experiments, *not* in the laboratory!), he found that the measuring device (a beam of light) would upset the experiment and make precise measurements impossible. For example, if one wanted to determine the position and momentum of an electron, one could bombard it with a photon and observe the recoil of the photon. But if the photon used were too energetic, it would change the momentum of the electron, and if a low-energy photon were used, its wavelength would be so long that the position of the electron would be indeterminate. Heisenberg generalized this into the following rule: In an experiment, the position and momentum of a given particle cannot be simultaneously and accurately measured. The error in measuring the position, Δq, multiplied by the error in measuring the momentum, Δp, must exceed or equal‡ Planck's constant; thus

$$\Delta p \, \Delta q \geq h \tag{1.18}$$

Heisenberg's argument does not stop here, however. If one cannot measure p and q simultaneously, then is it proper to have a physics which implies that p and q are independent? The answer of modern physics is a strong negative; one of the most fundamental principles of physics is that no theory should imply the possibility of the existence of information which cannot in principle be determined experimentally. The indistinguishability principle of statistical mechanics is based on this precept, as are the basic ideas behind both special and general relativity theory and some of the more recent theories of gravitation. Einstein, when he later argued against the Heisenberg principle, tried to find cases in which it would not hold; he did not question the way it was to be applied to the theory of quantum mechanics.

What then is a particle if one cannot measure its position and momentum? Is it really a particle, or is it a wave? Is the electron a point charge, or is it some kind of smeared-out charge cloud? The answer is that the electron *is* a particle, of indeterminate position and/or momentum, but a particle nevertheless. In many experiments it may behave as a wave because the uncertainty principle prevents it from possessing a definite location; for that matter, radiation, which consists of waves, may act as a particle in certain instances. Since, in many kinds of experiments, the momentum is known exactly (the hydrogen atom is a good example; the energy levels are proportional to the momentum of the

‡ Hameka (1967, p. 57) has shown that in fact the lower uncertainty limit is $\frac{1}{2}\hbar$.

electron, and the energy levels are observed to be quantized), there must be *complete* uncertainty in the position of the electron. The most one can then say about the electron is where it is *likely* to be, and in this way the wave function and wave equation of Schrödinger can be interpreted as a statement about probabilities. Furthermore, the interpretation of wave functions as probability maps‡ goes a long way toward an explanation of the statistical behavior of nature. It is known, for example, that radioactive decay occurs spontaneously and apparently randomly. The Heisenberg uncertainty principle helps us understand this randomness by insisting on a statistical interpretation of position for systems of fixed momentum. Cause and effect, in the classical sense, are thus forced to the wayside, and indeterminacy becomes a requirement of physics.

This, in rough form, is the Copenhagen interpretation of quantum mechanics. This interpretation excludes certain kinds of questions from physics; there are some things we apparently cannot know, or to go more deeply, there are some things which do not happen because we cannot know them. This is perhaps a somewhat bizarre point of view, and as we shall see in the following chapter, it leads to a rather unusual way of regarding the relationship of experiment and observation. The indeterminacy can be handled through either the wave-mechanical point of view of Schrödinger or the matrix mechanics of Heisenberg; fortunately, both points of view have been united in the work of Dirac, and so we shall follow his approach. This latter method of treating quantum mechanics has the advantage that all the laws, including Schrödinger's equation, can be derived from a small number of basic postulates. The postulates are justified both on the basis of "intrinsic reasonableness" and by appeal to experiment; naturally, the ultimate justification lies in the agreement the derived laws give with experiment.§ There is also a practical reason for taking the Dirac point of view: much of the current literature uses his notation and formalism.

PROBLEMS

1.1 Show that the Planck radiation formula and the Rayleigh-Jeans formula agree with each other in the limit of very small frequencies.

1.2 The Wien displacement law states that the frequency at which black-body radiation intensity is at a maximum, at a given temperature, is proportional to the temperature. Prove this using the Planck formula.

1.3 An electron moves with energies 10 ev and 200 ev. Find the de Broglie wavelength in each case.

‡ For example, the well-known charge-cloud pictures of the electronic distribution in atoms.
§ What seems intrinsically reasonable to theoreticians may well appear outlandish to the beginning student.

1.4 What is the ionization potential of the hydrogen atom? How would you obtain it experimentally?

1.5 In the *atomic* system of units, $\hbar = 1$, $e = 1$, and $m = 1$. What is the ionization potential of the hydrogen atom in these units? What is the value of the first Bohr radius (a_0) in atomic units? In angstroms?

1.6 Find the value of kT at room temperature in ergs, electron volts, cm^{-1}, and atomic units. What is the repulsive energy between a pair of electrons 1 Å apart in each of these units? See also Appendix 1.

1.7 Using the Bohr method, solve for the energy of a hydrogenic atom with arbitrary nuclear charge Z.

SUGGESTIONS FOR FURTHER READING

History is a far more complicated business than science. Scientific truth is almost by definition simple and elegant; historical truth, equally by definition, is complex. One cannot summarize the events of 30 years or even 1 year in a few pages and hope to achieve any genuine accuracy. For simplicity, I have deliberately overemphasized the role of Niels Bohr in the development of quantum theory; on the other hand, I have not even mentioned his principle of complementarity in the belief that it is better for the modern student to make his own interpretation of the wave-particle duality. Those who are interested in gaining a deeper understanding of the origins and philosophical basis of quantum mechanics will do well to read the books listed below.

1. Jammer, Max: "The Conceptual Development of Quantum Mechanics," McGraw-Hill Book Company, New York, 1966.

   This book is unquestionably *the* definitive study of the scientific history of quantum mechanics. The author is skilled in both the methods of history and of science and has painstakingly read almost the entire scientific literature of quantum mechanics. The history of quantum mechanics is not as orderly as either scientists or historians would like to think, and thus Jammer's book is at times heavy reading as it wades through the often contradictory and illogical reasoning of the early days of quantum mechanics. However, Jammer has probably come closer to the truth than anyone else, and this is the supreme virtue of his book.

2. Cline, Barbara Lovett: "The Questioners," Thomas Y. Crowell Company, New York, 1965.

   This is an excellent account of the history of quantum mechanics and the physicists who created it. Written for laymen, the book can be read by a scientist in a short time and so makes excellent bedtime reading.

3. Moore, Ruth: "Niels Bohr, The Man, His Science, and the World They Changed," Alfred A. Knopf, Inc., New York, 1966.

   A personal, scientific, and political biography of the Great Dane. Not perhaps as clear on the scientific aspects as *The Questioners*, this book nevertheless is an excellent history of the development of quantum mechanics from the Bohr atom to Hiroshima and Nagasaki.

4. Hoffmann, Banesh: "The Strange Story of the Quantum," Dover Publications, Inc., New York, 1959.

An unusually lucid account of quantum theory for laymen. Although overly elementary for anyone knowing any mathemetics at all, the exposition of physical principles is superb.

5. Gamow, George: "Thirty Years that Shook Physics, The Story of Quantum Theory," Doubleday & Company, Inc., Garden City, N.Y., 1966.

An anecdotal account of the history of quantum mechanics, but not written as well as such an account ought to be.

6. Slater, John C.: "Quantum Theory of Atomic Structure," vol. I, chaps. 1 and 2, McGraw-Hill Book Company, New York, 1960.

One of the best accounts, in a standard text on quantum mechanics, of the scientific history of the subject. Chapter 2 contains an excellent bibliography of earlier references on quantum mechanics, including several histories.

7. Petersen, Aage: "Quantum Physics and the Philosophical Tradition," M.I.T. Press, Cambridge, Massachusetts, 1968.

A very deep, terse, and yet lucid account of the philosophical implications of quantum mechanics and of the unsolved epistemological and ontological problems posed by present quantum formalism. The author is a former assistant to Niels Bohr, and his account is of great interest to scientists because it suggests that many of the difficulties lie within quantum theory itself. For those who find philosophy irrelevant or misleading in questions of science, Petersen's treatment will be a productive revelation.

2

THE DIRAC FORMULATION
OF QUANTUM MECHANICS

There is a limit to the fineness of
our powers of observation and
the smallness of the accompany-
ing disturbance—a limit which is
inherent in the nature of things
and which can never be sur-
passed by improved technique or
increased skill on the part of the
observer.
*P. A. M. Dirac, "The Principles of
Quantum Mechanics"*

2.1 SUPERPOSITION AND INDETERMINACY

Consider an isolated hydrogen atom in the center of an empty box.
Let us suppose we wish to determine what electronic state the atom is
in. What experiments can we propose which will yield this information?

The first experiment we might try is to surround the atom by
photodetectors and then simply wait. If the atom is in the ground state,
nothing will happen; if the atom is in an excited state, it will eventually
emit a photon and drop to a lower state. From the wavelength of the
photon we can determine something about the state the system *was* in;
of course, after the observation, the state is changed.

Another possible experiment is to illuminate the atom with photons
of infrared light and then allow a progressive decrease in the wavelength
of the light. When the energy of the photons becomes sufficiently large,
the atom will absorp light and go to the next highest energy state; again,

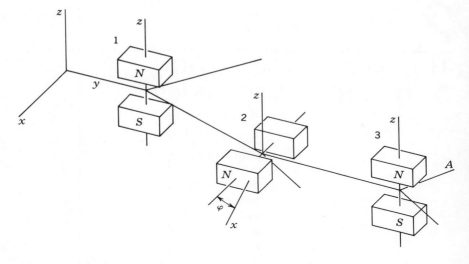

Fig. 2.1 Modified Stern-Gerlach experiment. The second magnetic field may be rotated from the x axis by an angle φ (in the xz plane). When $\varphi = 90°$, all three fields are parallel.

we can determine the initial state only by changing that state to something else.

Thus we see that one cannot ordinarily make an observation on a system in a state without disturbing the system and destroying the state.‡ In constructing a theory for such a system, we should take this rule into account. We also should investigate how measurements are made and what limit can be placed on the number of measurements which can be carried out on a given system in a given state before the state is destroyed. Our theory should also reflect the discrete nature of the states of many systems.

There are other physical systems which do not show quite the kind of behavior exemplified by the simple hydrogen atom above. For example, consider a beam of noninteracting hydrogen atoms, presumed to be in their ground electronic states. The z component of the electronic spin of these atoms can be either $+\frac{1}{2}\hbar$ or $-\frac{1}{2}\hbar$. If we apply a magnetic field perpendicular to the direction of the beam, as in Fig. 2.1, the beam will divide sharply into two segments, one containing the hydrogen atoms with positive spin component and one containing the atoms with nega-

‡ If we suspected or knew the hydrogen atom was in a certain state, we could *verify* this by illuminating the atom with all wavelengths of light except those which could be absorbed.

tive spin component. If we then observe one of these two segments in a magnetic field which is perpendicular to both the beam direction and the direction of the first field, our beam segment will again be split. If, finally, we observe one of these new components in a magnetic field placed in the same manner as the first field, the beam will once again be divided, which is certainly curious. However, if the second magnetic field is turned off, no splitting will occur in the third magnetic field; the entire beam will be deflected.

This experiment, patterned after the Stern-Gerlach experiment, in which the magnetic moment of the electron was first measured, can be interpreted as follows. The hydrogen beam, moving in the y direction, passes through the first magnetic field, oriented in the z direction. This field separates the beam into components with $s_z = \frac{1}{2}\hbar$ and $s_z = -\frac{1}{2}\hbar$. However, the second magnetic field, in creating two new states with $s_x = \frac{1}{2}\hbar$ and $-\frac{1}{2}\hbar$, *destroys* the s_z polarization (here $\varphi = 0$). It is to be noted that the original beam of hydrogen was not in any definite electronic spin state, that the beam between the first and second magnets does not have any definite value of s_x, and that the beam between the second and third magnets (with the second magnet turned on) has no definite value of s_z. Each magnet *prepares* a polarized state.

The spin state of the original beam is indeterminate. Any experiment we perform destroys this indeterminacy and throws a given atom into a definite state. Therefore a question about the spin state of a given electron cannot be answered unless a *prior* experiment has forced it into a fixed state. In particular, it is not possible to make any conclusion about the value of s_z for an electron with fixed s_x. We are forced to assume that the state of s_z in this case is a *superposition* of states of s_x; sometimes we will refer to this as an *impure state* of s_z.

The concept of superposition may be made somewhat clearer if we change the preceding experimental arrangement somewhat. Suppose the second magnetic field is rotated by an angle φ so that the field is no longer in the x direction but is between the x and z axes. Let us send a *single* hydrogen atom along the y axis, and let us further assume that the first magnetic field polarizes it so that it passes through the second field. Then the direction the atom takes when passing through the third field is still indeterminate, although as φ approaches 90°, it becomes more and more unlikely that the atom will appear at position A. Thus, although the result of the series of observations of a *single* atom remains indeterminate, in a large number of experiments, the fraction of atoms reaching A becomes smaller as φ approaches 90°. If our conception of superposition is to be of any use, it must express the probability of a given observation on a single atom; an impure state must be a relative concept, in the sense that an impure state can be made as nearly pure as is desired.

In still another kind of experiment, we might try to measure the position and momentum of a moving particle. Here we have a somewhat different kind of indeterminacy—the two measurements directly interfere. If we were to measure only the position of the particle, we would use light of very short wavelength; according to classical optics, the error in such a measurement is on the order of $\Delta q \sim \lambda$, where λ is the wavelength of the incident radiation. On the other hand, to accurately measure the momentum of the particle, long-wavelength radiation is necessary since the momentum in the photon, $p = h/\lambda$, can change the momentum of the particle. The two requirements conflict, and we find that the product of the uncertainties in p and q is of order h; an exact analysis shows that $\Delta p \, \Delta q \geq \frac{1}{2}\hbar$, which is one form of Heisenberg's uncertainty principle. This interference between measurements of p and q is fundamental, as we shall see, to an understanding of all other kinds of interference between measurements; the consequent indeterminacy in p and q is the basis of all indeterminacy in time-independent quantum mechanics.

2.2 STATE VECTORS

In order to give mathematical expression to the superposition principle and to our earlier discussion of the electronic states of the hydrogen atom, it is first necessary to develop a more precise concept of the state of a system. In classical mechanics, a state is a specification of the momenta and positions of all particles. A convenient mathematical definition for a system of n particles can be made in terms of a vector in a phase space of $6n$ dimensions, 3 for the position and 3 for the momentum of each particle. A vector in this space will specify the values of $6n$ coordinates, and thus the classical state. In quantum mechanics, we may still specify and define a state by a vector in a suitable phase space, but our space will have less than $6n$ dimensions. Thus, one of the problems in quantum mechanics is to define just how many dimensions the quantum mechanical phase space can have, but for the moment we can only say that the space should have as many dimensions "as are theoretically possible without mutual interference or contradiction" in the sense of the preceding discussion and the Heisenberg uncertainty principle. An impure state may then be regarded as a linear combination of other state vectors; we shall shortly see how this idea may be made precise.

In the Dirac system of nomenclature, the quantum-mechanical state vectors are called *kets*. A general ket will be denoted by the symbol $| \ \rangle$; a specific ket by, say, $|\Psi\rangle$. Such a ket is analogous to a *wave function* Ψ, but as yet our ket has no association with any special interpretation.

A vector has the properties of both length and direction. It can be multiplied by a scalar, added to another vector, etc.; in order to define the properties of our ket we must specify all these operations. However, there is a major difference between a classical and a quantum-mechanical phase space—in quantum mechanics the length of a ket has no physical meaning. For example, in the two-dimensional *classical* phase space of a single particle confined to the x axis, a vector specifies both the x coordinate and the x component of the momentum of the particle. The motion of such a particle can be described by a vector of changing direction and length. However, in small systems where the uncertainty principle must be applied, our phase space cannot include both the position and the momentum of the particle, and it turns out to be possible to define our kets solely in terms of their direction.

Consider the example of the beam of hydrogen atoms in the previous section. Let $|s_z = \frac{1}{2}\hbar\rangle$ and $|s_z = -\frac{1}{2}\hbar\rangle$ represent pure states, such that a measurement of the z component of the spin of an electron in a hydrogen atom in the state $|s_z = \frac{1}{2}\hbar\rangle$ is sure to give the result $\frac{1}{2}\hbar$. After the beam of atoms has passed the second magnet (with field in the x direction), the pure state of s_z is destroyed. The new state, which is a pure state of s_x, can be represented as a *superposition* of the states $|s_z = \frac{1}{2}\hbar\rangle$ and $|s_z = -\frac{1}{2}\hbar\rangle$. Writing this new state as $|\Psi\rangle$, we have

$$|\Psi\rangle = c_1|s_z = \tfrac{1}{2}\hbar\rangle + c_2|s_z = -\tfrac{1}{2}\hbar\rangle \tag{2.1}$$

where c_1 and c_2 are, in general, complex numbers, usually simply called scalars. *If this is the way we are to formulate the principle of superposition,* then the length of a vector cannot have meaning, since the superposition of $|s_z = \frac{1}{2}\hbar\rangle$ with itself must not result in a new state; for example, if

$$|\Psi\rangle = c_1|s_z = \tfrac{1}{2}\hbar\rangle + c_2|s_z = \tfrac{1}{2}\hbar\rangle = (c_1 + c_2)|s_z = \tfrac{1}{2}\hbar\rangle \tag{2.2}$$

then $|\Psi\rangle$ must correspond to the same state as $|s_z = \frac{1}{2}\hbar\rangle$. Of course, $c_1 + c_2$ must not equal zero. It is of interest to note that just as the direction of the second magnetic field in Fig. 2.1 can be rotated, the values of c_1 and c_2 can be varied continuously. In this way, in (2.1), the state $|\Psi\rangle$ can be varied from a pure state of s_x to a pure state of s_z. As we shall show in Sec. 3.5,

$$|s_x = \tfrac{1}{2}\hbar\rangle = \frac{1}{\sqrt{2}}|s_z = \tfrac{1}{2}\hbar\rangle + \frac{1}{\sqrt{2}}|s_z = -\tfrac{1}{2}\hbar\rangle$$

The kets may often be thought of as column vectors, that is, as $n \times 1$ matrices. Corresponding to any matrix $\mathbf{A}$, there is the adjoint matrix $\mathbf{A}\dagger = \tilde{\mathbf{A}}^*$. $\mathbf{A}\dagger$ and $\tilde{\mathbf{A}}$ are row vectors; $\mathbf{A}\dagger$ is the complex conjugate of $\tilde{\mathbf{A}}$. Without pushing this analogy too far, we may define a second set of vectors $\langle \ |$, called bras (the terms *bra* and *ket* come from the word *bracket*). Together, a bra and a ket define a scalar $\langle\Psi|\Psi\rangle$; this is similar to the dot product of two vectors in ordinary space or the matrix product of a $1 \times n$ row vector with an $n \times 1$ column vector to yield a (1×1) scalar. The scalar dot product is defined to be a linear relation, so

$$\langle \Psi_1 | (|\Psi_2\rangle + |\Psi_3\rangle) = \langle \Psi_1 | \Psi_2 \rangle + \langle \Psi_1 | \Psi_3 \rangle$$
$$((\langle \Psi_1 | + \langle \Psi_2 |)|\Psi_3\rangle = \langle \Psi_1 | \Psi_3 \rangle + \langle \Psi_2 | \Psi_3 \rangle \qquad (2.3)$$
$$\langle \Psi_1 | (c|\Psi_2\rangle) = ((\langle \Psi_1 | c)|\Psi_2\rangle = c\langle \Psi_1 | \Psi_2 \rangle$$

where c is any scalar. In the language of ordinary wave mechanics, a dot product such as $\langle \Psi_1 | \Psi_2 \rangle$ corresponds to the integral $\int \Psi_1^* \Psi_2 \, d\tau$ where the integration extends over the entire space spanned by Ψ_1 and Ψ_2. The nature of this correspondence will be clarified in Sec. 2.7.

It is convenient to make the relation between bras and kets somewhat more precise. If we let the bra $\langle \Psi |$ correspond to the ket $|\Psi\rangle$, then we shall also let $\bar{c}\langle \Psi | = \langle \Psi | \bar{c}$ correspond to $c|\Psi\rangle$, where $\bar{c} = c^*$ is the complex conjugate of the scalar c. It should be stressed that this particular correspondence is *not* an equality, for the same reason that a row vector and a column vector cannot be *equal*. However, $|\Psi\rangle$ and $\langle \Psi |$ define the same state, and in this sense they are equivalent. For the present we shall only define the effect of the bar operation on a scalar, in which case it is equivalent to the starring operation of taking the complex conjugate. Furthermore, *as a fundamental rule*, we define

$$\langle \Psi_1 | \Psi_2 \rangle^* = \overline{\langle \Psi_1 | \Psi_2 \rangle} = \langle \Psi_2 | \Psi_1 \rangle \qquad \blacktriangleright (2.4)$$

for all bras and kets. In wave-mechanical language,

$$(\int \Psi_1^* \Psi_2 \, d\tau)^* = \int \Psi_2^* \Psi_1 \, d\tau \qquad (2.5)$$

In the language of quantum chemistry, the overlap matrix is said to be Hermitian, and $S_{21}^* = S_{21}$.

We have not yet defined the integral of $|\Psi\rangle$ with its corresponding bra $\langle \Psi |$. First, we note from Eq. (2.4) that $\langle \Psi | \Psi \rangle$ is real. It is *convenient* throughout quantum mechanics to take this number as equal to unity, in which case $|\Psi\rangle$ is said to be normalized; that is,

$$\langle \Psi | \Psi \rangle = \overline{\langle \Psi | \Psi \rangle} = 1 \qquad (2.6)$$

This is not always possible, however, and in the most general case we can only assume that $\langle \Psi | \Psi \rangle$ is positive. By analogy with vector operations in ordinary space, $\langle \Psi | \Psi \rangle$ is said to be the square of the length of $|\Psi\rangle$. However, even with the length and direction of $|\Psi\rangle$ completely specified, there is still some ambiguity, since

$$\langle \Psi | \overline{e^{ik}} e^{ik} | \Psi \rangle = \langle \Psi | \Psi \rangle \qquad (2.7)$$

The number e^{ik} is called a phase factor; any ket $|\Psi\rangle$ may be multiplied by such a factor without changing its direction or length. (Note the discussion in the paragraph following Eq. (2.3); $\langle \Psi | \overline{e^{ik}}$ *corresponds* to $e^{ik}|\Psi\rangle$.) The reader should distinguish three levels of importance in discussing the length of a ket. On the one hand, we have shown that the length of a ket is not *physically* meaningful. On the other hand, we will find it mathematically convenient to take the length as unity whenever possible. Finally, we will also find it convenient to specify the phase of the ket, although, once again, the phase is not significant.

2.3 LINEAR HERMITIAN OPERATORS

In the preceding section we developed the concept of a ket vector as a description of a state. However, we have not shown in what manner the ket numerically specifies the state. In order to do this, we must first introduce the concept of a linear Hermitian operator.

We define a general operator $\mathbf{O}$ by the equation

$$|\Psi_1\rangle = \mathbf{O}|\Psi\rangle \tag{2.8}$$

By *linear* operator we imply

$$\mathbf{O}(|\Psi_1\rangle + |\Psi_2\rangle) = \mathbf{O}|\Psi_1\rangle + \mathbf{O}|\Psi_2\rangle \tag{2.9}$$

and we further assume

$$\mathbf{O}(c|\Psi\rangle) = c\mathbf{O}|\Psi\rangle$$
$$(\mathbf{O}_1 + \mathbf{O}_2)|\Psi\rangle = \mathbf{O}_1|\Psi\rangle + \mathbf{O}_2|\Psi\rangle \tag{2.10}$$

In classical vector spaces, there are many analogs of the effect of an operator on a ket. $\mathbf{O}$ may be viewed as a rotation operator, taking one vector into another. Alternatively, if we view the ket $|\Psi\rangle$ as a kind of $n \times 1$ column vector, then $\mathbf{O}$ can be considered as an $n \times n$ transformation matrix. In either case, the product of the two linear operators $\mathbf{O}_1$ and $\mathbf{O}_2$ is also a linear operator. Thus, we define

$$(\mathbf{O}_1\mathbf{O}_2)|\Psi\rangle = \mathbf{O}_1(\mathbf{O}_2|\Psi\rangle) = \mathbf{O}_1\mathbf{O}_2|\Psi\rangle \tag{2.11}$$

However, it is not necessary to assume that the operators commute, so if

$$|\Psi_2\rangle = \mathbf{O}_1\mathbf{O}_2|\Psi\rangle$$
$$|\Psi_1\rangle = \mathbf{O}_2\mathbf{O}_1|\Psi\rangle \tag{2.12}$$

we can take $|\Psi_1\rangle \neq |\Psi_2\rangle$ in the most general case.

Linear operators can also act on bra vectors. Just as $\langle\Psi|\bar{c}$ is the bra corresponding to $c|\Psi\rangle$, so we take $\langle\Psi|\bar{\mathbf{O}} = \langle\Psi_1|$ as the bra corresponding to $|\Psi_1\rangle = \mathbf{O}|\Psi\rangle$. The operator $\bar{\mathbf{O}}$ is called the adjoint of $\mathbf{O}$. Its action is not restricted to bra vectors; the ket $\bar{\mathbf{O}}|\Psi\rangle$ may also be defined. Finally, since we may write $|\Psi_1\rangle = \mathbf{O}|\Psi\rangle$, we can also form the scalar quantity

$$\langle\Psi_2|\Psi_1\rangle = \langle\Psi_2|\mathbf{O}|\Psi\rangle \tag{2.13}$$

which corresponds to the usual wave-mechanical expression $\int\Psi_2^*\mathbf{O}\Psi \, d\tau$.

As noted above, $\langle \Psi_2 | \Psi_1 \rangle = \overline{\langle \Psi_1 | \Psi_2 \rangle}$. Writing $| \Psi_1 \rangle = \mathbf{O} | \Psi \rangle$, we obtain

$$\langle \Psi | \bar{\mathbf{O}} | \Psi_2 \rangle^* \equiv \overline{\langle \Psi | \bar{\mathbf{O}} | \Psi_2 \rangle} = \langle \Psi_2 | \mathbf{O} | \Psi \rangle \qquad \blacktriangleright (2.14)$$

or in the language of wave mechanics,

$$\left(\int \Psi^* \bar{\mathbf{O}} \Psi_2 \, d\tau \right)^* = \int \Psi_2^* \mathbf{O} \Psi \, d\tau \qquad (2.15)$$

Equations (2.14) and (2.15) express a general requirement on all operators.

In order to proceed further with the development of the theory, we must attach some physical significance to the operators. We take as a fundamental postulate that the linear operators correspond to observable properties of a system. Thus, just as we have constructed a correspondence between states and kets (or bras), so we also set up a one-to-one correspondence between operators and observables.

If a quantity to be observed in a measurement is complex, it can be broken down, in general, into a real and an imaginary part. Thus, the observation of a complex quantity involves two measurements. As noted earlier, it is possible that an observation on a system so disturbs it that a second observation of some other property is not possible on the *original* state. We must therefore restrict ourselves to the measurement of real quantities in the most general case; if an observation of a complex quantity can be made, one may consider it as *two separate real measurements*. Thus, we are only interested in quantities which are in some sense real, and we shall assume that the necessary relation is $\bar{\mathbf{O}} = \mathbf{O}$. A proof is given as an exercise at the end of the next section. When the relation $\bar{\mathbf{O}} = \mathbf{O}$ is true for a given operator, the operator is called *self-adjoint*, or *Hermitian*. Note that for a Hermitian operator, $\langle \Psi_1 | \mathbf{O} | \Psi_2 \rangle^* = \langle \Psi_2 | \mathbf{O} | \Psi_1 \rangle$, or, in an obvious notation, $O_{12}^* = O_{21}$.

To complete the development of the theory, we need to know not only the correspondence between states and kets and between observables and operators, but also the correspondence between *results* of observations and some symbolic quantity. In the next section this development will be considered.

PROBLEMS

2.1 Prove that $\mathbf{O} = \bar{\bar{\mathbf{O}}}$ for any general linear operator, not necessarily Hermitian.

2.2 Prove that $\bar{\mathbf{O}}_1 \bar{\mathbf{O}}_2 = \overline{\mathbf{O}_2 \mathbf{O}_1}$ for any pair of linear operators.

2.4 EIGENVECTORS, EIGENVALUES, AND OBSERVABLES

The most interesting special case of a linear Hermitian operator acting on a ket is expressed by the equation

$$\mathbf{O} | \Psi \rangle = c | \Psi \rangle \qquad \blacktriangleright (2.16)$$

This type of relation is known as an eigenvalue equation. As we shall show later, an important theoretical problem in quantum mechanics is to find kets $|\Psi\rangle$ and scalars c such that this equation is satisfied for some particular operator $\mathbf{O}$. The ket $|\Psi\rangle$ is known as an eigenvector or eigenket or eigenfunction of the operator $\mathbf{O}$; c is known as the corresponding eigenvalue. It is frequently convenient to label the eigenkets of an operator according to their eigenvalues; thus,

$$\mathbf{O}|O_1\rangle = O_1|O_1\rangle \tag{2.17}$$

where O_1 is the (scalar) eigenvalue. As the notation implies, there may be more than one eigenvalue and corresponding eigenvector satisfying (2.16) for a given operator $\mathbf{O}$.

We now make an assumption which may be considered the cornerstone postulate of the theory. We assume that the eigenvalues correspond to observations, that is, to the numerical values of results of observations. Furthermore, we assume that if a system is in a state described by the ket $|O_n\rangle$ and if an observable corresponds to the operator $\mathbf{O}$, then the result of an observation of the property O in the state $|O_n\rangle$ will always give the experimental result O_n, *if and only if*

$$\mathbf{O}|O_n\rangle = O_n|O_n\rangle \tag{2.18}$$

This equation expresses in mathematical form the quantization of states in small systems.

The eigenvalue equation formalism may be illustrated by reference to our experiment on the beam of hydrogen atoms. As noted earlier, an individual atom is prepared in a fixed state of the z component of the electronic spin in the first magnetic field. There are two such states, and these may be written $|s_z = \tfrac{1}{2}\hbar\rangle$ and $|s_z = -\tfrac{1}{2}\hbar\rangle$ (in cases in which it is clear that it is the z component which is implied, we may abbreviate the two states as $|\tfrac{1}{2}\hbar\rangle$ and $|-\tfrac{1}{2}\hbar\rangle$ or even as $|\tfrac{1}{2}\rangle$ and $|-\tfrac{1}{2}\rangle$). Observation of s_z in a state corresponds to operating on the ket corresponding to the state with the appropriate operator corresponding to s_z. Without specifying this operator further, we shall call it $\mathbf{s}_z$. Then

$$\mathbf{s}_z|s_z = \tfrac{1}{2}\hbar\rangle = \tfrac{1}{2}\hbar|s_z = \tfrac{1}{2}\hbar\rangle \tag{2.19}$$

which simply and concisely expresses the fact than an observation of s_z in a state with s_z equal to $\tfrac{1}{2}\hbar$ is certain to give the result $\tfrac{1}{2}\hbar$.

One can also ask for the result of a measurement of s_z in an impure state, or in a pure state of s_x. For example, if a general impure state $|\Psi\rangle$ is written

$$|\Psi\rangle = c_1|s_z = \tfrac{1}{2}\hbar\rangle + c_2|s_z = -\tfrac{1}{2}\hbar\rangle \tag{2.20}$$

then

$$\mathbf{s}_z|\Psi\rangle = \tfrac{1}{2}\hbar c_1|s_z = \tfrac{1}{2}\hbar\rangle - \tfrac{1}{2}\hbar c_2|s_z = -\tfrac{1}{2}\hbar\rangle \neq c|\Psi\rangle \tag{2.21}$$

The state $|\Psi\rangle$ is clearly not an eigenstate of the operator $\mathbf{s}_z$. A measurement of s_z on an atom in the state $|\Psi\rangle$ is therefore indeterminate. Note that in using the word "indeterminate," we do not mean that we know nothing about the result of a measurement of s_z in the state $|\Psi\rangle$. For example, we know from experiment that the result of a measurement on a single atom will yield either the result $\frac{1}{2}\hbar$ or the result $-\frac{1}{2}\hbar$, and not something in between. We also know that if the state is a (pure) eigenstate of s_z, the result $s_z = \frac{1}{2}\hbar$ will occur as often as the result $s_z = -\frac{1}{2}\hbar$ in a series of experiments on a large number of atoms. Furthermore, if $|\Psi\rangle$ is not an eigenstate of $\mathbf{s}_z$, we can predict the relative probabilities of observing $s_z = \frac{1}{2}\hbar$ and $s_z = -\frac{1}{2}\hbar$ in such a series of measurements; the probabilities will not always be equal.

How can we determine the relative probabilities of these two results? Assume that $|\Psi\rangle$, $|s_z = \frac{1}{2}\hbar\rangle$, and $|s_z = -\frac{1}{2}\hbar\rangle$ are all normalized, and assume further that the latter two kets are orthogonal (i.e., that $\langle s_z = \frac{1}{2}\hbar | s_z = -\frac{1}{2}\hbar\rangle = 0$; see Prob. 2.4 for the proof). Then

$$\langle\Psi|\Psi\rangle = |c_1|^2 + |c_2|^2 = 1 \tag{2.22}$$

and we may *postulate* that $|c_1|^2$ is equal to the probability of observing $s_z = \frac{1}{2}\hbar$, whereas $|c_2|^2$ is the probability of observing $s_z = -\frac{1}{2}\hbar$ in the state $|\Psi\rangle$. With this postulate, we see that

$$\langle\Psi|\mathbf{s}_z|\Psi\rangle = \frac{1}{2}\hbar(|c_1|^2 - |c_2|^2) \tag{2.23}$$

so the integral $\langle\Psi|\mathbf{s}_z|\Psi\rangle$ represents the *average value* of s_z in the state $|\Psi\rangle$. By average value we mean the average which would be observed in a series of measurements on a large number of individual cases; we do *not* mean a time average or some other kind of average on a single atom. In a pure state of s_x, $\langle\Psi|\mathbf{s}_z|\Psi\rangle = 0$, and we have $|c_1|^2 = |c_2|^2$; the individual complex values of c_1 and c_2 are not so easily obtained.

PROBLEMS

Prove the following theorems:

2.3 The eigenvalues of a Hermitian operator are real. (Hermitian operators are sometimes referred to as *real* operators because of this theorem. It should be clear that because we are only interested in real observations and hence real eigenvalues, we can restrict our theory to Hermitian operators.)

2.4 Two eigenvectors belonging to different eigenvalues of a given Hermitian operator are orthogonal. [This theorem is of great importance throughout all of quantum chemistry, and its proof should be thoroughly understood. In mathematical form, if

$$\mathbf{O}|\Psi_1\rangle = c_1|\Psi_1\rangle$$
$$\mathbf{O}|\Psi_2\rangle = c_2|\Psi_2\rangle \qquad c_2 \neq c_1, \ \mathbf{O} = \bar{\mathbf{O}}$$

then we must prove

$$\langle \Psi_1 | \Psi_2 \rangle = 0$$

Since the scalar product $\langle \Psi_1 | \Psi_2 \rangle$ is analogous to the dot product of two vectors in vector algebra, we call $|\Psi_1\rangle$ and $|\Psi_2\rangle$ *orthogonal* when the bracket $\langle \Psi_1 | \Psi_2 \rangle$ (or $\langle \Psi_2 | \Psi_1 \rangle$) is zero. In wave-mechanical language, the wavefunctions Ψ_1 and Ψ_2 are orthogonal when their integral, taken over a suitable interval, vanishes, that is, when $\int \Psi_1^* \Psi_2 \, d\tau = 0.$]

2.5 COMMUTATION OF OPERATORS: COMPLETE COMMUTING SETS

We have established thus far that the result of the measurement of an observable in a given system is an eigenvalue. Thus the set of eigenvalues (and corresponding eigenkets) is just the set of possible results of measurements. In this sense the set of eigenkets is *complete*, since any observation can be found in the set.

The complete set of eigenvectors of an operator is not the most general complete set which can be constructed. Consider the case in which a ket $|\Psi\rangle$ is an eigenket of two different operators $\mathbf{O}_A$ and $\mathbf{O}_B$. Then

$$\mathbf{O}_A|\Psi\rangle = O_{A_i}|\Psi\rangle$$
$$\mathbf{O}_B|\Psi\rangle = O_{B_j}|\Psi\rangle \tag{2.24}$$

where O_{A_i} and O_{B_j} are the eigenvalues of $\mathbf{O}_A$ and $\mathbf{O}_B$, respectively. Since $|\Psi\rangle$ is an eigenket of both operators,

$$\mathbf{O}_A\mathbf{O}_B|\Psi\rangle = \mathbf{O}_A O_{B_j}|\Psi\rangle = O_{B_j}\mathbf{O}_A|\Psi\rangle = O_{B_j}O_{A_i}|\Psi\rangle$$
$$\mathbf{O}_B\mathbf{O}_A|\Psi\rangle = \mathbf{O}_B O_{A_i}|\Psi\rangle = O_{A_i}\mathbf{O}_B|\Psi\rangle = O_{A_i}O_{B_j}|\Psi\rangle$$
$$\mathbf{O}_A\mathbf{O}_B|\Psi\rangle = \mathbf{O}_B\mathbf{O}_A|\Psi\rangle \tag{2.25}$$
$$(\mathbf{O}_A\mathbf{O}_B - \mathbf{O}_B\mathbf{O}_A)|\Psi\rangle = 0$$

The first of Eqs. (2.24) can be read as a statement about a measurement of the properties corresponding to $\mathbf{O}_A$ and $\mathbf{O}_B$. If the property O_B is measured in the state $|\Psi\rangle$, the state is unchanged since it is an eigenstate of $\mathbf{O}_B$. Thus, the measurement of O_A, made after the measurement of O_B, gives the same result for O_A as when O_A is measured first. To put it another way, if a series of measurements, alternately of O_A and O_B, are made on the state $|\Psi\rangle$, the measurement of O_A will always give the same result, and the measurement of O_B will always give the same result; that is,

$$\mathbf{O}_A(\mathbf{O}_B\mathbf{O}_A\mathbf{O}_B\mathbf{O}_A \cdots)|\Psi\rangle = O_{A_i}(\mathbf{O}_B\mathbf{O}_A\mathbf{O}_B\mathbf{O}_A \cdots)|\Psi\rangle \tag{2.26}$$

This kind of relation only holds when the ket $|\Psi\rangle$ is an eigenket of both operators. For example, consider the case in which $|\Psi\rangle$ is an eigenket of s_x, with eigenvalue $s_x = \frac{1}{2}\hbar$. Then, by (2.19) and (2.20),

$$
\begin{aligned}
|\Psi\rangle &= c_1|s_z = \tfrac{1}{2}\hbar\rangle + c_2|s_z = -\tfrac{1}{2}\hbar\rangle \\
\mathbf{s}_x|\Psi\rangle &= \tfrac{1}{2}\hbar|\Psi\rangle = c_1\mathbf{s}_x|s_z = \tfrac{1}{2}\hbar\rangle + c_2\mathbf{s}_x|s_z = -\tfrac{1}{2}\hbar\rangle \\
\mathbf{s}_z\mathbf{s}_x|\Psi\rangle &= \tfrac{1}{2}\hbar\mathbf{s}_z|\Psi\rangle = \tfrac{1}{2}\hbar(c_1\mathbf{s}_z|s_z = \tfrac{1}{2}\hbar\rangle \\
&\hspace{5.5cm} + c_2\mathbf{s}_z|s_z = -\tfrac{1}{2}\hbar\rangle) \quad (2.27) \\
&= \tfrac{1}{4}\hbar^2(c_1|s_z = \tfrac{1}{2}\hbar\rangle - c_2|s_z = -\tfrac{1}{2}\hbar\rangle) \\
\mathbf{s}_x\mathbf{s}_z|\Psi\rangle &= \mathbf{s}_x\tfrac{1}{2}\hbar(c_1|s_z = \tfrac{1}{2}\hbar\rangle - c_2|s_z = -\tfrac{1}{2}\hbar\rangle)
\end{aligned}
$$

and since $(c_1|s_z = \frac{1}{2}\hbar\rangle - c_2|s_z = -\frac{1}{2}\hbar\rangle)$ is not the eigenfunction of $\mathbf{s}_x$ with the eigenvalue $\frac{1}{2}\hbar$,

$$\mathbf{s}_z\mathbf{s}_x|\Psi\rangle \neq \mathbf{s}_x\mathbf{s}_z|\Psi\rangle \tag{2.28}$$

This discussion illustrates two kinds of measurements on a state, measurements which do not interfere with each other and measurements which do interfere with each other. As mentioned in the first section of this chapter, one of the tasks of the theory of quantum mechanics is to predict how many different kinds of measurements can be made on a system without mutual interference, and (2.25) and (2.28) suggest the following possible condition: observations of the properties O_A and O_B will not interfere provided the operators corresponding to them commute, i.e., provided $\mathbf{O}_A\mathbf{O}_B = \mathbf{O}_B\mathbf{O}_A$.

The proof of this theorem is straightforward. Suppose $\mathbf{O}_A$ and $\mathbf{O}_B$ commute. Let their eigenfunctions be labeled by the appropriate eigenvalues, so that

$$
\begin{aligned}
\mathbf{O}_A|O_{A_i}\rangle &= O_{A_i}|O_{A_i}\rangle \\
\mathbf{O}_B|O_{B_j}\rangle &= O_{B_j}|O_{B_j}\rangle
\end{aligned} \tag{2.29}
$$

Then

$$\mathbf{O}_B\mathbf{O}_A|O_{B_j}\rangle = \mathbf{O}_A\mathbf{O}_B|O_{B_j}\rangle = O_{B_j}\mathbf{O}_A|O_{B_j}\rangle \tag{2.30}$$

Therefore the ket $\mathbf{O}_A|O_{B_j}\rangle$ is an eigenket of the operator $\mathbf{O}_B$ with eigenvalue O_{B_j}. However, the eigenkets of $\mathbf{O}_B$ form a complete set in the sense that if a ket is an eigenfunction of $\mathbf{O}_B$ with eigenvalue O_{B_j}, it must be equal to or proportional to the ket $|O_{B_j}\rangle$. Therefore

$$\mathbf{O}_A|O_{B_j}\rangle = c|O_{B_j}\rangle \tag{2.31}$$

and the ket $|O_{B_j}\rangle$ is also an eigenket of $\mathbf{O}_A$.

It should be noted that the eigenvalue of $\mathbf{O}_A$ in the ket $|O_{B_j}\rangle$ is not specified, and clearly c can be *any* eigenvalue O_{A_i}. This indefiniteness can be used to *construct* a new set of kets which we label $|O_{A_i}O_{B_j}\rangle$, such that

$$
\begin{aligned}
\mathbf{O}_A|O_{A_i}O_{B_j}\rangle &= O_{A_i}|O_{A_i}O_{B_j}\rangle \\
\mathbf{O}_B|O_{A_i}O_{B_j}\rangle &= O_{B_j}|O_{A_i}O_{B_j}\rangle
\end{aligned} \qquad \blacktriangleright (2.32)
$$

Thus the new kets $|O_{A_i}O_{B_j}\rangle$ are simultaneously eigenkets of both operators and form a complete set in the sense that any observation of $\mathbf{O}_A$ and/or $\mathbf{O}_B$ can be found in this set.

The above construction is valid only if the functions $|O_{A_i}\rangle$ and $|O_{B_j}\rangle$ belong to *different* degrees of freedom. For example, the operators $\mathbf{O}$ and $\mathbf{O}^2$ commute, but it may not always be possible to construct a set of eigenfunctions which are simultaneously eigenkets of both.

We have shown, therefore, that if two operators commute, measurements of the properties corresponding to the operators do not interfere with each other. The general problem of finding the maximum number of different measurements that can be made on a state without disturbing or destroying it reduces to finding a set of operators which all mutually commute; such a set is usually referred to as a *complete set of commuting observables*. The following two major problems remain in the general theory: first, to discover how individual operators may be constructed, given a certain property of interest, and second, to construct eigenfunctions and their eigenvalues, given the operators. The first problem will be treated in the following two sections; the second problem will be, in effect, the subject of all the remaining chapters.

One of the properties of a complete set of commuting observables is that if one has a general ket $|\Psi\rangle$, it can always be expanded in terms of the complete set; that is,

$$|\Psi\rangle = \sum_{i,j,k,\dots} c_{i,j,k,\dots}|O_{A_i}O_{B_j}O_{C_k}\cdots\rangle \tag{2.33}$$

where i, j, and k label the individual eigenvalues of the commuting operators $\mathbf{O}_A$, $\mathbf{O}_B$, and $\mathbf{O}_C$. In a simpler notation,

$$|\Psi\rangle = \sum_i c_i|\chi_i\rangle \tag{2.34}$$

The sum is over all possible combinations of eigenvalues. We can deduce some special properties of operators constructed from this set of kets. Consider the expression

$$|\chi_i\rangle\langle\chi_j| = \mathbf{O} \tag{2.35}$$

$\mathbf{O}$ is clearly an operator since

$$\mathbf{O}|\Psi\rangle = (|\chi_i\rangle\langle\chi_j|)|\Psi\rangle = \langle\chi_j|\Psi\rangle|\chi_i\rangle = \left(\sum_k c_k\langle\chi_j|\chi_k\rangle\right)|\chi_i\rangle$$
$$= c_j|\chi_i\rangle \tag{2.36}$$

and this equation can be used to determine the expansion coefficients in (2.34). (The last step above follows from the proof of Prob. 2.4.) In addition, the theorem

$$\sum_i |\chi_i\rangle\langle\chi_i| = 1 \qquad\qquad \blacktriangleright (2.37)$$

can be proved, provided the sum over i extends over all eigenstates of the complete set of commuting observables. The proof is reasonably easy and will be left as an exercise. The theorem can be used in the expansion of a general ket $|\Psi\rangle$ since

$$
\begin{aligned}
|\Psi\rangle &= \left(\sum_i |\chi_i\rangle\langle\chi_i|\right) |\Psi\rangle \\
&= \sum_i \langle\chi_i|\Psi\rangle|\chi_i\rangle \\
&= \sum_i c_i|\chi_i\rangle
\end{aligned}
\qquad (2.38)
$$

Furthermore,

$$
\begin{aligned}
\langle\Psi|\Psi\rangle &= \sum_j \sum_i c_j^* c_i \langle\chi_j|\chi_i\rangle \\
&= \sum_i c_i^* c_i = \sum_i |c_i|^2
\end{aligned}
\qquad (2.39)
$$

and thus

$$\sum_i |c_i|^2 = 1 \qquad\qquad (2.40)$$

2.6 POISSON BRACKETS

According to (2.28), the operators $\mathbf{s}_x$ and $\mathbf{s}_z$ do not commute, and measurements of s_x and s_z will interfere. Similarly, we know, according to Heisenberg's uncertainty principle, that measurements of momentum and position interfere; presumably, then, the operators corresponding to these quantities do not commute. We therefore require as part of our theory some rule for determining which operators commute. As it develops, we shall also need the value of the commutator $\mathbf{O}_A\mathbf{O}_B - \mathbf{O}_B\mathbf{O}_A$ for those operators which do not. Curiously enough, there are several ways of accomplishing this end, and none of them is without defect: there appears to be no unique way of determining the value of the commutator in the most general case. Fortunately, this seems to cause no difficulty in applying the theory, and one can choose any one of the published methods.

We shall follow the Poisson bracket formalism of Dirac, primarily because it is the most widely used (see Shewell, 1959, for a discussion of other possibilities). One virtue of the Dirac approach is that it follows classical mechanics rather closely, so that one may see immediately how a quantum-mechanical system behaves classically in certain limiting cases. This is an advantage in quantum chemistry since chemical systems are often "nearly classical." It may be a disadvantage in areas of nuclear and high-energy physics where the systems one works with have no classical analog.

In classical mechanics one can define the Poisson bracket $\{O_A,O_B\}$ by the expression

$$\{O_A,O_B\} \equiv \sum_{i=1}^{3n} \left[\frac{\partial O_A}{\partial q_i} \frac{\partial O_B}{\partial p_i} - \frac{\partial O_A}{\partial p_i} \frac{\partial O_B}{\partial q_i} \right] \qquad \blacktriangleright (2.41)$$

O_A and O_B are two measureable quantities, not the corresponding operators, which are always indicated by boldface type. The sum over i extends over the n particles of the system, and p_i and q_i represent components of the momentum and position vectors, respectively. We reserve the symbol $\{ \ \}$ for the Poisson bracket of O_A and O_B and use the symbol

$$[\mathbf{O}_A,\mathbf{O}_B] \equiv \mathbf{O}_A\mathbf{O}_B - \mathbf{O}_B\mathbf{O}_A \qquad \blacktriangleright (2.42)$$

for the commutator of the corresponding operators. Note that

$$\begin{aligned} [\mathbf{O}_A,\mathbf{O}_B] &= -[\mathbf{O}_B,\mathbf{O}_A] \\ \{O_A,O_B\} &= -\{O_B,O_A\} \\ \{O_A,c\} &= 0 = [\mathbf{O}_A,c] \end{aligned} \qquad (2.43)$$

Dirac (p. 87) has postulated a relationship between the classical Poisson bracket and the commutator; namely,

$$[\mathbf{O}_A,\mathbf{O}_B] = i\hbar\{O_A,O_B\} \qquad \blacktriangleright (2.44)$$

This equation may be taken as a fundamental postulate of quantum mechanics, although, as noted by Shewell and others before him, it is inadequate to handle *all* possible operators. Ignoring this difficulty (see Prob. 2.8), we may at once determine the commutator for the most important quantities in quantum mechanics, the position and momentum operators. Using (2.41) and (2.44), we find

$$\begin{aligned} [\mathbf{p}_i,\mathbf{p}_j] &= 0 \\ [\mathbf{q}_i,\mathbf{q}_j] &= 0 \\ [\mathbf{q}_i,\mathbf{p}_j] &= i\hbar\delta_{ij} \end{aligned} \qquad \blacktriangleright (2.45)$$

The last equation takes the special form

$$\mathbf{qp} - \mathbf{pq} = i\hbar \qquad (2.46)$$

when $i = j$ and is somewhat reminiscent of Heisenberg's uncertainty principle. Note that any pair $\mathbf{p}_i$ and $\mathbf{q}_j$ commute when they refer to *different* degrees of freedom; this implies, for example, that a measurement of the x component of the position of a particle does not interfere with a measurement of its momentum in the z direction. When p and q refer to the same degree of freedom, the measurements do interfere, although in the classical limit we may consider $\hbar$ to be so small that the interference may be neglected.

Another consequence of (2.44) arises from a classical theorem on Poisson brackets. If a function f depends explicitly only on the p's and q's, and not on the time, it is possible to show that (Goldstein, p. 256)

$$\frac{df}{dt} = \{f,H\} \tag{2.47}$$

where H is the classical Hamiltonian for the system. Therefore, the function f will be a *constant of the motion* if $\{f,H\}$ is zero. In quantum theory, (2.44) guarantees that any constant of the motion will commute with the quantum-mechanical Hamiltonian (which we write as $\mathfrak{IC}$, not $\mathbf{H}$). Conversely, any quantity which commutes with $\mathfrak{IC}$ will be a constant of the motion, and in this way we shall be able to verify that the classical laws of conservation of momentum, energy, and angular momentum (for isolated systems) hold in quantum theory.

PROBLEMS

2.5 Prove that $(\mathbf{O}_A\mathbf{O}_B - \mathbf{O}_B\mathbf{O}_A)$ is pure imaginary, assuming that $\mathbf{O}_A$ and $\mathbf{O}_B$ are Hermitian. Confirm that the constant $\hbar$ in (2.44) is thus real.

2.6 Prove that $(\mathbf{O}_A\mathbf{O}_B + \mathbf{O}_B\mathbf{O}_A)$ is Hermitian, again assuming that $\mathbf{O}_A$ and $\mathbf{O}_B$ are Hermitian.

2.7 Define $\mathbf{O}_{pq} = \frac{1}{2}(\mathbf{pq} + \mathbf{qp})$. Show that $\mathbf{O}_{pq} = \mathbf{qp} - \frac{1}{2}i\hbar$.

◆ **2.8** Expand $\mathbf{O}_{p^2q^2} \equiv \frac{1}{2}(\mathbf{p}^2\mathbf{q}^2 + \mathbf{q}^2\mathbf{p}^2)$ using (2.45), and show that $\mathbf{O}_{p^2q^2} = \mathbf{q}^2\mathbf{p}^2 - 2i\hbar\mathbf{qp} - \hbar^2$. Repeat the process using (2.44) with $\mathbf{O}_A = \mathbf{p}^2$ and $\mathbf{O}_B = \mathbf{q}^2$, and show that $\mathbf{O}_{p^2q^2} = \mathbf{q}^2\mathbf{p}^2 - 2i\hbar\mathbf{qp} - 2\hbar^2$. What does this contradiction imply?

2.7 THE SCHRÖDINGER REPRESENTATION

Since measurements of position and momentum interfere, it is impossible to define quantum-mechanical states in terms of the coordinates and the momenta of all particles in the system, as is done in classical mechanics. However, since the operators corresponding to all the components of the positions of the particles in the system commute, i.e., since

$$[\mathbf{q}_i,\mathbf{q}_j] = 0 \tag{2.48}$$

it is reasonable to assume that the maximum number of observable

◆ This symbol denotes problems of greater than average difficulty.

quantities in a system is $3n$, where n is the number of particles.‡ Therefore we may take as our *complete* set of commuting observables the coordinates of all the particles in the system.

It should be equally possible to take as the complete set of observables the momenta of all the particles. However, although such an approach is possible, it seems to be not really useful, at least in problems of chemical interest, and so we shall not pursue it further. All conventional approaches to quantum mechanics take the coordinates as the basic observables. This may seem puzzling at first, since in practice, measurements of the positions of particles are rarely carried out, and it is the momenta which are truly the quantities of experimental interest. Nevertheless, we choose the coordinates as the basic observables because of certain resultant simplifications in the basic state vectors.

Suppose we have a complete set of commuting observables q_1, q_2, . . . , q_{3n}. The most important property of such a set is that any ket $|\Psi\rangle$ can be expanded in terms of eigenkets of the complete set; i.e., in the notation of the previous section,

$$|\Psi\rangle = \sum_i c_i |\chi_i\rangle \tag{2.49}$$

where $|\chi_i\rangle$ is an eigenket of all the commuting observables q_j simultaneously. However, a difficulty arises in the theory at this point. From experiment we know that all possible values of the coordinates can be observed, so the eigenvalues of the operators q_j are not discrete, but form a multiply infinite set. Thus, there are an infinite number of kets $|\chi_i\rangle$, and the infinity is not countable. Consequently, we can no longer write Eq. (2.49), but we must take instead the more general expression

$$|\Psi\rangle = \int c(q_1, q_2, q_3, \ldots, q_{3n}) |\chi_i\rangle \, dq_1 \, dq_2 \, dq_3 \cdots dq_{3n} \tag{2.50}$$

At first glance (2.50) would appear to be but a trivial extension of (2.49), but unfortunately this is not true. Consider the special case of a system with a single operator q. Then

$$\begin{aligned}
|\Psi\rangle &= \int c(q) |q\rangle \, dq \\
\langle\Psi| &= \int c^*(q) \langle q| \, dq \\
\langle\Psi|\Psi\rangle &= \iint c^*(q) c(q') \langle q|q'\rangle \, dq \, dq'
\end{aligned} \tag{2.51}$$

If we apply the orthogonality condition, $\langle q|q'\rangle$ is zero except when $\langle q|$ and $|q'\rangle$ refer to the same point on the q axis. Thus the integral over dq' will be zero since the integrand is zero except at a single point. Therefore $\langle\Psi|\Psi\rangle = 0$, contradicting our fundamental assumptions.

‡ As will be shown later, this assumption is incorrect; there are other coordinates, connected with the spin of the particles, which we must add to our set of $3n$. However, this does not affect the validity of the present argument, which is based on a complete set of observables formed of coordinates (including spin coordinates if necessary) and omitting the conjugate momenta.

Clearly, the root of the difficulty is in the strict application of the orthogonality theorem. We have tried to compose an infinite number of mutually orthogonal kets from a single coordinate, yet the orthogonal kets are to have eigenvalues in a continuous range, infinitely close together. The difficulty can be circumvented by requiring $\langle q|q'\rangle$ to be infinite at $q = q'$ and by using the Dirac δ function to obtain the integral $\langle\Psi|\Psi\rangle$. However, this approach has fairly severe conceptual and mathematical difficulties, and we will not consider it further. At this point we may choose to attempt to be rigorous at the expense of the conceptual problems, or we may abandon all attempts to be rigorous and concentrate instead on the practical aspects of the theory. As chemists, it is natural to take the latter course; those who are interested in going into the basic theory more deeply are referred to the work of Dirac or to other sources in quantum theory.

Following a heuristic approach, we first note that $|\Psi\rangle$, in (2.51), is completely determined by the function $c(\mathbf{q})$. We may therefore *define*

$$\langle\Psi|\Psi\rangle = \int |c(\mathbf{q})|^2 \, d\mathbf{q} = 1 \qquad\qquad \blacktriangleright (2.52)$$

More generally, when $|\Psi\rangle$ is a function of all the coordinates q_1, q_2, $\ldots$, q_{3n},

$$\langle\Psi|\Psi\rangle = \int \cdots \int |c(\mathbf{q}_1, \ldots ,\mathbf{q}_{3n})|^2 \, d\mathbf{q}_1 \, d\mathbf{q}_2 \cdots d\mathbf{q}_{3n} \qquad (2.53)$$

It is frequently convenient to abbreviate the volume element by the symbol $d\tau$ and to abbreviate $c(\mathbf{q}_1, \ldots ,\mathbf{q}_{3n})$ by Ψ, so that

$$\langle\Psi|\Psi\rangle = \int \Psi^*\Psi \, d\tau \qquad\qquad \blacktriangleright (2.54)$$

In this sense, then,

$$|\Psi\rangle = \Psi(\mathbf{q}_1,\mathbf{q}_2, \ldots ,\mathbf{q}_{3n}) \qquad (2.55)$$

where Ψ is, of course, a continuous function of all the coordinates of the system. Multiplying a ket $|\Psi\rangle$ by a bra, say $\langle\Psi_1|$, implies that we integrate over the appropriate coordinate space; similarly,

$$\langle\Psi_1|O|\Psi_2\rangle = \int \Psi_1^*O\Psi_2 \, d\tau \qquad\qquad \blacktriangleright (2.56)$$

These definitions are not without certain difficulties. For example, the operator

$$O_1 = |\Psi_1\rangle\langle\Psi_1| \qquad (2.57)$$

can still be defined so that

$$O_1|\Psi\rangle = (\int \Psi_1^*\Psi \, d\tau)|\Psi_1\rangle \qquad (2.58)$$

but it is no longer clear in what sense (2.37) is true. Fortunately, this ambiguity, and others connected with it, will not cause trouble in the applications treated in the chapters which follow, and so it will be ignored.‡

We have still not determined the specific forms of the operators $\mathbf{q}_i$ and $\mathbf{p}_i$. The only requirements which we have thus far placed on them are the commutation relations (2.45) and the general requirement that all operators be Hermitian. Therefore,

$$\langle\Psi_1|\mathbf{p}|\Psi_2\rangle = \langle\Psi_2|\mathbf{p}|\Psi_1\rangle^*$$
$$\langle\Psi_1|\mathbf{q}|\Psi_2\rangle = \langle\Psi_2|\mathbf{q}|\Psi_1\rangle^* \qquad\qquad \blacktriangleright(2.59)$$

or, since we must reinterpret the brackets as integrals,

$$\int\Psi_1^*\mathbf{p}\Psi_2\,d\tau = (\int\Psi_2^*\mathbf{p}\Psi_1\,d\tau)^*$$
$$\int\Psi_1^*\mathbf{q}\Psi_2\,d\tau = (\int\Psi_2^*\mathbf{q}\Psi_1\,d\tau)^* \qquad\qquad (2.60)$$

It is natural to take the operators $\mathbf{q}_i$ as the corresponding coordinates q_i; operation of a ket $|\chi\rangle$ by the operator $\mathbf{q}_i$ is thus equivalent to multiplying by q_i. This is possible because of the continuity of the eigenvalues of $\mathbf{q}_i$; the eigenvalue equation

$$\mathbf{q}_i|\chi_j\rangle = q_i|\chi_j\rangle \qquad\qquad (2.61)$$

is an identity for all operators $\mathbf{q}_i$. Similarly, if the ket $|\Psi\rangle$ is a function of the kets $|\chi_j\rangle$, then the integral

$$\langle q_i\rangle \equiv \int\Psi^*q_i\Psi\,d\tau \qquad\qquad \blacktriangleright(2.62)$$

represents the *average value* or *expectation value* of q in the state $|\Psi\rangle$.

If the bracket $\langle\Psi|\Psi\rangle$ is redefined as the integral $\int\Psi^*\Psi\,d\tau$, then it is desirable to attach some physical significance to the function $\Psi^*\Psi$. Comparison with (2.39) and (2.53) leads us to the conclusion that $\Psi^*\Psi\,d\tau$, *evaluated at the point* $q_1, q_2, \ldots, q_{3n}$, represents the *probability* that the coordinates all have values between $q_1, q_2, \ldots, q_{3n}$ and $q_1 + dq_1, q_2 + dq_2, \ldots, q_{3n} + dq_{3n}$. In general, of course, $d\tau$ may be as small as desired, so $\Psi^*\Psi\,d\tau$ is essentially zero. We therefore define the function $\Psi^*\Psi$ as the *probability density* of the coordinates; $\Psi^*\Psi$ expresses the relative probability of the coordinates having a particular value. Thus, the requirement that $\langle\Psi|\Psi\rangle = 1$, that is, that $|\Psi\rangle$ be normalized, is equivalent to the statement that the coordinates must have some values within the space over which we integrate. Similarly, Eq. (2.62) expresses again the concept that $\Psi^*\Psi$ is a probability density which weights the different possible values of q_i.

We have yet to prove that $\mathbf{q}_i = q_i$ is consistent with the requirement that $\mathbf{q}_i$ be Hermitian. However, the proof is quite obvious; since q_i

‡ We shall later be constructing other complete sets of commuting observables in which the operators have discrete eigenvalues. The problem of finding a complete set of commuting operators has no unique solution, and the set of the $\mathbf{q}$'s is only one possible solution, useful only for limited purposes.

is no longer an operator but merely a (real) function,

$$\begin{aligned}(\int \Psi_1^* q \Psi_2 \, d\tau)^* &= \int \Psi_1 q^* \Psi_2^* \, d\tau^* = \int \Psi_1 q \Psi_2^* \, d\tau \\ &= \int \Psi_2^* q \Psi_1 \, d\tau = \int \Psi_2^* \Psi_1 q \, d\tau\end{aligned} \tag{2.63}$$

Only one major task remains in the construction of the theory—we must find operators $\mathbf{p}_i$ which are Hermitian and which satisfy the commutation relations. It is clear that the operator

$$\mathbf{p}_i = -i\hbar \frac{\partial}{\partial q_i} \tag{2.64}$$

satisfies at least the commutation relations. For example,

$$\begin{aligned}\mathbf{p}_i \mathbf{p}_j &= -\hbar^2 \frac{\partial^2}{\partial q_i \, \partial q_j} = -\hbar^2 \frac{\partial^2}{\partial q_j \, \partial q_i} = \mathbf{p}_j \mathbf{p}_i \\ \mathbf{p}_i \mathbf{q}_j &= -i\hbar \frac{\partial}{\partial q_i} q_j = -i\hbar q_j \frac{\partial}{\partial q_i} = \mathbf{q}_j \mathbf{p}_i \qquad i \neq j \\ \mathbf{p}_i \mathbf{q}_i &= -i\hbar \frac{\partial}{\partial q_i} q_i = -i\hbar - i\hbar q_i \frac{\partial}{\partial q_i} = -i\hbar + \mathbf{q}_i \mathbf{p}_i\end{aligned} \tag{2.65}$$

so $[\mathbf{p}_i, \mathbf{p}_j] = 0$ and $[\mathbf{q}_i, \mathbf{p}_j] = i\hbar \delta_{ij}$ as required.‡ However, this form of $\mathbf{p}_i$ is not the only form which satisfies the commutation requirements. For example, the operator $\mathbf{p}_i' = -i\hbar q_i^{-n}(\partial/\partial q_i) q_i^n$ also satisfies the requirements, since it clearly commutes with all $\mathbf{p}_j'$ and $\mathbf{q}_j$ and since

$$\mathbf{p}_i' \mathbf{q}_i = -i\hbar(n+1) - i\hbar q_i \frac{\partial}{\partial q_i}$$

$$\mathbf{q}_i \mathbf{p}_i' = -i\hbar(n) - \hbar q_i \frac{\partial}{\partial q_i} \tag{2.66}$$

$$[\mathbf{q}_i, \mathbf{p}_i'] = i\hbar$$

Furthermore, the expression for $\mathbf{p}_i'$ is not the most general operator obeying the commutation relations; in fact, it is clear that the more general operator

$$\mathbf{p}_i'' = -i\hbar \frac{\partial}{\partial q_i} + f(q_i) \tag{2.67}$$

can also be shown to obey the commutation relations, provided $f(q_i)$ commutes with the q_i.

Faced with this wide choice of operators $\mathbf{p}_i$, we turn next to the

‡ It must be remembered that $\mathbf{p}_i \mathbf{q}_i$ is an *operator* expression; written together with a general function f, the last of Eqs. (2.65) would read

$$\mathbf{p}_i \mathbf{q}_i f = -i\hbar \frac{\partial}{\partial q_i} q_i f = -i\hbar f - i\hbar q_i \frac{\partial f}{\partial q_i} = -i\hbar f + \mathbf{q}_i \mathbf{p}_i f$$

requirement that all operators be Hermitian. As we shall show, the form of $\mathbf{p}_i$ depends on the kind of coordinate we are dealing with. Consider the case of a single coordinate, say $q = x$, whose range is $-\infty$ to ∞. Then, integrating by parts‡ and letting $u = \Psi_1^*$ and $dv = d\Psi_2$, we have

$$\int_{-\infty}^{\infty} \Psi_1^*(x) \left(-i\hbar \frac{\partial}{\partial x} \right) \Psi_2(x) \, dx = -i\hbar \Psi_1^*(x) \Psi_2(x) \Big|_{-\infty}^{\infty}$$

$$- \int_{-\infty}^{\infty} \Psi_2(x) \left(-i\hbar \frac{\partial}{\partial x} \right) \Psi_1^*(x) \, dx \quad (2.68)$$

$$\left[\int_{-\infty}^{\infty} \Psi_2^*(x) \left(-i\hbar \frac{\partial}{\partial x} \right) \Psi_1(x) \, dx \right]^*$$

$$= - \int_{-\infty}^{\infty} \Psi_2(x) \left(-i\hbar \frac{\partial}{\partial x} \right) \Psi_1^*(x) \, dx$$

Therefore, the requirement that $-i\hbar(\partial/\partial x)$ be Hermitian is equivalent to the requirement that the product $\Psi_1^*(x)\Psi_2(x)$ vanish§ at $\pm \infty$. This requirement is reasonable on physical grounds as well since we wish our probability densities to go to zero at $\pm \infty$, so our system is physically confined.

It is clear that the form of the operator $\mathbf{p}_i$ is dependent on the kinds of boundary conditions we place on our problem, and that when the coordinates are cartesian, the form $\mathbf{p}_i = -i\hbar(\partial/\partial q_i)$ satisfies all the conditions enumerated. In other coordinate systems, as we shall see, this form is incorrect; in particular, when the range of Ψ is restricted to the positive real axis, the present form for $\mathbf{p}_i$ cannot be applied.

We have still not shown that in a cartesian coordinate system the operator $\mathbf{p}_i = -i\hbar(\partial/\partial q_i)$ is the *only* Hermitian operator satisfying the commutation relations. A proof has been given by Dirac; we shall ordinarily assume in future work that any operator which is Hermitian and which satisfies the commutation relations is in fact unique. Alternatively, operators in a noncartesian coordinate system will be derived from the cartesian operators by appropriate transformations which preserve their Hermitian properties.

The course we have been following in this section, that is, choosing a particular set of commuting operators (the $\mathbf{q}$'s) as the complete set of observables and forming from them a complete set of commuting eigenkets, is called forming a *representation* of the operators. The particular one we have obtained is called the Schrödinger representation because Schrödinger's wave mechanics is but a simple extension of it. In the following section we shall obtain the Schrödinger equation from the

‡ $\int u \, dv = uv - \int v \, du$.
§ The weaker condition that the product not depend on the sign of x may alternatively be used.

present theory. In Chap. 8, on group theory, the concept of a representation will be considerably amplified, and in particular it will be shown how a representation can be formed using matrices. As Dirac has shown, the Schrödinger representation can be viewed as a kind of matrix formulation as well, although this approach has its difficulties since the resulting matrices are neither finite nor discrete.

2.8 THE SCHRÖDINGER EQUATION AND THE POSTULATES OF QUANTUM MECHANICS

Before proceeding to applications of quantum mechanics, we shall first summarize the discussions of the previous sections in a series of postulates. These postulates may be taken, by themselves, as the entire basis of the theory of quantum mechanics, and in this light the preceding sections may be viewed as background material which serves to provide an intuitive understanding of the theoretical and logical basis of the postulates.

POSTULATE 1

From our discussion of state vectors and kets, and from the development of the Schrödinger representation, we know that the state of a system is described as completely as possible by a vector $|\Psi\rangle$ which is only a function of the coordinates (including, as we shall see, spin coordinates) of the particles in the system. Thus, for a system of n particles,

$$|\Psi\rangle = \Psi(q_1, q_2, \ldots, q_{3n}, \xi_1, \xi_2, \ldots, \xi_n) \tag{2.69}$$

where the ξ_i are the (undefined) spin coordinates of the particles. In the most general case, $|\Psi\rangle$ is also a function of the time, but in this book we are only concerned with time-independent theory.

Since the function $|\Psi\rangle$ specifies the system as completely as is possible, it contains all possible information about any given state. The rest of the postulates are concerned with obtaining this information.

POSTULATE 2

To every observable property of a system there corresponds a linear Hermitian operator. In a cartesian coordinate system we may determine these operators using the following rules:

1. Take the classical mathematical expression for the observable, and express it solely in terms of the coordinates and momenta of the system.

2. Replace all components of the momenta p_i by $-i\hbar(\partial/\partial q_i)$ to obtain the quantum-mechanical operator. The resulting operator must be Hermitian *if it is to be valid. Thus, since the classical expression for the kinetic energy of a system of n particles in cartesian space is*

$$T = \frac{1}{2} \sum_{i=1}^{n} m_i(\mathbf{v}_i \cdot \mathbf{v}_i) = \frac{1}{2} \sum_{i=1}^{n} \frac{(\mathbf{p}_i \cdot \mathbf{p}_i)}{m_i}$$

$$= \frac{1}{2} \sum_{i=1}^{n} \frac{(p_{x_i}^{\,2} + p_{y_i}^{\,2} + p_{z_i}^{\,2})}{m_i} \quad (2.70)$$

then

$$\mathbf{T} = -\frac{1}{2} \sum_{i=1}^{n} \frac{\hbar^2}{m_i} \left(\frac{\partial^2}{\partial x_i{}^2} + \frac{\partial^2}{\partial y_i{}^2} + \frac{\partial^2}{\partial z_i{}^2} \right) = -\frac{\hbar^2}{2} \sum_{i=1}^{n} \frac{1}{m_i} \nabla_i{}^2 \quad \blacktriangleright (2.71)$$

where

$$\nabla_i{}^2 = \nabla_i \cdot \nabla_i \quad \text{and} \quad \nabla_i = \hat{\mathbf{i}} \frac{\partial}{\partial x_i} + \hat{\mathbf{j}} \frac{\partial}{\partial y_i} + \hat{\mathbf{k}} \frac{\partial}{\partial z_i} \quad (2.72)$$

It is easily verified that $\nabla_i{}^2$ is Hermitian (Prob. 2.11).

POSTULATE 3

Results of measurements of observable properties of a system correspond to the result of operating with the appropriate operator on the vector describing the state of the system. This operation may have several effects; in general, the result of the measurement is to force the system into a new state in which the observable has a definite fixed value. Subsequent measurements of the observable will yield the same result unless the system is otherwise disturbed.

There are two fundamentally different kinds of initial states $|\Psi\rangle$. In one case, $|\Psi\rangle$ is a pure eigenstate of the operator $\mathbf{O}$, and if we prepare a large number of systems, all with state $|\Psi\rangle$, the result of the observations will always yield the same result, that is,

$$\mathbf{O}|\Psi\rangle = c|\Psi\rangle \quad (2.73)$$

Here $|\Psi\rangle$ is an eigenket of $\mathbf{O}$ with eigenvalue c, and a measurement of $\mathbf{O}$ in the state $|\Psi\rangle$ is sure to give the result c. On the other hand, in the more general case, $|\Psi\rangle$ is not an eigenket of $\mathbf{O}$. In this instance a large number of measurements of $\mathbf{O}$ in the state $|\Psi\rangle$ will give rise to a variety of different results. Although we cannot predict the result of any particular measurement, we can say that the average value of $\mathbf{O}$ in the state $|\Psi\rangle$ is

$$\langle \mathbf{O} \rangle = \frac{\langle \Psi | \mathbf{O} | \Psi \rangle}{\langle \Psi | \Psi \rangle} \quad \blacktriangleright (2.74)$$

Usually, $|\Psi\rangle$ is presumed to be normalized to unity, so the integral $\langle \Psi | \Psi \rangle$ can be left out of the denominator. It should be carefully noted that although *only the average value $\langle O \rangle$ can be predicted when $|\Psi\rangle$ is an impure state of O, this does not imply that any value for O can be observed. As in the example experiment outlined in Fig. 2.1, the possible values for O may be discrete and limited.*

POSTULATE 4

For a system in an impure state $|\Psi\rangle$ of an operator O, we can, in general, expand $|\Psi\rangle$ in terms of pure states of O, so that

$$|\Psi\rangle = \sum_i c_i |\chi_i\rangle \tag{2.75}$$

where

$$O|\chi_i\rangle = k_i|\chi_i\rangle \tag{2.76}$$

In this case $|c_i|^2$ is the probability that a measurement of O in the state $|\Psi\rangle$ will throw the system into the new state $|\chi_i\rangle$ and give k_i as the result of the measurement.

Furthermore, when $|\Psi\rangle$ is explicitly considered as a function of the coordinates $|\Psi\rangle = \Psi(q_1, q_2, \ldots, q_{3n}, \xi_1, \xi_2, \ldots, \xi_n)$, then the function $\Psi^*\Psi \, d\tau$ is the probability that the coordinate q_1 is between q_1 and $q_1 + dq_1$, that q_2 is between q_2 and $q_2 + dq_2$, etc. Since $d\tau = dq_1 \, dq_2 \cdots dq_{3n} \, d\xi_1 \, d\xi_2 \cdots d\xi_n$ may be as small as desired, $\Psi^*\Psi$ alone is often referred to as the probability density of the values $q_1, \ldots, \xi_n$. Usually the spin coordinates are omitted entirely since we often work in systems in which spin is not explicitly considered.

Of course, the above postulates are not a unique formulation of quantum mechanics; different authors choose differing forms of the statements and take a variety of rules as being most fundamental. All formulations lead, however, to the same theorems and the same working rules, and, of course, this is what is crucial.

Schrödinger's equation is quite easily derived from the postulates. From the experimental evidence we may assume that the total energy of a conservative system is an observable. The classical expression for the energy is $H = T + V$, where T is given by (2.70) and V is a function only of the coordinates. The operator corresponding to the total energy is thus

$$\mathcal{H} \equiv \mathbf{H} = -\frac{\hbar^2}{2} \sum_{i=1}^{n} \frac{\nabla_i^2}{m_i} + V(q_1, \ldots, q_{3n}, \xi_1, \ldots, \xi_n) \qquad \blacktriangleright (2.77)$$

This is, of course, the quantum-mechanical Hamiltonian.

From Postulate 3 it must be possible to define states which are eigenstates of the operator $\mathcal{H}$. Thus

$$\mathcal{H}|\Psi\rangle = E|\Psi\rangle$$

or, using (2.77),

$$-\frac{\hbar^2}{2} \sum_{i=1}^{n} \frac{1}{m_i} \nabla_i^2 \Psi - (E - V)\Psi = 0 \qquad \blacktriangleright (2.78)$$

This is Schrödinger's equation for stationary states. For a single particle it may be written as

$$-\frac{\hbar^2}{2m} \left(\frac{\partial^2}{\partial x^2} + \frac{\partial^2}{\partial y^2} + \frac{\partial^2}{\partial z^2} \right) \Psi - (E - V)\Psi = 0 \qquad (2.79)$$

which is a common form of the equation. As noted earlier, Schrödinger's equation is most easily written in cartesian coordinates because it is only in this system that the momentum operators and the Hamiltonian are straightforwardly obtained. In other systems of coordinates, the operators corresponding to $\mathbf{p}_i$ are often difficult to write, and we can best obtain the Hamiltonian by direct transformation of the cartesian form into the new coordinate system.

A variety of other consequences of the postulates are easily derived. In Sec. 2.5 we showed that if two operators commute, then one may construct a set of kets which are simultaneous eigenkets of both operators. Several theorems have also been given as problems; the most important is Prob. 2.4, concerning the orthogonality of eigenvectors belonging to different eigenvalues of an operator. Equation (2.37) is also of great utility, although it must be used carefully since it only applies to cases in which the complete commuting set is discrete.

PROBLEMS

2.9 Consider the spherical-coordinate system $x = r \cos \varphi \sin \theta$, $y = r \sin \varphi \sin \theta$, $z = r \cos \theta$. Show that $\nabla^2 = (1/r^2)(\partial/\partial r)[r^2(\partial/\partial r)] + (1/r^2 \sin \theta)(\partial/\partial \theta)[\sin \theta (\partial/\partial \theta)] + (1/r^2 \sin^2 \theta)(\partial^2/\partial \varphi^2)$.
Hint: Obtain the partial derivatives $\partial r/\partial x$, $\partial \varphi/\partial y$, etc.

2.10 Find the Hamiltonian for a particle confined to move on a regular helical wire.
Hint: Let the independent coordinate be a nonperiodic angle.

2.11 If $\mathbf{O}$ is a Hermitian operator, show that $\mathbf{O}^n$ is also Hermitian. Use this theorem to prove that the Hamiltonian operator of (2.77) is Hermitian.

SUGGESTIONS FOR FURTHER READING

Students learning quantum mechanics for the first time find two very different types of difficulties. Either they find the physical principles hazy, or else they are unable to follow the mathematical steps linking the principles together. No two students have the same difficulties, and so I cannot recommend a single text suitable for all. For the mathematically inclined, the following three sources may be helpful; the reader should bear in mind, however, that this is a severely limited list, and a brief search of the quantum-theory shelves in a good library may be equally useful:

1. Dirac, P. A. M.: "The Principles of Quantum Mechanics," 4th ed., Oxford University Press, London, 1958.

   Sections 1–22 provide an extraordinarily detailed exposition of the fundamental postulates. Dirac's writing and approach are deceptively clear; the material requires careful and repeated study before even partial mastery can be achieved.

2. Margenau, H.: chaps. 1 and 2, in D. R. Bates (ed.), "Quantum Theory," vol. I, Academic Press, Inc., New York, 1961.

   The material here duplicates that of Dirac almost exactly, but a different approach to the subject may be of use.

3. Margenau, H., and G. M. Murphy: "The Mathematics of Physics and Chemistry," 2d ed., vol. I, D. Van Nostrand, New York, 1956.

   For the mathematically well-prepared student, the first few sections of Chap. 2 provide a clear and terse introduction to quantum mechanics. For the student who is not so well prepared, the other chapters provide what is needed.

For the student who does *not* find the references above particularly helpful, two other approaches are possible. The first three texts listed below are intended specifically for chemists, and although they are somewhat dated, in them the reader will at least find less emphasis on the mathematical details. On the other hand, a more elementary introduction to quantum theory from the physicist's point of view may be more useful, and the last two books listed are among the most popular undergraduate texts. Once again, the selection given is but a sampling of what is available.

4. Kauzman, W.: "Quantum Chemistry," Academic Press, Inc., New York, 1957.

5. Pauling, L., and E. B. Wilson: "Introduction to Quantum Mechanics," McGraw-Hill Book Company, New York, 1935.

6. Eyring, H., J. Walter, and G. E. Kimball: "Quantum Chemistry," John Wiley and Sons, New York, 1944.

7. Hameka, H. F.: "Introduction to Quantum Theory," Harper and Row, New York, 1967.

8. Dicke, R. H., and J. P. Wittke: "Introduction to Quantum Mechanics," Addison-Wesley Publishing Co., Inc., Cambridge, Mass., 1960.

I also warmly recommend a new text written by an undergraduate classmate. The approach is quite similar to that of the present text, but with greater emphasis on spectroscopy and a considerable development of time-dependent theory. The zeroth chapter contains a good description of the Feynman two-slit *gedanken* experiment which beautifully demonstrates the nonclassical behavior of small particles and the inadequacy of classical concepts.

9. Anderson, J. M.: "Introduction to Quantum Chemistry," W. A. Benjamin, Inc., New York, 1969.

3

ANGULAR MOMENTUM

The angular momentum of the
electron round the nucleus in a
stationary state of the system is
equal to an entire multiple of a
universal value, independent of
the charge on the nucleus.
Niels Bohr, Philosophical Magazine
(1913)

3.1 ORDINARY ANGULAR MOMENTUM

The most important theorem in classical mechanics is the mass-energy conservation law. It is this theorem which allows us to write down the Hamiltonian operator in quantum mechanics and obtain stationary functions which are its eigenstates. Classically, angular momentum is also conserved. However, we cannot immediately assume that both energy and angular momentum are conserved in quantum mechanics. To do so would imply that states can be defined which are simultaneously eigenstates of the Hamiltonian and the appropriate angular momentum operator, and this in turn implies that the two operators commute. In order to *prove* that both energy and angular momentum can be conserved in quantum mechanics, we must therefore show that the appropriate operators commute.‡ Before doing so, we shall first develop

‡ Here we do not assume the truth of Eq. (2.47).

some of the classical relations concerning the angular momentum of a system of particles.

For a single particle, the angular momentum is defined as

$$\mathbf{l} = \mathbf{r} \times \mathbf{p} \tag{3.1}$$

where $\mathbf{p} = m\mathbf{v}$ is the (linear) momentum of the particle. For a collection of n particles, we define $\mathbf{L} = \sum_{i=1}^{n} \mathbf{l}_i = \sum_{i=1}^{n} \mathbf{r}_i \times \mathbf{p}_i$ as the total angular momentum of the system. In a classical system it is straightforward to show that

$$\frac{d}{dt}\mathbf{L} = \sum_i \mathbf{r}_i \times \mathbf{F}_i \tag{3.2}$$

where $\mathbf{F}_i$ is the force on the ith particle. Clearly, for an isolated system with no forces, angular momentum is conserved; furthermore, it is conserved in a system of particles acted upon by *central forces* in the direction of the vectors $\mathbf{r}_i$. The motion of the planets is one example; electron motion in atoms is another.

Let us expand the cross product in (3.1). We obtain the components‡

$$\begin{aligned}
l_x &= yp_z - zp_y \\
l_y &= zp_x - xp_z \\
l_z &= xp_y - yp_x
\end{aligned} \tag{3.3}$$

Equations (3.3) provide the means of obtaining the quantum-mechanical operator corresponding to the angular momentum. Replacing the components of $\mathbf{p}$ by the corresponding operators (see Sec. 2.8), we obtain

$$\begin{aligned}
\mathbf{l}_x &= -i\hbar \left(y\frac{\partial}{\partial z} - z\frac{\partial}{\partial y} \right) \\
\mathbf{l}_y &= -i\hbar \left(z\frac{\partial}{\partial x} - x\frac{\partial}{\partial z} \right) \\
\mathbf{l}_z &= -i\hbar \left(x\frac{\partial}{\partial y} - y\frac{\partial}{\partial x} \right)
\end{aligned} \tag{3.4}$$

which, again, are only valid for a single particle. Using either (3.3) or (3.4), it is possible to prove a number of valuable commutation relations. As in Chap. 2, we write $[\mathbf{O}_1,\mathbf{O}_2] = \mathbf{O}_1\mathbf{O}_2 - \mathbf{O}_2\mathbf{O}_1$ for the commutator of two operators and find

$$\begin{aligned}
[\mathbf{l}_x,\mathbf{l}_y] &= i\hbar\mathbf{l}_z \\
[\mathbf{l}_z,\mathbf{l}_x] &= i\hbar\mathbf{l}_y \\
[\mathbf{l}_y,\mathbf{l}_z] &= i\hbar\mathbf{l}_x
\end{aligned} \tag{3.5}$$

‡ It is not possible to write the components of the total angular momentum vector $\mathbf{L}$ in such a simple form.

The second and third relations in (3.5) follow from the first by cyclic permutation of x, y, and z; the first relation is most easily proved using (3.3) and the commutation properties of the coordinates and momenta.

For a system of n particles, the commutation relations for the components of $\mathbf{L}$ still hold. For example,

$$
\begin{aligned}
[\mathbf{L}_x, \mathbf{L}_y] &= \left(\sum_{i=1}^{n} 1_{x_i} \right) \left(\sum_{j=1}^{n} 1_{y_j} \right) - \left(\sum_{j=1}^{n} 1_{y_j} \right) \left(\sum_{i=1}^{n} 1_{x_i} \right) \\
&= \sum_{i=1}^{n} (1_{x_i} 1_{y_i} - 1_{y_i} 1_{x_i}) \\
&= i\hbar \sum_{i=1}^{n} 1_{z_i} = i\hbar \mathbf{L}_z
\end{aligned}
\tag{3.6}
$$

and likewise for the other two commutators. The second line of (3.6) follows from the fact that the angular momenta of different particles commute, as do the coordinates and momenta of the different particles. As will be shown in the next section, most of the important properties of angular momentum can be derived solely from the commutation relations, without recourse to the classical definition. In particular, whereas the angular behavior of a single particle will be derived primarily from the classical definition (3.3) and the operator form (3.4), the relations for a system of particles will be developed almost entirely from the commutation relations.

Consider the operator 1_z. Since we are dealing with an *angular* quantity, it is convenient to reexpress 1_z in spherical coordinates. By transforming the differential operators $\partial/\partial x$, $\partial/\partial y$, and $\partial/\partial z$, (see Prob. 2.9) we find

$$
1_z = -i\hbar \frac{\partial}{\partial \varphi}
\tag{3.7}
$$

which is a considerably simpler form than (3.4). The expressions for 1_x and 1_y are not so simple, but fortunately we shall not need them. We *will* need the operator $1^2 = 1 \cdot 1 = 1_x{}^2 + 1_y{}^2 + 1_z{}^2$, however. The transformation to spherical coordinates gives

$$
1^2 = -\hbar^2 \left[\frac{1}{\sin \theta} \frac{\partial}{\partial \theta} \left(\sin \theta \frac{\partial}{\partial \theta} \right) + \frac{1}{\sin^2 \theta} \frac{\partial^2}{\partial \varphi^2} \right].
\tag{3.8}
$$

As is easily shown, 1^2 and 1_z commute. Using the spherical polar form developed for ∇^2 in the previous section, we can also show that 1_z and ∇^2 commute; hence, by symmetry, 1_x, 1_y, and 1^2 all commute with ∇^2.

We thus have a set of three commuting quantities, ∇^2, 1^2, and (say) 1_z. We cannot add 1_x and 1_y to this set since neither one commutes with

1_z; of course, we can replace 1_z in our set by 1_x or 1_y. Unfortunately, a commuting set including ∇^2 is not especially useful since the kinetic energy is not necessarily a constant of the motion. In many cases (for example, in atoms), 1^2 and 1_z commute with the potential energy term V of the Hamiltonian operator as well, so $\mathcal{3C}$, 1^2, and 1_z form a commuting set, and we can construct states which are simultaneously eigenkets of all three operators. The energy, the square of the total angular momentum and the z component of the angular momentum may thus all be constants of the motion. The specific eigenvalues for the angular momentum will be obtained in the next section without further use of the classical definition. However, the eigenkets cannot be obtained without explicitly solving the eigenvalue equations

$$
\begin{aligned}
1^2 |Y_{lm}\rangle &= c |Y_{lm}\rangle \\
1_z |Y_{lm}\rangle &= k |Y_{lm}\rangle
\end{aligned}
\tag{3.9}
$$

for the eigenfunctions Y_{lm}. This will be done in Sec. 3.3.

PROBLEMS

3.1 Verify the angular-momentum commutator relations (3.5).

3.2 Prove that 1^2 and 1_z commute.

3.3 Show that both 1^2 and 1_z commute with ∇^2.

3.4 Verify (3.7) and (3.8).

3.2 GENERALIZED ANGULAR MOMENTA

Consider a set of operators, which we shall denote $\mathbf{L}_x$, $\mathbf{L}_y$, and $\mathbf{L}_z$, obeying the commutation relations (3.5). We shall call such a set a generalized angular-momentum vector; it need not have a definition in terms of classical mechanics. As noted earlier, $\mathbf{L}^2$ and $\mathbf{L}_z$ can be shown to commute, and so they possess a set of simultaneous eigenkets. We write this set as $|ck\rangle$, such that

$$
\begin{aligned}
\mathbf{L}^2 |ck\rangle &= c |ck\rangle \\
\mathbf{L}_z |ck\rangle &= k |ck\rangle
\end{aligned}
\tag{3.10}
$$

The eigenkets have been labeled according to their eigenvalues. We may also define the two operators

$$
\begin{aligned}
\mathbf{L}^+ &= \mathbf{L}_x + i\mathbf{L}_y \\
\mathbf{L}^- &= \mathbf{L}_x - i\mathbf{L}_y
\end{aligned}
\tag{3.11}
$$

which are variously referred to as shift operators, ladder operators, or raising and lowering operators. As we shall show, it is possible to determine the eigenvalues c and k by using the shift operators. Writing $\mathbf{L}^{\pm}$

for the raising and lowering operators, we see that the proof follows in five major steps.

Lemma 1 $[\mathbf{L}_z, \mathbf{L}^\pm] = \pm \hbar \mathbf{L}^\pm$ $\qquad\qquad$ (3.12)

Proof: $\mathbf{L}_z \mathbf{L}^\pm = \mathbf{L}_z \mathbf{L}_x \pm i\mathbf{L}_z \mathbf{L}_y$, and $\mathbf{L}^\pm \mathbf{L}_z = \mathbf{L}_x \mathbf{L}_z \pm i\mathbf{L}_y \mathbf{L}_z$, so

$$[\mathbf{L}_z, \mathbf{L}^\pm] = [\mathbf{L}_z, \mathbf{L}_x] \pm i[\mathbf{L}_z, \mathbf{L}_y] = i\hbar \mathbf{L}_y \pm i(-i\hbar \mathbf{L}_x) = \pm \hbar \mathbf{L}^\pm$$

Lemma 2 $\mathbf{L}^\pm |ck\rangle = N^\pm |c, k \pm \hbar\rangle$ where $|c, k \pm \hbar\rangle$ are eigenkets of $\mathbf{L}^2$ and $\mathbf{L}_z$ with eigenvalues c and $k \pm \hbar$, respectively, and $N^\pm$ are numerical factors to be determined.

Proof: Using Lemma 1, we may write

$$\mathbf{L}_z \mathbf{L}^\pm |ck\rangle = \mathbf{L}^\pm \mathbf{L}_z |ck\rangle \pm \hbar \mathbf{L}^\pm |ck\rangle = (k \pm \hbar)\mathbf{L}^\pm |ck\rangle$$

so

$$\mathbf{L}_z(\mathbf{L}^\pm |ck\rangle) = (k \pm \hbar)(\mathbf{L}^\pm |ck\rangle)$$

This shows that $(\mathbf{L}^\pm |ck\rangle)$ is an eigenket of $\mathbf{L}_z$ with eigenvalue $k \pm \hbar$. But the ket $|c, k \pm \hbar\rangle$ is just an eigenket with this eigenvalue, $\mathbf{L}_z |c, k \pm \hbar\rangle = (k \pm \hbar)|c, k \pm \hbar\rangle$. Therefore the two kets must be *proportional;* that is

$$\mathbf{L}^\pm |ck\rangle = N^\pm |c, k \pm \hbar\rangle \qquad\qquad (3.13)$$

Lemma 3 $\mathbf{L}^\mp \mathbf{L}^\pm = \mathbf{L}^2 - \mathbf{L}_z{}^2 \mp \hbar \mathbf{L}_z$ $\qquad\qquad$ (3.14)

Proof: $(\mathbf{L}_x \mp i\mathbf{L}_y)(\mathbf{L}_x \pm i\mathbf{L}_y) = \mathbf{L}_x{}^2 + \mathbf{L}_y{}^2 \pm i(\mathbf{L}_x \mathbf{L}_y - \mathbf{L}_y \mathbf{L}_x)$
$$= \mathbf{L}^2 - \mathbf{L}_z{}^2 \mp \hbar \mathbf{L}_z$$

Theorem 1 $|k| \leq \sqrt{c}$

Proof: $\langle ck|\mathbf{L}^\mp \mathbf{L}^\pm |ck\rangle = c - k^2 \mp \hbar k \geq 0$ by Lemma 3 and because the *length* of the vector $\mathbf{L}^\pm |ck\rangle$ is positive. Here we recall from Sec. 2.2 that $\langle \psi|\overline{\mathbf{O}}$ corresponds to $\mathbf{O}|\psi\rangle$. The proof that $\overline{\mathbf{L}}^\pm = \mathbf{L}^\mp$ is given as an exercise in Prob. 3.5. We now obtain

$$c - k^2 - \hbar k \geq 0$$
$$c - k^2 + \hbar k \geq 0$$

Adding, we obtain

$$c - k^2 \geq 0$$

According to this proof, the range of values of k is limited. This is only possible if there exists some pair of kets such that

$$\mathbf{L}^+|ck_{max}\rangle = 0 \qquad \text{with } |ck_{max}\rangle \neq 0$$
$$\mathbf{L}^-|ck_{min}\rangle = 0 \qquad \text{with } |ck_{min}\rangle \neq 0 \tag{3.15}$$

Using Theorem 1, we obtain

$$\langle ck_{max}|\mathbf{L}^-\mathbf{L}^+|ck_{max}\rangle = 0 = c - k_{max}^2 - \hbar k_{max}$$
$$\langle ck_{min}|\mathbf{L}^+\mathbf{L}^-|ck_{min}\rangle = 0 = c - k_{min}^2 + \hbar k_{min} \tag{3.16}$$

Subtracting the two equations, we have

$$(k_{max} + k_{min})(k_{max} - k_{min} + \hbar) = 0 \tag{3.17}$$

The use of the shift operators in Lemma 2 shows that $k_{max} = k_{min} + n\hbar$, where n is an integer. Therefore the second factor in (3.17) cannot be zero, and

$$k_{max} = -k_{min}$$
$$2k_{max} = n\hbar \tag{3.18}$$

It is conventional to write

$$k_{max} = L\hbar \tag{3.19}$$

where $2L$ is an integer. From (3.16) we obtain

$$c = L(L + 1)\hbar^2 \tag{3.20}$$

The variable k is conventionally denoted $M_L\hbar$, so Theorem 1 becomes, in this notation,

$$-L \leq M_L \leq L \tag{3.21}$$

Since $2L$ is an integer, either

$$|M_L| = \tfrac{1}{2}, \tfrac{3}{2}, \tfrac{5}{2}, \ldots, L$$

or

$$|M_L| = 0, 1, 2, 3, \ldots, L \tag{3.22}$$

For ordinary angular momentum, as discussed in Sec. 3.1, the second series holds; this will be proved explicitly in the next section. The first series in (3.22) holds for spin and for a special kind of angular momentum to be discussed in Sec. 3.4.

The preceding discussion may be summarized in a theorem.

Theorem 2 $\mathbf{L}^2|LM_L\rangle = L(L + 1)\hbar^2|LM_L\rangle$ ▶(3.23)

$$\mathbf{L}_z|LM_L\rangle = M_L\hbar|LM_L\rangle \qquad\qquad ▶(3.24)$$

where $|M_L| \leq L$. We have relabeled the kets using L and M_L for clarity; the kets remain the same, of course. The constant L is either an integer or half an integer.

Expressing Lemma 2 in the new notation, we have

$$\mathbf{L}^{\pm}|LM_L\rangle = N^{\pm}|L,M_L \pm 1\rangle$$

Using Theorem 1, we have at once

$$\langle LM_L|\mathbf{L}^{\mp}\mathbf{L}^{\pm}|LM_L\rangle = (N^{\pm})^2 = \hbar^2[L(L+1) - M_L^2 \mp M_L]$$

so

$$\mathbf{L}^{\pm}|LM_L\rangle = \hbar[(L \mp M_L)(L \pm M_L + 1)]^{1/2}|L,M_L \pm 1\rangle \qquad \blacktriangleright (3.25)$$

Equations (3.23), (3.24), and (3.25) hold for any generalized angular momentum satisfying commutation relations of the type (3.5).

It is worthwhile pointing out that (3.21) implies that there are $2L + 1$ kets $|LM_L\rangle$ for a given L. As we shall show in Chap. 4, there are many systems in which the energy depends on L, but not M_L. Such systems therefore have at least $2L + 1$ degeneracy; that is, there are at least $2L + 1$ independent states with the same energy.

PROBLEMS

3.5 Prove that $\overline{\mathbf{L}^{\pm}} = \mathbf{L}^{\mp}$.
Hint: Are $\mathbf{L}_x$ and $\mathbf{L}_y$ self-adjoint?
3.6 Verify Eq. (3.25). Show that indeed $\mathbf{L}^+|LL\rangle = 0$ and $\mathbf{L}^-|L, -L\rangle = 0$.

3.3 EIGENFUNCTIONS FOR ANGULAR MOMENTUM

In most cases, we need only know the eigenvalues of the operators; the specific form of the eigenfunctions (known generally as spherical harmonics) need not concern us. In certain applications, however (to be treated in later chapters), it is of interest to be able to determine the eigenfunctions and sketch their forms. In order to find the eigenfunctions for a single electron, we must set up and solve the equations

$$\begin{aligned}
\mathbf{1}^2|lm_l\rangle &= \hbar^2 l(l+1)|lm_l\rangle \\
\mathbf{1}_z|lm_l\rangle &= \hbar m_l|lm_l\rangle
\end{aligned} \tag{3.26}$$

using (3.7) and (3.8) for $\mathbf{1}_z$ and $\mathbf{1}^2$. The eigenfunctions for $\mathbf{1}_z$ are easy to find. Writing $|lm_l\rangle = Y_{lm} = \Theta_{lm}\Phi_m$, that is, assuming the ket $|lm_l\rangle$ can be written as a product of two functions, we obtain

$$-i\hbar \frac{\partial}{\partial\varphi} \Phi_m = \hbar m_l \Phi_m \tag{3.27}$$

where we have assumed further that the function Θ_{lm} is independent of φ. The solution to (3.27) is

$$\Phi_m(\varphi) = Ne^{im\varphi} \tag{3.28}$$

and if Φ_m is to be a single-valued function of φ, then m must be an integer. In this way we have proved that only the second series in (3.22) is valid for ordinary angular momentum. The constant N is obtained by normalizing the function Φ_m to unity;‡ thus

$$\int_0^{2\pi} \Phi_m^*(\varphi)\Phi_{m'}(\varphi)\,d\varphi = \delta_{mm'} = 2\pi N^2 \qquad \text{for } m = m' \tag{3.29}$$

Substituting for N, we have

$$\Phi_m(\varphi) = \frac{1}{\sqrt{2\pi}}\,e^{im\varphi} \tag{3.30}$$

Finding the eigenfunctions for the first of Eqs. (3.26) is more difficult. Writing $|lm_l\rangle = Y_{lm} = \Theta_{lm}\dfrac{1}{\sqrt{2\pi}}e^{im\varphi}$ and substituting in (3.26), we obtain

$$\frac{1}{\sin\theta}\frac{d}{d\theta}\left(\sin\theta\frac{d\Theta_{lm}}{d\theta}\right) + \left[l(l+1) - \frac{m^2}{\sin^2\theta}\right]\Theta_{lm} = 0 \tag{3.31}$$

The solutions to this differential equation are the normalized associated Legendre polynomials. Their explicit forms are

$$\Theta_{lm} = (-1)^m \sqrt{\frac{(2l+1)}{2}\frac{(l-m)!}{(l+m)!}}\,P_l^m(\cos\theta)$$

$$\text{for } m \text{ positive} \tag{3.32}$$

The unnormalized associated Legendre polynomials P_l^m are defined as

$$P_l^m(x) = \frac{(1-x^2)^{m/2}}{2^l l!}\frac{d^{l+m}}{dx^{l+m}}(x^2-1)^l \tag{3.33}$$

For negative values of m, we take the phase of Θ_{lm} as always positive, so for positive m,

$$\Theta_{lm} = (-1)^m\Theta_{l-m} \qquad m > 0$$

Of course, the choice of phase is arbitrary (see Sec. 2.8). This particular choice has been taken from Condon and Shortley (p. 52) and is extensively used throughout the literature. The choice arises from the way in which the kets are connected by the shift operators. Some authors prefer to take the phase always positive, and one must be careful in comparing results from different sources to see which phase has been chosen.

A list of the normalized functions $Y_{lm}(\theta,\varphi) = \Theta_{lm}(\theta)\Phi_m(\varphi)$ is given in Table 3.1. The related real functions are given as well; the reader will

‡ It is of interest to note that eigenfunctions belonging to different eigenvalues are orthogonal, as anticipated in Chap. 2.

Table 3.1 The Y_{lm} and the corresponding real functions for $l \leq 2$

l	m	Y_{lm}	Real forms	Orbital names
0	0	$\sqrt{\dfrac{1}{4\pi}}$	$\sqrt{\dfrac{1}{4\pi}}$	(s)
1	± 1	$\mp\sqrt{\dfrac{3}{8\pi}}\,\sin\theta\,e^{\pm i\varphi}$	$\sqrt{\dfrac{3}{4\pi}}\,\sin\theta\,\sin\varphi;\ \sqrt{\dfrac{3}{4\pi}}\,\sin\theta\,\cos\varphi$	(p_x, p_y)
1	0	$\sqrt{\dfrac{3}{4\pi}}\,\cos\theta$	$\sqrt{\dfrac{3}{4\pi}}\,\cos\theta$	(p_z)
2	± 2	$\sqrt{\dfrac{15}{32\pi}}\,\sin^2\theta\,e^{\pm 2i\varphi}$	$\sqrt{\dfrac{15}{16\pi}}\,\sin^2\theta\,\sin 2\varphi;$ $\sqrt{\dfrac{15}{16\pi}}\,\sin^2\theta\,\cos 2\varphi$	$(d_{xy}, d_{x^2-y^2})$
2	± 1	$\mp\sqrt{\dfrac{15}{8\pi}}\,\sin\theta\,\cos\theta\,e^{\pm i\varphi}$	$\sqrt{\dfrac{15}{4\pi}}\,\sin\theta\,\cos\theta\,\sin\varphi;$ $\sqrt{\dfrac{15}{4\pi}}\,\sin\theta\,\cos\theta\,\cos\varphi$	(d_{yz}, d_{xz})
2	0	$\sqrt{\dfrac{5}{16\pi}}\,(3\cos^2\theta - 1)$	$\sqrt{\dfrac{5}{16\pi}}\,(3\cos^2\theta - 1)$	(d_{z^2})

note the similarity to the angular part of the hydrogen atom wave functions. Sketches of the real functions (weighted by a radial factor) are given in this connection in Fig. 4.8 (see also the discussion in Sec. 4.6).

PROBLEMS

3.7 Using (3.32) and (3.33), verify the entries in Table 3.1 for $l \leq 1$.

3.8 Show that the normalized linear combinations $1/\sqrt{2}\ (Y_{lm} \pm Y_{l-m})$ are eigenfunctions of $\mathbf{L}^2$. The combinations are either purely real or purely imaginary, and so they can all be made real by multiplication by $\sqrt{-1}$ where necessary.

3.4 ADDITION OF ANGULAR MOMENTA

In 3.1, we briefly considered the angular momentum of a system of particles and showed that the total angular-momentum vector

$$\mathbf{L} = \sum_{i=1}^{n} \mathbf{l}_i \tag{3.34}$$

also satisfied the commutation relations. Therefore $\mathbf{L}$ is also a proper generalized angular momentum, according to the discussion of Sec. 3.2.

We now wish to find the allowed values of L and M_L for such a system of particles and to relate them to the possible values of the l_i and m_{l_i}. For simplicity, let us consider a two-particle system, with eigenvalues l_1, l_2, m_{l_1}, and m_{l_2}. Offhand, we cannot say what the value of L is for this system since the angular momentum vectors for the two particles need not be parallel. They could have equal magnitude and opposite direction, for example, in which case the total angular momentum would be zero. However, we do know the components of both vectors in the z direction, that is, m_{l_1} and m_{l_2}. If both m_{l_1} and m_{l_2} have their maximum values l_1 and l_2, we can be sure that the two vectors are parallel‡ with $M_L = m_{l_1} + m_{l_2} = L$. One might also suppose that the vectors could add in other ways, and we seek restrictions on the relative orientations of $\mathbf{l}_1$ and $\mathbf{l}_2$ in order to find the possible values of L. Clearly, L must be integral since $M_L = m_{l_1} + m_{l_2}$ is integral. Furthermore, since a ket with $L = M_L$ implies the existence of $2L$ other kets with M_L ranging from $L - 1$ to $-L$, the maximum value of L is $l_1 + l_2$.

We might at first suppose that L can have all values ranging from $l_1 + l_2$ to 0. In general, this is not possible unless $l_1 = l_2$, since so many values of L imply

$$\sum_{L=0}^{l_1+l_2} (2L + 1) = (l_1 + l_2 + 1)^2$$

independent kets, whereas in fact there are only $(2l_1 + 1)(2l_2 + 1)$ kets, one for each possible pair of values m_{l_1}, m_{l_2}. Equivalently, each possible value of L implies the existence of a ket with $M_L = 0$, and there would be $l_1 + l_2 + 1$ such kets if all values of L were allowed. However, $M_L = 0$ implies $m_{l_1} = -m_{l_2}$, and there are only $(2l_< + 1)$ ways that this can be achieved, where $l_<$ is the lesser of l_1 and l_2. A detailed analysis of this type of restriction shows that the minimum value of L is $|l_1 - l_2|$ and that all values of L between $|l_1 - l_2|$ and $l_1 + l_2$ are permitted.

To be specific, consider the case $l_1 = 2$, $l_2 = 1$. There are 15 pairs of values m_{l_1}, m_{l_2}. The case $L = 3$ uses up only seven of them. The two vectors can add to give $L = 2$, which uses up five more, and finally the two vectors can add to give $L = 1$, which requires the last three kets. Table 3.2 shows that the $M_L = m_{l_1} + m_{l_2}$ values are consistent with this scheme; there is only one way of achieving $M_L = 3$ (that is, with $L = 3$), two ways of accomplishing $M_L = 2$ ($L = 3$, $L = 2$), three ways for $M_L = 1$ ($L = 3$, $L = 2$, and $L = 1$), and so forth. Thus the possible values for the vector addition of $\mathbf{l}_1$ and $\mathbf{l}_2$ can be determined solely from an examination of the z components $\mathbf{l}_{z_1}$ and $\mathbf{l}_{z_2}$.

‡ The two vectors are not actually *parallel;* they only have a maximal z component. The actual length of each vector is $\sqrt{l(l + 1)}$, so although the z component is at its maximum, the vector does not lie along the z axis.

Table 3.2 Total values of M_L for a two-particle system with $l_1 = 2$ and $l_2 = 1$

		$l_2 = 1$		
		$m_{l_2} = 1$	0	-1
	$m_{l_1} = 2$	3	2	1
	1	2	1	0
$l_1 = 2$	0	1	0	-1
	-1	0	-1	-2
	-2	-1	-2	-3

The above discussion suggests two different ways of forming eigenkets for a two-particle system. Either we may write the set of commuting operators as $l_1{}^2$, $l_2{}^2$, l_{z_1}, and l_{z_2}, or we may take the set $l_1{}^2$, $l_2{}^2$, $\mathbf{L}^2$, and $\mathbf{L}_z$. It is easily shown that $l_1{}^2$ and $\mathbf{L}^2$, for example, commute; we must also prove that the two sets are equivalent, i.e., that they span the same vector space.

Consider the kets $|l_1 l_2 m_{l_1} m_{l_2}\rangle$, defined so that

$$l_1{}^2 |l_1 l_2 m_{l_1} m_{l_2}\rangle = \hbar^2 l_1 (l_1 + 1) |l_1 l_2 m_{l_1} m_{l_2}\rangle$$
$$l_{z_1} |l_1 l_2 m_{l_1} m_{l_2}\rangle = \hbar m_{l_1} |l_1 l_2 m_{l_1} m_{l_2}\rangle \tag{3.35}$$

and so forth. These kets are clearly eigenkets of $\mathbf{L}_z$, with eigenvalues $\hbar(m_{l_1} + m_{l_2})$; thus

$$\begin{aligned} \mathbf{L}_z |l_1 l_2 m_{l_1} m_{l_2}\rangle &= (1_{z_1} + 1_{z_2}) |l_1 l_2 m_{l_1} m_{l_2}\rangle \\ &= \hbar(m_{l_1} + m_{l_2}) |l_1 l_2 m_{l_1} m_{l_2}\rangle \\ &= \hbar M_L |l_1 l_2 m_{l_1} m_{l_2}\rangle \end{aligned} \tag{3.36}$$

Taking the ket with $m_{l_1} = l_1$ and $m_{l_2} = l_2$ and operating with $\mathbf{L}^+ = 1_1^+ + 1_2^+$, we have

$$\mathbf{L}^+ |l_1 l_2 l_1 l_2\rangle = 0 \tag{3.37}$$

since the values of m_{l_1} and m_{l_2} are at a maximum. Operating on the left with $\mathbf{L}^-$ and using Lemma 3, we obtain

$$(\mathbf{L}^2 - \mathbf{L}_z{}^2 - \hbar \mathbf{L}_z) |l_1 l_2 l_1 l_2\rangle = 0$$
$$[\mathbf{L}^2 - \hbar^2 (l_1 + l_2)^2 - \hbar^2 (l_1 + l_2)] |l_1 l_2 l_1 l_2\rangle = 0 \tag{3.38}$$

which implies

$$\begin{aligned} \mathbf{L}^2 |l_1 l_2 l_1 l_2\rangle &= \hbar^2 (l_1 + l_2)(l_1 + l_2 + 1) |l_1 l_2 l_1 l_2\rangle \\ &= \hbar^2 L(L + 1) |l_1 l_2 l_1 l_2\rangle \end{aligned} \tag{3.39}$$

so $|l_1l_2l_1l_2\rangle$ is an eigenket of $\mathbf{L}^2$ and $\mathbf{L}_z$ as well as of $\mathbf{l}_1{}^2$, $\mathbf{l}_2{}^2$, $\mathbf{l}_{z_1}$, and $\mathbf{l}_{z_2}$. If we apply $\mathbf{L}^-$ to this ket, we obtain

$$
\begin{aligned}
\mathbf{L}^-|l_1l_2l_1l_2\rangle &= (\mathbf{l}_1^- + \mathbf{l}_2^-)|l_1l_2l_1l_2\rangle \\
&= \mathbf{L}^-|l_1l_2LL\rangle = \hbar\sqrt{2}\,(\sqrt{l_1}\,|l_1l_2,l_1-1,l_2\rangle + \sqrt{l_2}\,|l_1l_2l_1,l_2-1\rangle) \\
&= \hbar\sqrt{2L}\,|l_1l_2L,L-1\rangle
\end{aligned}
\tag{3.40}
$$

so

$$
|l_1l_2L,L-1\rangle = \sqrt{\frac{l_1}{L}}\,|l_1l_2,l_1-1,l_2\rangle + \sqrt{\frac{l_2}{L}}\,|l_1l_2l_1,l_2-1\rangle
\tag{3.41}
$$

In this way we can generate a new set of kets $|l_1l_2LM_L\rangle$ with $L = l_1 + l_2$ and $M_L \leq L$, and we can find the explicit relation between kets in this set of commuting observables and in the old set $|l_1l_2m_{l_1}m_{l_2}\rangle$. The reader should carefully note the use of the notation $|l_1l_2LM_L\rangle$ for the simultaneous eigenkets of the four mutually commuting operators $\mathbf{l}_1{}^2$, $\mathbf{l}_2{}^2$, $\mathbf{L}^2$ and $\mathbf{L}_z$.

As mentioned above, the value $L = l_1 + l_2$ is not the only way the two vectors can add. In fact, all possible values of L from $l_1 + l_2$ to $|l_1 - l_2|$ can occur. The proof is rather complicated, however, and so it will not be given here (see, for example, Griffith, 1961, p. 15). One may find the linear combinations of kets for $L \neq l_1 + l_2$ by repeated application of the shift operator $\mathbf{L}^-$ and by the orthogonality of all the kets in any given scheme. For example, from (3.41) we obtain

$$
|l_1l_2,L-1,L-1\rangle = -\sqrt{\frac{l_2}{L}}\,|l_1l_2,l_1-1,l_2\rangle + \sqrt{\frac{l_1}{L}}\,|l_1l_2l_1,l_2-1\rangle
\tag{3.42}
$$

since this is the only ket (aside from a phase factor) with $M_L = l_1 + l_2 - 1$ which is orthogonal to $|l_1l_2L, L-1\rangle$. In this way all kets in the new scheme may be generated. The number of linearly independent kets is the same in both schemes, of course; in fact, the equation

$$
\sum_{L=|l_1-l_2|}^{l_1+l_2} (2L+1) = (2l_1+1)(2l_2+1)
$$

is an identity.

In systems with more than two particles, the addition of angular momenta is an even more complicated business. Fortunately, however, in the systems of interest, the values of l_1, l_2, . . . will be small, usually 2 or less, since we are not ordinarily interested in electrons outside the d shell of atoms (see Chap. 6 for further discussion).

The addition formula we have developed in this section may be summarized in the equation

$$|l_1 l_2 L M_L\rangle = \sum_{m_{l_1}, m_{l_2}} \langle l_1 l_2 m_{l_1} m_{l_2} | l_1 l_2 L M_L\rangle | l_1 l_2 m_{l_1} m_{l_2}\rangle \qquad \blacktriangleright (3.43)$$

The coefficients $\langle l_1 l_2 m_{l_1} m_{l_2} | l_1 l_2 L M_L\rangle$, which are commonly abbreviated as $\langle m_{l_1} m_{l_2} | L M_L\rangle$, are known as the Wigner coefficients, and (3.43) is referred to as Wigner's formula. Tables of Wigner coefficients are available from many sources; for example, see Condon and Shortley, p. 76. In elementary situations, the coefficients may be obtained solely from the shift operators and the orthogonality relations.

Equation (3.43) and the methods we have developed in this section are not restricted to the addition of the angular momenta of separate particles. Our proofs depended only on the fact that $l_1{}^2$ and $l_2{}^2$ were commuting generalized angular momenta. As will be shown in the next section, l_1 and l_2 may refer to two different kinds of angular momentum on the same particle, classical angular momentum and spin. The resulting vector is then called **j**.

Equation (3.43) contains an arbitrary phase choice. If we take the phase of the ket with highest L and M_L as positive

$$(|l_1 l_2 L M_L\rangle = |l_1 l_2 m_{l_1} m_{l_2}\rangle \quad \text{when } m_{l_1} = l_1,\ m_{l_2} = l_2,\ \text{and } M_L = L)$$

then the shift operators will take care of the phases of the kets with this L but lower M_L. However, in constructing the kets with next highest L by orthogonality, there is a choice to be made. For example, we have made such a choice in (3.42). It is conventional to take the phase of the Wigner coefficient $\langle m_{l_1} m_{l_2} | L L\rangle$ as $(-1)^{l_1 - m_{l_1}}$. This indirectly determines the phases of all coefficients.

PROBLEMS

3.9 We have argued in this section that the kets $|l_1 l_2 m_{l_1} m_{l_2}\rangle$ and the kets $|l_1 l_2 L M_L\rangle$ span the same vector space. Assuming that they do, verify (3.43).

3.10 For a system with $l_1 = 2$ and $l_2 = 1$, find the Wigner coefficients by using the shift operators, the orthogonality relations, and the Wigner phase convention.

3.5 SPIN

Every elementary particle possesses a quantity known as spin. If the particles are in fact point masses they cannot, of course, literally "spin"; the word comes from the experimentally observed fact that the elementary particles possess an extra magnetic moment which cannot be explained on the basis of ordinary angular momentum. This magnetic moment is easily detected, either spectroscopically, as in the anomolous Zeeman effect, or by direct magnetic measurement, as in the Stern-Gerlach experiment outlined in Chap. 2.

The existence of spin cannot be predicted on the basis of the quantum mechanics which we have developed thus far. However, it is possible to develop a more rigorous form of quantum mechanics which is relativistically correct; this leads to the Dirac equation instead of the Schrödinger equation, and it can be shown (Dirac, p. 263, or Griffith, 1961, p. 116) that the Dirac equation predicts the existence of electronic spin.‡

Spin is thus a consequence of the marriage of relativity theory and quantum mechanics, but since we treat only nonrelativistic theory in this book, we will introduce spin from a phenomenological point of view. The method is consistent with the Dirac equation.

Assume, therefore, that every electron possesses an intrinsic angular momentum called spin and denoted s. Assume further that the vector s and its components obey the commutation relations (3.5) and that the eigenvalue of s^2 is $\hbar^2 s(s+1) = \frac{3}{4}\hbar^2$. Since $s = \frac{1}{2}$ the only allowed eigenvalues of s_z are $\frac{1}{2}\hbar$ and $-\frac{1}{2}\hbar$, according to (3.22). Thus $m_s = \pm\frac{1}{2}$. We finally assume that s commutes with all the r and p vectors, and hence with all the l. Following the notation of the previous section, we have, from (3.25),

$$
\begin{aligned}
s^2|\tfrac{1}{2}\ \tfrac{1}{2}\rangle &= \tfrac{3}{4}\hbar^2|\tfrac{1}{2}\ \tfrac{1}{2}\rangle & s^2|\tfrac{1}{2}\ -\tfrac{1}{2}\rangle &= \tfrac{3}{4}\hbar^2|\tfrac{1}{2}\ -\tfrac{1}{2}\rangle \\
s_z|\tfrac{1}{2}\ \tfrac{1}{2}\rangle &= \tfrac{1}{2}\hbar|\tfrac{1}{2}\ \tfrac{1}{2}\rangle & s_z|\tfrac{1}{2}\ -\tfrac{1}{2}\rangle &= -\tfrac{1}{2}\hbar|\tfrac{1}{2}\ -\tfrac{1}{2}\rangle \\
s^+|\tfrac{1}{2}\ \tfrac{1}{2}\rangle &= 0 & s^+|\tfrac{1}{2}\ -\tfrac{1}{2}\rangle &= \hbar|\tfrac{1}{2}\ \tfrac{1}{2}\rangle \\
s^-|\tfrac{1}{2}\ \tfrac{1}{2}\rangle &= \hbar|\tfrac{1}{2}\ -\tfrac{1}{2}\rangle & s^-|\tfrac{1}{2}\ -\tfrac{1}{2}\rangle &= 0
\end{aligned}
\qquad \blacktriangleright (3.44)
$$

The kets $|\tfrac{1}{2}\ \tfrac{1}{2}\rangle$ and $|\tfrac{1}{2}\ -\tfrac{1}{2}\rangle$ are often abbreviated $|\alpha\rangle$ and $|\beta\rangle$, respectively, since the reminder $s = \frac{1}{2}$ is superfluous. Since s commutes with l, and in many cases, with $\mathcal{H}$, we may add it to our set of commuting operators, forming the set (for a single particle) $\mathcal{H}$, l^2, l_z, s^2, and s_z.

For a system of n particles, we may also define the quantity

$$
S = \sum_{i=1}^{n} s_i \tag{3.45}
$$

and clearly, from the discussion of the previous sections, S is also an angular momentum. The results obtained in Sec. 3.3 for the total momentum of two particles apply to S as well as to L, so in a two-particle system we can have either $S = 1$ or $S = 0$. Explicitly, the eigenkets are, in the

‡ It also predicts the existence of the positron.

$|s_1s_2SM_S\rangle$ notation scheme (but leaving off the redundant values of s_1 and s_2),

$$
\begin{aligned}
|1\ 1\rangle &= |\alpha_1\alpha_2\rangle \\
|1\ 0\rangle &= \sqrt{\tfrac{1}{2}}\,|\alpha_1\beta_2\rangle + \sqrt{\tfrac{1}{2}}\,|\beta_1\alpha_2\rangle \\
|1\ -1\rangle &= |\beta_1\beta_2\rangle \\
|0\ 0\rangle &= \sqrt{\tfrac{1}{2}}\,|\alpha_1\beta_2\rangle - \sqrt{\tfrac{1}{2}}\,|\beta_1\alpha_2\rangle
\end{aligned}
\tag{3.46}
$$

where the kets on the right-hand side are in the scheme $|s_1s_2m_{s_1}m_{s_2}\rangle$, abbreviated as noted above.

The *total*‡ angular momentum of a particle is defined as

$$
\mathbf{j} = \mathbf{l} + \mathbf{s} \qquad\qquad \blacktriangleright(3.47)
$$

and as

$$
\mathbf{J} = \mathbf{L} + \mathbf{S} = \sum_{i=1}^{n} \mathbf{j}_i \qquad\qquad \blacktriangleright(3.48)
$$

for a system of n particles. Of course, the quantities $\mathbf{J}$ and $\mathbf{j}$ are proper generalized angular momenta. We can use the equations concerning the addition of angular momenta to discuss the sum $\mathbf{J} = \mathbf{L} + \mathbf{S}$ as well, and it turns out to be possible to define the kets for a single particle in terms of $\mathbf{l}^2$, $\mathbf{s}^2$, $\mathbf{j}^2$, and $\mathbf{j}_z$ instead of $\mathbf{l}^2$, $\mathbf{s}^2$, $\mathbf{l}_z$, and $\mathbf{s}_z$. The generalization for two and more particles will be developed in Chap. 6.

PROBLEM

3.11 The angular momentum of a single (p) electron with $l = 1$ can be classified either according to the scheme $|lsm_lm_s\rangle$ or the scheme $|lsjm_j\rangle$. Show that

$$
\begin{aligned}
|1\ \tfrac{1}{2}\ \tfrac{3}{2}\ \tfrac{3}{2}\rangle &= |1\ \tfrac{1}{2}\ 1\ \tfrac{1}{2}\rangle \\
|1\ \tfrac{1}{2}\ \tfrac{3}{2}\ \tfrac{1}{2}\rangle &= \sqrt{\tfrac{2}{3}}\,|1\ \tfrac{1}{2}\ 0\ \tfrac{1}{2}\rangle + \sqrt{\tfrac{1}{3}}\,|1\ \tfrac{1}{2}\ 1\ -\tfrac{1}{2}\rangle \\
|1\ \tfrac{1}{2}\ \tfrac{3}{2}\ -\tfrac{1}{2}\rangle &= \sqrt{\tfrac{2}{3}}\,|1\ \tfrac{1}{2}\ 0\ -\tfrac{1}{2}\rangle + \sqrt{\tfrac{1}{3}}\,|1\ \tfrac{1}{2}\ -1\ \tfrac{1}{2}\rangle \\
|1\ \tfrac{1}{2}\ \tfrac{3}{2}\ -\tfrac{3}{2}\rangle &= |1\ \tfrac{1}{2}\ -1\ -\tfrac{1}{2}\rangle \\
|1\ \tfrac{1}{2}\ \tfrac{1}{2}\ \tfrac{1}{2}\rangle &= \sqrt{\tfrac{2}{3}}\,|1\ \tfrac{1}{2}\ 1\ -\tfrac{1}{2}\rangle - \sqrt{\tfrac{1}{3}}\,|1\ \tfrac{1}{2}\ 0\ \tfrac{1}{2}\rangle \\
|1\ \tfrac{1}{2}\ \tfrac{1}{2}\ -\tfrac{1}{2}\rangle &= \sqrt{\tfrac{1}{3}}\,|1\ \tfrac{1}{2}\ 0\ -\tfrac{1}{2}\rangle - \sqrt{\tfrac{2}{3}}\,|1\ \tfrac{1}{2}\ -1\ \tfrac{1}{2}\rangle
\end{aligned}
$$

where the $|lsjm_j\rangle$ scheme is written on the left-hand side and the $|lsm_lm_s\rangle$ scheme on the right.

Hint: Operate on both sides with shift operator $\mathbf{j}^-$, and use (3.25) and (3.44). According to the discussion of the last section, it is conventional to take the phase of the Wigner coefficients $\langle m_lm_s|jj\rangle$ as $(-1)^{l-m_l}$.

‡ The quantity $\mathbf{l}$ is usually called the *angular* momentum or more frequently the *orbital* angular momentum.

3.6 ANTISYMMETRY AND THE PAULI EXCLUSION PRINCIPLE

Pauli has proposed that for a system of identical particles of half-integral spin, the ket describing any state must reverse sign upon permutation of the coordinates of any two particles. This principle seems to be connected with the physical indistinguishability of identical particles. According to the discussion in Sec. 1.3, if no experiment can be performed which will distinguish two particles, then our equations must not imply such distinguishability. Thus a ket describing a system of n particles must not contain explicit labels, and questions concerning the behavior of a particular particle are beyond the scope of both theory and experiment. If P_{ij} is the permutation which interchanges the coordinates of two electrons, one described by the spatial and spin coordinates $\mathbf{r}_i$ and ξ_i, the other described by $\mathbf{r}_j$ and ξ_j, then

$$P_{ij}|\Psi\rangle = -|\Psi\rangle \qquad \blacktriangleright(3.49)$$

according to the Pauli principle.‡ Note that in keeping with the indistinguishability principle, we can reverse either the coordinates or the particles themselves provided $\mathbf{r}_i$ and ξ_i describe *some* particle, not a specific labeled one.

Throughout the latter part of this book we shall be assuming that the total wavefunction for a system of n particles can be written as a product of one-particle functions. This assumption is by no means necessarily true, but it forms the basis of both atomic and molecular orbital theory—an orbital is specifically defined as a one-electron function. Let us then write

$$|\Psi\rangle = |u_1 u_2 \cdots u_n\rangle \qquad (3.50)$$

where $u_1, u_2, \ldots, u_n$ are one-electron functions, all different from each other. We require

$$P_{ij}|u_1 u_2 \cdots u_n\rangle = -|u_1 u_2 \cdots u_n\rangle \qquad (3.51)$$

but the meaning of P_{ij} on $|\Psi\rangle$ is not entirely clear if $|\Psi\rangle$ contains no labels for the particles. The requirements of the Pauli principle and the principle of indistinguishability seem to conflict. We avoid this difficulty by labeling the electrons but at the same time preserving their indistinguishability. We write $|u_1(1)u_2(2) \cdots u_n(n)\rangle$ to indicate that electron 1 has coordinates $\mathbf{r}_1$, ξ_1; electron 2 has coordinates $\mathbf{r}_2$, ξ_2; and so forth. The ket $|u_1(2)u_2(1) \cdots u_n(n)\rangle$ similarly indicates that electron 1 has coordi-

---

‡ Particles which obey the Pauli exclusion principle are called fermions. There exists a second class of particles called bosons which satisfy the requirement

$$P_{ij}|\Psi\rangle = |\Psi\rangle$$

Such particles have integral spin; photons are a familiar example.

nates u_2, etc. Of course, both kets violate the principle of indistinguishability, but the linear combination

$$|\Psi'\rangle = \sum_P (-1)^P P |u_1(1)u_2(2) \cdots u_n(n)\rangle \qquad (3.52)$$

is suitable. The sum extends over all possible distinct permutations of the electrons; the symbolic multiplier $(-1)^P$ is positive when an *even* number of two-electron permutations are required to bring the electrons back into some standard order, and negative when an *odd* number of permutations are required. The many-electron permutation is accordingly called even or odd. We illustrate this form for $|\Psi'\rangle$ with the three-electron system as follows:

$$\begin{aligned}|\Psi'\rangle = {} &|u_1(1)u_2(2)u_3(3)\rangle - |u_1(2)u_2(1)u_3(3)\rangle \\ &+ |u_1(2)u_2(3)u_3(1)\rangle - |u_1(3)u_2(2)u_3(1)\rangle \\ &+ |u_1(3)u_2(1)u_3(2)\rangle - |u_1(1)u_2(3)u_3(2)\rangle \quad (3.53)\end{aligned}$$

Reversing the labels of any two electrons in $|\Psi'\rangle$ clearly causes the ket to change sign; such a ket is called an antisymmetric function of the electrons [the corresponding symmetric function would be obtained by requiring $(-1)^P$ to be always positive]. $|\Psi'\rangle$ also preserves the indistinguishability of the electrons since all possible forms of labeling are represented.

In general, there are $n!$ terms in the sum, which implies that the form (3.52) is not properly normalized. Thus, in the two-electron case,

$$\begin{aligned}\langle\Psi'|\Psi'\rangle = {} &\langle u_1(1)u_2(2)|u_1(1)u_2(2)\rangle - 2\langle u_1(1)u_2(2)|u_1(2)u_2(1)\rangle \\ &+ \langle u_1(2)u_2(1)|u_1(2)u_2(1)\rangle \quad (3.54)\end{aligned}$$

The first and last terms are equal to unity; the second term is zero, provided the one-electron orbitals are chosen to be orthogonal. Thus the normalization factor is, in general, $\sqrt{1/n!}$, and the ket

$$|\Psi\rangle = \sqrt{\frac{1}{n!}} \sum_P (-1)^P P |u_1(1)u_2(2) \cdots u_n(n)\rangle \qquad \blacktriangleright (3.55)$$

is the desired antisymmetric normalized many-electron function. For simplicity, we often abbreviate

$$|\Psi\rangle = \mathcal{Q} |u_1(1)u_2(2) \cdots u_n(n)\rangle \qquad (3.56)$$

The operator $\mathcal{Q}$ is known as the antisymmetrizer.

Another way of writing our many-electron function is as a determinant. The normalized determinant

$$|\Psi\rangle = \sqrt{\frac{1}{n!}} \begin{vmatrix} u_1(1) & u_2(1) & \cdots & u_n(1) \\ u_1(2) & u_2(2) & \cdots & u_n(2) \\ \cdot & \cdot & \cdots & \cdot \\ u_1(n) & u_2(n) & \cdots & u_n(n) \end{vmatrix} \qquad (3.57)$$

is in fact mathematically defined to be equal to (3.55). This way of writing the many-electron function was developed by J. C. Slater, and (3.57) is known as a Slater determinant. It is a somewhat cumbersome notation, but it can be abbreviated to

$$|\Psi\rangle = \sqrt{\frac{1}{n!}} \, |u_1(1)u_2(2) \, \cdots \, u_n(n)| \tag{3.58}$$

where the parallel lines indicate the formation of a determinant. Any of the four equivalent forms (3.55), (3.56), (3.57), or (3.58) may be used to represent a normalized antisymmetric function of n electrons, and all will be found in the literature. Occasionally the normalizing factor is not explicitly included, being understood as part of the definition. For that matter, the antisymmetrizer itself is often omitted, and all kets used in this text are implicitly antisymmetrized unless indicated otherwise.

The term *exclusion principle* arises from an alternative way of stating Pauli's theorem. The older form states that no two identical fermions can have the same spatial and spin coordinates. This form appears easily as a special case of the more general antisymmetry requirement; for example, using the Slater determinantal form for $|\Psi\rangle$, we see that if two electrons have the same spatial and spin coordinates, i.e., occupy the same orbital, then the determinant has two identical columns. Any determinant with identical columns or rows is zero, and so the many-electron function vanishes if any two electrons have the same coordinates.

The Pauli exclusion principle has an important effect on the energies of many-electron systems; we illustrate this with the two-electron system. We take

$$|\Psi\rangle = \frac{1}{\sqrt{2}} \, [u_1(1)u_2(2) \, - \, u_1(2)u_2(1)] = \frac{1}{\sqrt{2}} \, (u_1u_2 \, - \, u_2u_1) \tag{3.59}$$

where u_1 and u_2 are orthogonal orbitals; the arguments are assumed to be in the standard order when they are omitted. The orbitals specify both the spatial and spin coordinates of the electron which they describe, and so we can write, for example,

$$u_1 = \phi_1\xi_1 \tag{3.60}$$

where ϕ_1 describes only the spatial coordinates, and ξ_1 only the spin coordinate. ξ_1 can take only the two values α and β, so there are only four possible spin combinations, $\alpha(1)\alpha(2)$, $\alpha(1)\beta(2)$, $\beta(1)\alpha(2)$, and $\beta(1)\beta(2)$. Note that subscripts on α and β are not required since we only need to distinguish α from β. It is therefore conventional to replace the *arguments* of α and β by *subscripts*, which may cause some momentary confusion. We write‡ $\alpha_1 = \alpha(1)$, and so forth, and obtain the four different

---

‡ See also (3.46), in which the subscripts arose in a more natural way.

antisymmetric functions

$$|\Psi_1\rangle = \frac{1}{\sqrt{2}}(\phi_1\phi_2 - \phi_2\phi_1)\alpha_1\alpha_2$$

$$|\Psi_2\rangle = \frac{1}{\sqrt{2}}(\phi_1\phi_2\alpha_1\beta_2 - \phi_2\phi_1\beta_1\alpha_2)$$

$$|\Psi_3\rangle = \frac{1}{\sqrt{2}}(\phi_1\phi_2\beta_1\alpha_2 - \phi_2\phi_1\alpha_1\beta_2)$$

$$|\Psi_4\rangle = \frac{1}{\sqrt{2}}(\phi_1\phi_2 - \phi_2\phi_1)\beta_1\beta_2$$

(3.61)

All four functions are eigenfunctions of $\mathbf{S}_z = \mathbf{s}_{z_1} + \mathbf{s}_{z_2}$, with eigenvalues $\hbar$, 0, 0, and $-\hbar$, respectively. The first and last functions are eigenfunctions of $\mathbf{S}^2$ as well, with $S = 1$, since $\mathbf{S}^+|\Psi_1\rangle = \mathbf{S}^-|\Psi_4\rangle = 0$. According to our discussion of the addition of angular momenta, we should also have a function with $S = 0$, as well as a function with $S = 1$ and $M_S = 0$. The latter function may be generated by acting on $|\Psi_1\rangle$ with $\mathbf{S}^-$; we obtain

$$\begin{aligned}
\mathbf{S}^-|\Psi_1\rangle = \mathbf{S}^-|1\ 1\rangle &= \hbar\sqrt{2}|1\ 0\rangle \\
&= \mathbf{s}_1^-|\Psi_1\rangle + \mathbf{s}_2^-|\Psi_1\rangle \\
&= \hbar|\Psi_3\rangle + \hbar|\Psi_2\rangle
\end{aligned}$$

(3.62)

so

$$\begin{aligned}
|1\ 0\rangle &= \sqrt{\frac{1}{2}}\,[\phi_1(1)\phi_2(2) - \phi_1(2)\phi_2(1)](\alpha_1\beta_2 + \beta_1\alpha_2) \\
&= \frac{1}{\sqrt{2}}(|\Psi_2\rangle + |\Psi_3\rangle)
\end{aligned}$$

(3.63)

The function $|0\ 0\rangle$ is orthogonal to $|1\ 0\rangle$, and thus is given as

$$\begin{aligned}
|0\ 0\rangle &= \sqrt{\tfrac{1}{2}}\,[\phi_1(1)\phi_2(2) + \phi_1(2)\phi_2(1)](\alpha_1\beta_2 - \beta_1\alpha_2) \\
&= \frac{1}{\sqrt{2}}(|\Psi_2\rangle - |\Psi_3\rangle)
\end{aligned}$$

(3.64)

where we use the $|SM_S\rangle$ notation for the eigenfunctions of $\mathbf{S}^2$ and $\mathbf{S}_z$. It should be noted that the last line of (3.62) follows from the definitions

$$\mathbf{s}_1^-|\alpha_1\alpha_2\rangle = \hbar|\beta_1\alpha_2\rangle \qquad \mathbf{s}_2^-|\alpha_1\alpha_2\rangle = \hbar|\alpha_1\beta_2\rangle$$

(3.65)

The functions $|1\ 1\rangle = |\Psi_1\rangle$, $|1\ 0\rangle$, and $|1\ -1\rangle = |\Psi_4\rangle$ all have the same spatial part; the function $|0\ 0\rangle$ has a completely different spatial part. In fact, the spatial part in the first three cases is antisymmetric, whereas the spin functions are symmetric; in the case of the $|0\ 0\rangle$ function, the spin function is antisymmetric, whereas the space part is symmetric.

Naturally, in all instances the total function remains antisymmetric and obeys the Pauli principle. As it turns out, the Hamiltonians which we use are usually independent of the spin, and the energy of a function depends only on its spatial part. Nonetheless, spin does have an important effect on the energy since the symmetry behavior of the spatial part is determined by the value of S. As we shall show in Chap. 6, the three functions with $S = 1$, which comprise what is called a triplet, have the same energy; the function with $S = 0$ has an entirely different (and generally higher) energy.

Two points should be stressed. The Pauli principle applies to any many-electron function, whether it is composed of orbitals or not. The orbital form has been introduced only for convenience. Secondly, when using the orbital form, the orbitals u_i must always be chosen to be orthogonal. In fact, it is not necessary that the spatial parts be themselves orthogonal, provided that the spin parts are orthogonal. For example, the functions

$$u_1 = \phi_1\alpha \qquad u_2 = \phi_2\beta \qquad \phi_1 = \phi_2 \tag{3.66}$$

are acceptable. The orbitals u_i are often called spin orbitals, since they include a specification of the spin coordinate. Some authors write a bar over the orbital to indicate β spin and leave the orbital unmarked to indicate α spin; for example,

$$\phi_1 = \phi_1\alpha \qquad \bar{\phi}_1 = \phi_1\beta \tag{3.67}$$

Another notation common in the literature is

$$\phi_1^+ = \phi_1\alpha \qquad \phi_1^- = \phi_1\beta \tag{3.68}$$

This notation, especially that in (3.67), can cause some confusion, so one must be sure whether a given orbital contains a spin specification.

For compactness in notation, we do not always use different symbols for spatial orbitals and spin orbitals in the chapters which follow. Thus, we may often use the single symbol φ (instead of u) to represent a spin orbital. Furthermore, we may use the same symbol to abbreviate an integral over spatial orbitals and an integral over spin orbitals; this is especially true in Chap. 6 in which K represents in one case an exchange integral over spatial orbitals and in another case an exchange integral over spin orbitals. This allows for the possibility of some confusion in nomenclature, but I have not been able to discover a suitable alternative which completely avoids all difficulties. Instead I have tried to point out carefully those instances in which confusion can occur, and if the reader bears in mind the *possibility* of ambiguity he should encounter no difficulties.

PROBLEMS

3.12 For a two-electron system, show that $S^2 = {s_1}^2 + {s_2}^2 + s_1^+ s_2^- + s_1^- s_2^+ + 2s_{z_1} s_{z_2}$.

3.13 Using the equation developed in the problem above, operate on the ket $|0\ 0\rangle$ as given by (3.64), and show that $S = 0$.

3.14 Consider an n-electron system with spin coordinates $\alpha_1 \alpha_2 \alpha_3 \cdots \alpha_n$, described by the ket $|\Psi\rangle$. Show that $S^2|\Psi\rangle = \hbar^2 S(S + 1)|\Psi\rangle$, with $S = n/2$, and that $S_z|\Psi\rangle = \hbar M_S|\Psi\rangle$, $M_S = S$.

3.15 Show that $|s_x = \frac{1}{2}\hbar\rangle = \sqrt{1/2}(|s_z = \frac{1}{2}\hbar\rangle + |s_z = -\frac{1}{2}\hbar\rangle)$

SUGGESTIONS FOR FURTHER READING

1. Griffith, J. S.: "The Theory of Transition Metal Ions," chap. 2, Cambridge University Press, London, 1961.

2. Dirac, P. A. M.: "Quantum Mechanics," 4th ed., secs. 35, 36, and 37, Oxford University Press, London, 1958.

3. Kauzmann, W.: "Quantum Chemistry," chaps. 8 and 9D, Academic Press Inc., New York, 1957.

4. Condon, E. U., and G. H. Shortley: "The Theory of Atomic Spectra," chap. III, Cambridge University Press, London, 1935.

5. Pauling, L., and E. B. Wilson: "Introduction to Quantum Mechanics," chap. VIII, McGraw-Hill Book Company, New York, 1935.

4

ELEMENTARY APPLICATIONS
OF THE
SCHRÖDINGER EQUATION

Had I known that we were not
going to get rid of the damned
quantum jumping, I would never
have involved myself in this
business.
Erwin Schrödinger

4.1 THE PARTICLE IN A BOX

Some of the most instructive examples in quantum mechanics are the box problems. As noted in Sec. 2.8, the various Hamiltonians we use differ only in the form of the potential term V. The box problems have an especially simple form, with the potential term always a constant. In the most sophisticated problems, V may have several different constant values in various regions of space; in the simplest boxes, the potential term is zero inside and infinite outside.

Let us consider the most elementary case, that of a particle in a one-dimensional box. We define $V = 0$ inside the box, that is, from $x = 0$ to $x = a$, and take V infinite elsewhere. The form of V is sketched in Fig. 4.1. The Hamiltonian is given by

$$\mathcal{3C} = -\frac{\hbar^2}{2m}\frac{d^2}{dx^2} + V \qquad\qquad \blacktriangleright (4.1)$$

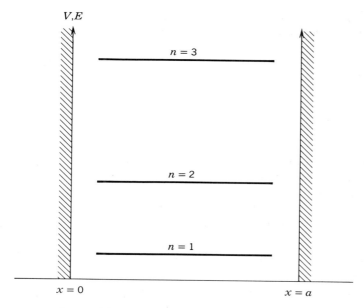

Fig. 4.1 Potential energy for a particle in a one-dimensional box. Inside the box, $V = 0$; outside the box, $V = \infty$. The lowest three energy levels are shown as horizontal lines.

so we must solve the Schrödinger equation

$$-\frac{\hbar^2}{2m}\frac{d^2\Psi}{dx^2} = (E - V)\Psi \tag{4.2}$$

The particle is not found outside the box where $V = \infty$, so $\Psi = 0$, and $d^2\Psi/dx^2 = 0$. Inside the box,

$$-\frac{\hbar^2}{2m}\frac{d^2\Psi}{dx^2} = E\Psi \tag{4.3}$$

This has the general solution

$$\Psi = A \sin kx + B \cos kx \tag{4.4}$$

where $k^2 = 2mE/\hbar^2$. The values of A, B, and k are to be determined by the boundary conditions. At $x = 0$, the potential undergoes a discontinuity from zero to infinity. If the function Ψ also were to undergo such a change, it would not be possible to properly normalize it; thus we define $\Psi = 0$ at $x = 0$ and similarly take $\Psi = 0$ at $x = a$. The first condition requires $B = 0$. The second condition can only be satisfied if $k = 0$ or $ka = n\pi$. If $k = 0$, the function Ψ completely vanishes, and the probability of the electron being somewhere in space is zero. We therefore reject this solution on physical grounds and take $ka = n\pi$, $n = \pm 1$, ± 2, ± 3, ... Thus, the possible energy levels are restricted to

$$E = \frac{\pi^2 \hbar^2 n^2}{2ma^2} \qquad \blacktriangleright (4.5)$$

with eigenfunctions

$$\Psi_n = A \sin \frac{n\pi x}{a} \qquad \blacktriangleright (4.6)$$

The solutions with negative n are identical to those with n positive, except for a phase factor, and so lead to no new functions. We can therefore restrict n to the set of positive integers.

We have yet to determine the constant A. Since, according to the postulates, Ψ must be normalized, we have

$$\int_{-\infty}^{\infty} \Psi^* \Psi \, dx = \int_0^a \Psi^* \Psi \, dx = A^2 \int_0^a \sin^2 \frac{n\pi x}{a} \, dx = 1 \qquad (4.7)$$

so $A^2 = 2/a$ and $A = \sqrt{2/a}$. The wavefunctions for several values for n are illustrated in Fig. 4.2. It should be noted that the functions are alternatively symmetric and antisymmetric about the center of the box; this will have important consequences when we discuss (in Sec. 4.5) the probability of the particle absorbing or emitting radiation and moving to a different energy level. It should also be noted that although the wavefunctions are continuous at the walls of the box, the first derivatives $d\Psi/dx$ are discontinuous. This is perfectly permissible, however, since the second derivatives are properly continuous as demanded by the form of the Schrödinger equation.

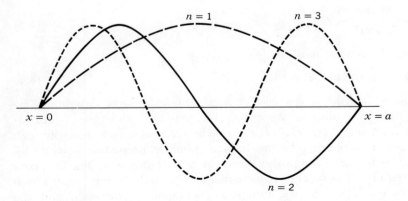

Fig. 4.2 Wavefunctions for the particle in a one-dimensional box.

We may now proceed to a somewhat more realistic case, that of a particle in a three-dimensional box. The Hamiltonian is given by Eq. (2.79). Taking $V = 0$ for $0 \leq x \leq a$, $0 \leq y \leq b$, and $0 \leq z \leq c$, we can easily derive

$$\Psi_{n_a, n_b, n_c} = \sqrt{\frac{8}{abc}} \sin \frac{n_a \pi x}{a} \sin \frac{n_b \pi y}{b} \sin \frac{n_c \pi z}{c} \tag{4.8}$$

where abc is the volume of the box and n_a, n_b, and n_c are positive integers. The energy levels are readily shown to be

$$E = \frac{\pi^2 \hbar^2}{2m} \left(\frac{n_a^2}{a^2} + \frac{n_b^2}{b^2} + \frac{n_c^2}{c^2} \right) \qquad \blacktriangleright (4.9)$$

The fact that the energy is a sum of three terms is a direct consequence of the fact that the wavefunction can be factored into three functions, each of which depends only on a single coordinate x, y, or z. This factoring is in turn due to the form of the Hamiltonian, which is a sum of independent terms each involving a single coordinate. It can be shown as a general theorem that when the Hamiltonian operator can be expressed as a sum of independent terms, the energy can always be expressed as a similar sum, and the wavefunction can be factored. Thus the three-dimensional differential equation reduces to three independent one-dimensional equations.

The solutions to the three-dimensional problem show several points of rather instructive behavior. We note, first of all, that a given energy level E for a cubical box may be achieved in more than one way. For example, the energy level $E = 6\pi^2 \hbar^2 / 2ma^2$ may be accomplished in three different ways: with $n_a = 1$, $n_b = 1$, and $n_c = 2$; with $n_a = 1$, $n_b = 2$, and $n_c = 1$; or with $n_a = 2$, $n_b = 1$, and $n_c = 1$. Similarly, the level $E = 14\pi^2 \hbar^2 / 2ma^2$ may be achieved in six ways. The number of linearly independent functions with a given energy is called the degeneracy of the energy level. A level is called nondegenerate if it has a degeneracy of unity; the lowest level of the particle in the three-dimensional box is nondegenerate.

The minimum energy that the particle can have is clearly nonzero. This is a direct consequence of the uncertainty principle; if the energy were zero, as all energy is in kinetic form, the momentum would also be zero, and the uncertainty in the momentum would be zero. This is impossible unless the uncertainty in position is infinite, which is the case only when the size of the box is infinite. Thus, even at absolute zero, the energy of a quantum-mechanical system can be finite; such energies are usually called zero-point energies. We will meet them again when we discuss the harmonic oscillator.

The box model can also be used to develop the equations for an ideal monatomic gas through the use of quantum-statistical mechanics.

If we assume that each particle is independent of every other, the Hamiltonian for the system will be a sum of one-particle Hamiltonians, and the wave function will be a product of one-particle functions, i.e., of orbitals. Of course, the energy levels for the system are the sum of the one-particle energies. Strictly speaking, one should apply Bose-Einstein or Fermi-Dirac statistics, depending on the spin of the particles, and the total wavefunction for the system should be either symmetric or antisymmetric on permutation of particles. However, in most cases of interest, the temperature is so high and the energy levels so close that few of the one-particle energy levels are occupied (see Prob. 4.2). In this case, ordinary Maxwell-Boltzmann statistics are sufficient.

Box models can be used to simulate many kinds of complex systems. We may, for example, consider an alpha particle constrained to a nucleus as a particle in a box, as illustrated in Fig. 4.3. If we assume that the most stable situation is obtained when the alpha particle is separated from the nucleus and that only a high potential-energy barrier prevents the particle from being ejected, it is possible to calculate the probability that the alpha will be emitted within a certain length of time and thus obtain the half-life of our model system. Such problems belong, however, in the realm of time-dependent theory, and so will not be treated here. The student who is interested will find that a study of similar box-barrier or "tunneling" problems is well worth the effort.

PROBLEMS

4.1 Find $\langle \mathbf{p}_x \rangle$, $\langle \mathbf{p}_x{}^2 \rangle$, $\langle \mathbf{x} \rangle$, and $\langle \mathbf{x}^2 \rangle$ for the particle in a one-dimensional box. What is the relation between $\langle \mathbf{p}_x{}^2 \rangle$ and E? Why is there no relation between $\langle \mathbf{p}_x \rangle$ and E?

4.2 Find the value of n_x necessary to give an oxygen atom in a one-dimensional box an energy equal to kT. Assume the box is 1 cm in length. How many different energy levels are there between $E = \frac{1}{2}kT$ and $E = \frac{3}{2}kT$?

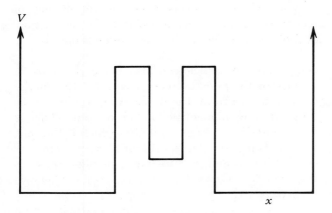

Fig. 4.3 Box model of an alpha particle in a nucleus.

4.3 In the lowest state of the one-dimensional system, find the probability that the particle is between $x = 0$ and $x = a/4$.

4.4 Derive the expression $PV = RT$ for an ideal monatomic gas, starting with the expression for the energy levels of the particle in a three-dimensional box.

4.2 THE HARMONIC OSCILLATOR

One of the more trivial problems in classical mechanics is the description of a mass point oscillating at the end of an ideal Hooke's law spring. In quantum-mechanical systems we have no springs, but the vibrations of molecules and the vibrations of atoms in a crystal often behave as if the controlling potentials were harmonic. Unfortunately, however, the quantum problem is a good deal more difficult than the classical one.

For simplicity, let us consider a pair of point masses restricted to move in a single dimension in a potential field $V = \frac{1}{2}k(x - x_0)^2$. The constant k is known as the Hooke's law force constant; x_0 could be, for example, the equilibrium internuclear distance between the two atoms of a diatomic molecule. The potential field is thus parabolic, or *harmonic*, and although this is not a completely accurate description of the forces found in molecules, it will do for the present.

The kinetic energy of the (nonrotating) system is the sum of the kinetic energies of the two particles. Referring the position of the two particles to their center of mass,

$$T = \frac{1}{2}(m_1 v_1{}^2 + m_2 v_2{}^2) = \frac{1}{2}(m_1 \dot{x}_1{}^2 + m_2 \dot{x}_2{}^2) = \frac{1}{2}\mu \dot{x}^2$$

where $x = x_1 - x_2$. It is then easy to show that the Hamiltonian is

$$\mathcal{H} = -\frac{\hbar^2}{2\mu}\frac{d^2}{dx^2} + \frac{1}{2}k(x - x_0)^2 \qquad \blacktriangleright (4.10)$$

but the wave functions are not so readily obtained. Making the substitutions $\alpha = 2\mu E/\hbar^2$, $\beta = \sqrt{\mu k}/\hbar$, $s = \sqrt{\beta}(x - x_0)$, and $y = e^{s^2/2}\Psi$, we see that the Schrödinger equation takes the form

$$\frac{d^2 y}{ds^2} - 2s\frac{dy}{ds} + \left(\frac{\alpha}{\beta} - 1\right)y = 0 \qquad (4.11)$$

This differential equation is known as Hermite's equation; it has solutions for any values of α and β. However, it is only for $\alpha/\beta - 1$ equal to twice an integer that the function $\Psi = e^{-s^2/2}y$ is zero at $x = \pm\infty$. Since this boundary condition is clearly necessary, the allowed energies of our system are limited to

$$E = \frac{\hbar}{2}\sqrt{\frac{k}{\mu}}\left(v + \frac{1}{2}\right) \qquad \blacktriangleright (4.12)$$

where v is a nonnegative integer. The corresponding eigenfunctions Ψ are given by

$$\Psi_v(s) = ye^{-s^2/2} = Ne^{-s^2/2}H_v(s) \tag{4.13}$$

where N is a normalizing factor and the functions $H_v(s)$, the solutions to (4.11), are known as the Hermite polynomials. The general expression for these polynomials is

$$H_v(s) = (-1)^v e^{s^2} \frac{d^v}{ds^v}(e^{-s^2}) \tag{4.14}$$

so that $H_0(s) = 1$, $H_1(s) = 2s$, $H_2(s) = 4s^2 - 2$, and so forth. The normalizing factor N is readily shown to be

$$N = \left(\frac{\sqrt{\beta/\pi}}{2^v v!}\right)^{1/2} \tag{4.15}$$

so

$$\Psi_v(s) = \left(\frac{\sqrt{\beta/\pi}}{2^v v!}\right)^{1/2} H_v(s)e^{-s^2/2} \qquad \blacktriangleright (4.16)$$

These functions, and their eigenvalues, show some close similarities to the functions for the particle in a one-dimensional box (compare Figs. 4.2 and 4.4). We again have a zero-point energy. This is a consequence of the shape of the potential well, which requires that a particle with zero energy must lie at the bottom of the parabola with fixed position and momentum. Of course, this is impossible. The corresponding eigenfunctions are again alternatively symmetric and antisymmetric about the center of the potential well, although since the harmonic oscillator does not have infinitely steep walls, the particle has finite probability of being "outside" the potential-energy curve. Classically, this would imply that the particle could have a negative kinetic energy in certain regions of space; in quantum mechanics, however, when we are discussing stationary states, it is not meaningful to speak of kinetic energy or potential energy except in reference to their expectation values (see the postulates in Sec. 2.8).

It is of some interest to further compare the classical harmonic oscillator with its quantum-mechanical analog since they are not as similar as might be imagined. The classical equation of motion is

$$\mu \frac{d^2x}{dt^2} = -k(x - x_0) \tag{4.17}$$

with solution

$$x - x_0 = A \cos 2\pi\nu_0 t \qquad \nu_0 = \frac{1}{2\pi}\sqrt{\frac{k}{\mu}} \tag{4.18}$$

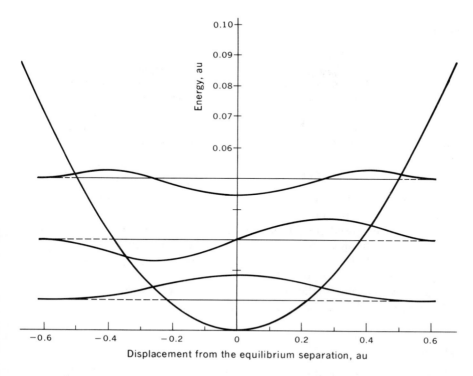

Fig. 4.4 Potential function for the harmonic oscillator. The allowed energy levels are indicated as horizontal lines, and the corresponding wavefunctions are sketched at each level. Compare Figs. 4.1 and 4.2. Data given is for $k = 0.4$ and $\beta = 20$, the approximate experimental constants for the hydrogen molecule.

The constant A is, of course, the amplitude of oscillation; ν_0 is the frequency. It is well known that the particle has maximum velocity and zero acceleration at $x = x_0$ and zero velocity and maximum acceleration at $x - x_0 = \pm A$, the turning points of the oscillation. For this reason the particle is most likely to be found at the turning points, whereas in the lowest quantum-mechanical state, the particle is most likely to be found near the point $x = x_0$. Clearly a quantum-mechanical vibration is quite different from a classical one. It can be shown,‡ however, that as v becomes large, the quantum-mechanical probability distribution function goes over into the classical, and the quantum-mechanical vibration becomes identical to the classical.

Since the classical frequency of oscillation is given by

$$\nu_0 = (1/2\pi) \sqrt{k/\mu}$$

it is customary to write the expression for the allowed quantum energy

‡ See Pauling and Wilson, p. 76.

levels as

$$E_v = (v + \tfrac{1}{2})h\nu_0 \tag{4.19}$$

It must be carefully noted that ν_0 has no real physical meaning in quantum mechanics, and is only used as a means of simplifying the energy formula. Further confusion can be caused by the similarity between ν_0 and the frequency of light obsorbed in a transition, ν. If a transition occurs between the levels E_v and E_{v+1}, then by the Planck-Einstein relation,

$$E_{v+1} - E_v = h\nu = h\nu_0 \qquad\qquad \blacktriangleright (4.20)$$

Thus, for a transition with $\Delta v = \pm 1$, the frequency of light absorbed or emitted is equal to the classical frequency of vibration. This identity seems to have caused considerable confusion in the early development of quantum mechanics, especially in the blackbody radiation problem, since it was not clear whether it was the radiation energy or the oscillator frequencies which were quantized. The modern point of view does not really recognize an oscillator frequency since this term implies simultaneous knowledge of position and momentum.

PROBLEMS

4.5 Compute the expectation values $\langle \mathbf{x} \rangle$, $\langle \mathbf{x}^2 \rangle$, $\langle \mathbf{p}_x \rangle$, and $\langle \mathbf{p}_x{}^2 \rangle$ for the harmonic oscillator.

4.6 Does the Hamiltonian, by itself, form a complete set of commuting observables for the harmonic oscillator?

4.3 CREATION AND ANNIHILATION OPERATORS

There is another technique for treating the harmonic oscillator, which, although elegant, is not widely used in orbital theory. The technique is purely algebraic and involves no differential equations; in a formal sense, the treatment in this section bears the same relation to the discussion of the previous section that the algebraic treatment of the angular-momentum problem (Sec. 3.2) bears to the usual solution by differential equations (Sec. 3.3). In the hope that the elegance of the method will attract as many advocates among chemists as it already has among physicists, we present an algebraic solution to the harmonic oscillator problem.

The Hamiltonian of (4.10) can be written in the form

$$\mathfrak{IC} = \frac{\mathbf{p}^2}{2\mu} + \tfrac{1}{2}k\mathbf{q}^2 \qquad\qquad \blacktriangleright (4.21)$$

where $\mathbf{q} = (x - x_0)$ is the displacement coordinate. A further simplification can be performed if we define

$$\mathbf{Q} = (k\mu)^{1/4}\mathbf{q} \qquad \mathbf{P} = \frac{\mathbf{p}}{(k\mu)^{1/4}} \tag{4.22}$$

so that

$$\mathcal{3C} = \frac{1}{2} \sqrt{\frac{k}{\mu}} (\mathbf{P}^2 + \mathbf{Q}^2) \tag{4.23}$$

We now introduce two new operators by the single equation

$$\mathbf{A}^{\pm} = \frac{1}{\sqrt{2}} (\mathbf{Q} \mp i\mathbf{P}) \qquad \blacktriangleright (4.24)$$

Note that $\overline{\mathbf{A}^{\pm}} = \mathbf{A}^{\mp}$ since both $\mathbf{P}$ and $\mathbf{Q}$ are Hermitian. $\mathbf{A}$ corresponds, in a rough sense, to the classical amplitude since

$$\mathbf{Q} = \frac{1}{\sqrt{2}} (\mathbf{A}^+ + \mathbf{A}^-) \tag{4.25}$$

The operators $\mathbf{A}^{\pm}$ have very simple commutation properties. Since

$$[\mathbf{q},\mathbf{p}] = i\hbar \tag{4.26}$$

we can derive

$$[\mathbf{Q},\mathbf{P}] = i\hbar \tag{4.27}$$

and hence

$$[\mathbf{A}^{\mp},\mathbf{A}^{\pm}] = \pm\hbar \qquad \blacktriangleright (4.28)$$

Using this last commutator, we can write the Hamiltonian in the form

$$\mathcal{3C} = \sqrt{\frac{k}{\mu}} (\mathbf{A}^{\pm}\mathbf{A}^{\mp} \pm \tfrac{1}{2}\hbar) \tag{4.29}$$

One last commutator is now readily obtained as follows:

$$[\mathcal{3C},\mathbf{A}^{\pm}] = \sqrt{\frac{k}{\mu}} [\mathbf{A}^{\pm}\mathbf{A}^{\mp},\mathbf{A}^{\pm}] = \pm\hbar \sqrt{\frac{k}{\mu}} \mathbf{A}^{\pm}$$

$$[\mathbf{A}^{\pm},\mathcal{3C}] = \sqrt{\frac{k}{\mu}} [\mathbf{A}^{\pm},\mathbf{A}^{\pm}\mathbf{A}^{\mp}] = \mp\hbar \sqrt{\frac{k}{\mu}} \mathbf{A}^{\pm} \qquad \blacktriangleright (4.30)$$

The operators $\mathbf{A}^{\pm}$ act very much like the shift operators of angular momentum theory. Let us assume there exists an eigenket $|\Psi_v\rangle$, such that

$$\mathcal{3C}|\Psi_v\rangle = E_v|\Psi_v\rangle \tag{4.31}$$

Then, from (4.30),

$$\mathfrak{IC}[\mathbf{A}^{\pm}|\Psi_v\rangle] = \left(E_v \pm \hbar \sqrt{\frac{k}{\mu}}\right)[\mathbf{A}^{\pm}|\Psi_v\rangle] \qquad (4.32)$$

It is thus clear that the kets $[\mathbf{A}^{\pm}|\Psi_v\rangle]$ are eigenkets of $\mathfrak{IC}$ with eigenvalues $E_v \pm \hbar \sqrt{k/\mu}$. Let us write

$$N_v^{\pm}|\Psi_{v\pm1}\rangle \equiv \mathbf{A}^{\pm}|\Psi_v\rangle \qquad (4.33)$$

where $N_v^{\pm}$ is a normalizing factor and where

$$\mathfrak{IC}|\Psi_{v\pm1}\rangle = \left(E_v \pm \hbar \sqrt{\frac{k}{\mu}}\right)|\Psi_{v\pm1}\rangle \qquad (4.34)$$

The operators $\mathbf{A}^{\pm}$ therefore generate a series of kets, differing from each other in energy by successive amounts $\hbar \sqrt{k/\mu}$. We now ask if this series is bounded. First, we note that

$$\langle\Psi_v|\mathbf{A}^{\mp}\mathbf{A}^{\pm}|\Psi_v\rangle \geq 0 \qquad (4.35)$$

This follows from the postulate that the length of the ket $\mathbf{A}^{\pm}|\Psi_v\rangle$ (where $\langle\Psi_v|\mathbf{A}^{\mp}$ is the corresponding bra) must be positive or zero. Combining (4.35) and (4.29), we may now write

$$\langle\Psi_v|\mathfrak{IC}|\Psi_v\rangle \pm \frac{1}{2}\hbar\sqrt{\frac{k}{\mu}} \geq 0$$
$$E_v \pm \frac{1}{2}\hbar\sqrt{\frac{k}{\mu}} \geq 0 \qquad (4.36)$$

This equation holds for all E_v, so the minimum possible value of E_v is $\frac{1}{2}\hbar\sqrt{k/\mu}$, that is, the zero-point energy. We call this energy E_0, and the corresponding eigenket $|\Psi_0\rangle$.

From (4.34) we see that the eigenkets have energies $\frac{1}{2}\hbar\sqrt{k/\mu}$, $\frac{3}{2}\hbar\sqrt{k/\mu}$, $\frac{5}{2}\hbar\sqrt{k/\mu}$, ad infinitum. We have therefore found all allowed energies. Additional information about the wavefunctions may be obtained by solving (4.33) for the $N_v^{\pm}$. From (4.35) and (4.36),

$$\mathbf{A}^-|\Psi_0\rangle = 0 \qquad (4.37)$$

(note that $\mathbf{A}^-$ lowers the energy). In addition,

$$\langle\Psi_0|\mathbf{A}^-\mathbf{A}^+|\Psi_0\rangle = \hbar \qquad (4.38)$$

Therefore,

$$\mathbf{A}^+|\Psi_0\rangle = \hbar^{1/2}|\Psi_1\rangle \qquad (4.39)$$

It is not difficult to show that, in general,

$$\mathbf{A}^-|\Psi_v\rangle = \hbar^{1/2}(v)^{1/2}|\Psi_{v-1}\rangle = N_v^-|\Psi_{v-1}\rangle$$
$$\mathbf{A}^+|\Psi_v\rangle = \hbar^{1/2}(v+1)^{1/2}|\Psi_{v+1}\rangle = N_v^+|\Psi_{v+1}\rangle$$

▶ (4.40)

The expectation values of the operators may be found without development of the wavefunctions. For example,

$$\langle \mathbf{q}^2 \rangle = \frac{1}{\sqrt{k\mu}} \langle \mathbf{Q}^2 \rangle \tag{4.41}$$

Expanding $\mathbf{Q}^2$ as $\frac{1}{2}(\mathbf{A}^+\mathbf{A}^+ + \mathbf{A}^-\mathbf{A}^- + \mathbf{A}^+\mathbf{A}^- + \mathbf{A}^-\mathbf{A}^+)$, we find

$$\langle \Psi_v|\mathbf{Q}^2|\Psi_v\rangle = \frac{1}{2}\langle\Psi_v|\mathbf{A}^+\mathbf{A}^- + \mathbf{A}^-\mathbf{A}^+|\Psi_v\rangle = \frac{1}{2}(N_v^-)^2 + \frac{1}{2}(N_v^+)^2$$

▶ (4.42)

$$= \hbar(v + \tfrac{1}{2})$$

$$\langle \mathbf{q}^2 \rangle = \frac{\hbar}{\sqrt{k\mu}}(v + \tfrac{1}{2})$$

Similarly,

$$\langle \mathbf{p}^2 \rangle = \frac{1}{2}\sqrt{k\mu}\,\langle\Psi_v|\mathbf{A}^+\mathbf{A}^- + \mathbf{A}^-\mathbf{A}^+|\Psi_v\rangle$$
$$= \hbar\sqrt{k\mu}\,(v + \tfrac{1}{2})$$

▶ (4.43)

Other expectation values may be found in the same way. Since a state is completely defined by its expectation values, there is no need to introduce explicit wavefunctions, and we avoid entirely having to solve any differential equations.

The operators $\mathbf{A}^+$ and $\mathbf{A}^-$ are *related* to the creation and annihilation operators used in quantum field theory. In such applications, $\mathbf{A}^+$ acting on an n-particle ket results in an $(n + 1)$-particle ket. In the present application $\mathbf{A}^+$ acts to increase the energy of the system. Both processes involve either matter or energy creation. Similarly, $\mathbf{A}^-$ annihilates particles or energy.

PROBLEMS

4.7 Verify (4.28), (4.29), and (4.30).

4.8 Verify (4.40).

4.9 Show that $\langle \mathbf{p} \rangle$ and $\langle \mathbf{q} \rangle$ are both zero.

4.10 Derive a general expression for the matrix elements
$\langle \Psi_v|\mathbf{q}|\Psi_{v'}\rangle$.

4.4 THE RIGID ROTOR

The theory of the rigid rotor is of primary value in the study of the rotation of diatomic molecules, and is thus outside the direct line of development of this book. However, a study of the theory can aid the student in understanding the practical applications of the angular-momentum theory, and the wavefunctions which we shall develop bear a useful resemblance to those which we shall later obtain for the hydrogen atom and for many-electron atoms. Furthermore, the later study of rotation-vibration spectroscopy will serve as a reasonably elementary introduction to some of the methods and problems involved in atomic and molecular spectroscopy.

It is convenient to take as our basic example of a rigid rotor an idealized diatomic molecule. Let the internuclear distance be the fixed length R, let the distance of the first atom from the center of mass be R_1, and let the distance of the second atom from the center of mass be R_2. We will also label the corresponding masses m_1 and m_2. The kinetic energy of this system is thus

$$T = \tfrac{1}{2}(m_1 v_1^2 + m_2 v_2^2) \tag{4.44}$$

where v_1 and v_2 are the velocities of the two particles. If we assume that the center of mass is fixed in space, the angular velocities $\omega = v/R$ of the two particles are equal; thus

$$\omega = \frac{|\mathbf{v}_1|}{R_1} = \frac{|\mathbf{v}_2|}{R_2} \tag{4.45}$$

Therefore

$$T = \tfrac{1}{2} I \omega^2 \tag{4.46}$$

where $I = m_1 R_1^2 + m_2 R_2^2$ is the moment of inertia of the molecule. The angular momentum of the molecule is given by

$$L = m_1 |\mathbf{v}_1| R_1 + m_2 |\mathbf{v}_2| R_2 = I \omega \tag{4.47}$$

so the kinetic energy of rotation may be written as

$$T = \frac{L^2}{2I} \tag{4.48}$$

The potential energy for unrestricted rotation is zero, of course, so the Schrödinger equation becomes

$$\mathfrak{IC}\Psi = E\Psi \tag{4.49}$$

or

$$\mathbf{L}^2\Psi = 2IE\Psi \tag{4.50}$$

where $\mathbf{L}^2$ is the angular-momentum operator.

We have already set up and solved (4.50) in Chap. 3. The allowed energies are

$$2IE = L(L + 1)\hbar^2$$
$$E = \frac{L(L + 1)\hbar^2}{2I} \qquad \blacktriangleright (4.51)$$

where L is an integer. The corresponding wavefunctions are given in Eqs. (3.30) and (3.32) and Table 3.1. It is of interest to note that the energy of the rigid rotor, unlike the energy of the harmonic oscillator or the particle in a box, can be zero. This is once again a consequence of the uncertainty principle; the energy and *angular* momentum of the rotor can be zero since the *angular* coordinate which controls the orientation of the rotor in space can have *any* value.

The eigenfunctions Ψ for the rigid rotor are simultaneously eigenfunctions of the Hamiltonian, of $\mathbf{L}^2$, and of $\mathbf{L}_z$. Thus the degeneracy of any given energy level is $2L + 1$, and only the ground state of the system is nondegenerate. The large degeneracy of the excited states will have important consequences when we discuss, in the next section, rotational and vibrational spectroscopy.

PROBLEMS

4.11 (*a*) Assuming that ν_0 for HCl^{35} is 2,886 cm^{-1} and that the internuclear separation is 1.28 Å, find the ratio of the energy separations between the first two rotational states and the first two vibrational states. Compare both separations to kT. (*b*) ν_0 for DCl is less than ν_0 for HCl by approximately the ratio $\sqrt{\frac{1}{2}}$. Explain why.

4.5 ROTATION–VIBRATION SPECTROSCOPY

A real diatomic molecule is neither a *rigid* rotor nor a *harmonic* oscillator nor, for that matter, are the motions of vibration and rotation strictly separate; note from Fig. 4.4 the large amplitudes of vibration in the ground state of H_2. Nevertheless, the approximation is useful, and can to a certain extent be justified. Let us then consider the case of two rotating mass points, connected by a spring of force constant k. This serves as a fairly accurate model for a diatomic molecule, and the method of finding the energy levels and wavefunctions for this model will prove instructive.

Let the first mass point have coordinates x_1, y_1, z_1 and mass m_1, and let the second mass point have similar coordinates and mass. The potential energy of the system is defined as $V = \frac{1}{2}k(r - r_e)^2$, where $r^2 = (x_2 - x_1)^2 + (y_2 - y_1)^2 + (z_2 - z_1)^2$ and r_e is the equilibrium value of r. The Hamiltonian is thus

$$\mathcal{H} = -\frac{\hbar^2}{2m_1}\left(\frac{\partial^2}{\partial x_1^2} + \frac{\partial^2}{\partial y_1^2} + \frac{\partial^2}{\partial z_1^2}\right) - \frac{\hbar^2}{2m_2}\left(\frac{\partial^2}{\partial x_2^2} + \frac{\partial^2}{\partial y_2^2} + \frac{\partial^2}{\partial z_2^2}\right)$$

$$+ \frac{k}{2}(r - r_e)^2 \quad (4.52)$$

In this form the Hamiltonian is quite inconvenient. Introducing the center-of-mass coordinates

$$x = \frac{m_1 x_1 + m_2 x_2}{m_1 + m_2} \tag{4.53}$$

(with similar definitions for y and z) and simultaneously introducing spherical polar coordinates

$$
\begin{aligned}
r \sin \theta \cos \varphi &= x_2 - x_1 \\
r \sin \theta \sin \varphi &= y_2 - y_1 \\
r \cos \theta &= z_2 - z_1
\end{aligned}
\tag{4.54}
$$

we obtain

$$
\begin{aligned}
\mathcal{H} = {}&- \frac{\hbar^2}{2(m_1 + m_2)} \left(\frac{\partial^2}{\partial x^2} + \frac{\partial^2}{\partial y^2} + \frac{\partial^2}{\partial z^2} \right) \\
&- \frac{\hbar^2}{2\mu} \left[\frac{1}{r^2} \frac{\partial}{\partial r} \left(r^2 \frac{\partial}{\partial r} \right) + \frac{1}{r^2 \sin^2 \theta} \frac{\partial^2}{\partial \varphi^2} + \frac{1}{r^2 \sin \theta} \frac{\partial}{\partial \theta} \left(\sin \theta \frac{\partial}{\partial \theta} \right) \right] \\
&+ \frac{k}{2} (r - r_e)^2 \quad (4.55)
\end{aligned}
$$

where $\mu = m_1 m_2 / (m_1 + m_2)$ is the reduced mass of the system (see Fig. 4.5). The first term above is just the kinetic-energy operator of a single particle of mass $m_1 + m_2$; the second term represents the kinetic energy of the two particles relative to the center of mass. Introducing a wavefunction Ψ and assuming it can be written as a product of a function of

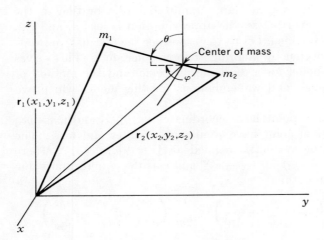

Fig. 4.5 Coordinate system for a diatomic molecule (rotation-vibration).

x, y, z and a function of r, θ, φ, we may separate the Schrödinger equation for this system into two parts,

$$\frac{-\hbar^2}{m_1 + m_2} \left(\frac{\partial^2 \Psi_t}{\partial x^2} + \frac{\partial^2 \Psi_t}{\partial y^2} + \frac{\partial^2 \Psi_t}{\partial z^2} \right) = E_t \Psi_t$$

and

$$\frac{-\hbar^2}{2\mu} \left[\frac{1}{r^2} \frac{\partial}{\partial r} \left(r^2 \frac{\partial \Psi_i}{\partial r} \right) + \frac{1}{r^2 \sin^2 \theta} \frac{\partial^2 \Psi_i}{\partial \varphi^2} + \frac{1}{r^2 \sin \theta} \frac{\partial}{\partial \theta} \left(\sin \theta \frac{\partial \Psi_i}{\partial \theta} \right) \right]$$

$$+ \frac{k}{2} (r - r_e)^2 \Psi_i = E_i \Psi_i \quad (4.56)$$

where

$$\Psi(x,y,z,r,\theta,\varphi) = \Psi_t(x,y,z) \Psi_i(r,\theta,\varphi)$$

and

$$E = E_t + E_i$$

Of course, the first equation is Schrödinger's equation for a particle of mass $m_1 + m_2$ with $V = 0$. Its solutions are similar to those for a particle in a box and indeed may be made identical to them by restricting our molecule to a limited region of space. We shall not investigate this particular problem further since we are primarily interested in the internal motions of the molecule which we observe in spectroscopic transitions and which are described by the second equation. It is of interest to note, however, that the separation of translational motion and internal motion is perfectly rigorous and can be accomplished for any many-particle system regardless of the form of the potential-energy function, provided it depends only on the internal coordinates.

The equation for internal motion may be further separated into two parts. Writing

$$\Psi_i = R(r) Y_{lm}(\theta,\varphi) \quad (4.57)$$

we obtain two differential equations. One of them is identical to (4.50). The second may be written as

$$\frac{1}{r^2} \frac{d}{dr} \left(r^2 \frac{dR}{dr} \right) + \left\{ -\frac{L(L + 1)}{r^2} + \frac{2\mu}{\hbar^2} \left[E_i - \frac{k}{2} (r - r_e)^2 \right] \right\} R = 0$$

$$(4.58)$$

This equation is not of the form of (3.10) or (3.11) and cannot be brought into such form. The vibrational part of the energy is therefore not the same as for a harmonic oscillator since the molecule is somewhat stretched by the rotational motion. The solution to (4.58) is quite difficult, but the energy can be written to a high degree of accuracy as (Pauling and Wilson, p. 271)

$$E_i = (v + \tfrac{1}{2})h\nu_0 + \frac{L(L+1)\hbar^2}{2I} - \frac{L^2(L+1)^2\hbar^4}{8\pi^2\nu_0^2 I^3} \qquad \blacktriangleright (4.59)$$

The first term is clearly the energy of an ordinary harmonic oscillator; the second term is that of a rigid rotor. The third term essentially corrects the energy for rotationally induced stretching; its numerical value, however, is quite small as the reader may wish to verify. Thus we see that the internal energy of a diatomic molecule (neglecting, for the present, the electronic energy) can be approximated as

$$E_i = (v + \tfrac{1}{2})h\nu_0 + \frac{J(J+1)\hbar^2}{2I} \qquad \blacktriangleright (4.60)$$

which is the sum of the energies of a harmonic oscillator and a rigid rotor. A given energy level is thus described by the two integers v and J; these labels are conventional in rotation-vibration theory. The degeneracy of a given level is of course $2J + 1$; this last expression is valid even when we consider the exact theory, and essentially it depends only on the kinds of degrees of freedom involved and not on the specific form of the wave equation. Let us now assume that the eigenfunctions of the diatomic molecule are given with sufficient accuracy by

$$\Psi_i = Y_{Jm}(\theta,\varphi)\Psi_v(s) \qquad \blacktriangleright (4.61)$$

where the Y_{Jm} are the normalized associated Legendre polynomials (Sec. 3.3) and the $\Psi_v(s)$ are the eigenfunctions of the harmonic oscillator [see (4.16)]. In the ground state of the system, $v = J = m = 0$; since the rotational levels are far more closely spaced than the vibrational levels, the first excited state has $v = 0$ and $J = 1$. A qualitative sketch of these levels is given in Fig. 4.6.

In spectroscopy, we are concerned not only with the location of energy levels but also with the probability that a system in a given state will absorb energy in the form of radiation and climb to a higher state. The intensity of absorption is, of course, determined by this probability, as is the intensity of spontaneous emission from a higher to a lower state. Another factor of importance is the number of states which absorb or emit light. This is essentially a question of concentration, i.e., the number of molecules making the transition. Furthermore, if there is a large concentration of molecules with the upper state occupied, light may induce an emission in such a way that the observed intensity of absorption is diminished. We shall simplify the situation by assuming that the number of molecules in the upper state is too small to cause a noticeable induced emission. This assumption is reasonable for rotation-vibration spectro-

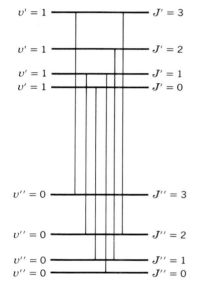

$v' = 1$ ——————— $J' = 3$

$v' = 1$ ——————— $J' = 2$

$v' = 1$ ——————— $J' = 1$
$v' = 1$ ——————— $J' = 0$

$v'' = 0$ ——————— $J'' = 3$

$v'' = 0$ ——————— $J'' = 2$

$v'' = 0$ ——————— $J'' = 1$
$v'' = 0$ ——————— $J'' = 0$

Fig. 4.6 Schematic rotation-vibration energy levels for a diatomic molecule. Only levels with $J \leq 3$ are shown. Vertical lines indicate allowed transitions. Relative separations between rotational levels and vibrational levels are not to scale since the lowest rotational separations are typically less than one-hundredth the vibrational separations.

scopy and electronic absorption spectroscopy; it will not work when discussing electron spin and nuclear magnetic resonance.

Through the use of time-dependent perturbation theory it is possible to show that the intensity of absorption is proportional to the square of the expectation value $\langle \mathbf{u} \rangle$ between the two states times the fraction f of the number of molecules in the lower state; that is,

$$I \propto f \langle \mathbf{u} \rangle^2 \qquad\qquad \blacktriangleright (4.62)$$

Here $\mathbf{u}$ is the dipole moment operator for the system. It is convenient to expand $\mathbf{u}$ in a power series with respect to some arbitrary coordinate r as follows:

$$\mathbf{u} = \mathbf{u}_0 + \left(\frac{\partial \mathbf{u}}{\partial r} \right)_0 r + \cdots \qquad\qquad (4.63)$$

The constant $\mathbf{\mu}_0$ is the permanent dipole moment of the system.

We may now apply (4.62) to our idealized diatomic molecule. We take r as the internuclear displacement distance. The operator $\mathbf{\mu}$ has three components, and we may write, for example,

$$\langle \mu_x \rangle = \langle \mu_{0x} \rangle + \left\langle \left(\frac{\partial \mu_x}{\partial r} \right) r \right\rangle$$

The partial derivatives cannot be obtained unless we know the *entire* molecular wavefunction (i.e., including electronic motion), but the directional properties are given by (4.54). Therefore, ignoring the contribution of $\mathbf{\mu}_0$ (see Prob. 4.16),

$$\langle \mu_x \rangle \propto \langle r \sin \theta \cos \varphi \rangle \tag{4.64}$$

Note, however, that $(\partial \mathbf{\mu} / \partial r)$ *can* be zero (Prob. 4.15). Similar expressions hold for the y and z components. Since the Y_{lm} are functions only of θ and φ, we may write (4.64) in the form

$$\langle \mu_x \rangle \propto \langle Y_{l'm'} | \sin \theta \cos \varphi | Y_{l''m''} \rangle \langle \Psi_{v'} | r | \Psi_{v''} \rangle \tag{4.65}$$

Let us first consider the matrix elements for the harmonic-oscillator part of the problem. Taking r along the internuclear axis and writing $s = q = \sqrt{\beta}\, r$, we find from Prob. 4.10 that

$$\langle \Psi_{v'} | r | \Psi_{v''} \rangle = \begin{cases} \sqrt{\dfrac{v_>}{2\beta}} & |v' - v''| = 1 \\ 0 & \text{otherwise} \end{cases} \qquad \blacktriangleright (4.66)$$

where $v_>$ is the larger of v' and v''. Alternatively, the integral may be evaluated using (4.16) and the well-known recursion formula for the Hermite polynomials

$$s H_v(s) = v H_{v-1}(s) + \tfrac{1}{2} H_{v+1}(s)$$

Equation (4.66) gives the following selection rule for v: all transitions are forbidden except those with $\Delta v = \pm 1$.

The selection rules for J and m are somewhat more difficult to obtain. One way to derive them is by using commutator relations (see for example Griffith, 1961, p. 33–39); alternatively, the explicit forms of the spherical harmonic functions (3.30) and (3.32) may be used (Herzberg, 1950, p. 119). The result, which is valid for any problem involving the associated Legendre polynomials, is the rule

$$\Delta J = \pm 1 \qquad \blacktriangleright (4.67)$$

(This rule holds for molecules with nondegenerate electronic ground states. If the electronic state is degenerate, a transition with $\Delta J = 0$ is also allowed.) The additional selection rules

$$\Delta m = 0$$
$$\Delta m = \pm 1$$

▶(4.68)

hold for light polarized in the z, x, or y directions, respectively.

In our simplified model, only the states with $v'' = 0$ are thermally occupied,‡ so only the transition from $v'' = 0$ to $v' = 1$ will occur in our model. However, states with $J'' \neq 0$ are thermally populated, with the fractional population f given by the Boltzmann formula (Sec. 1.1)

$$f = f_0(2J'' + 1)e^{-J''(J''+1)\hbar^2/2IkT}$$ (4.69)

The intensity of absorption of a system in the state $J = J''$ which makes a transition to the state $J = J' = J'' \pm 1$ is therefore given by

$$I \propto (2J'' + 1)e^{-J''(J''+1)\hbar^2/2IkT}\langle Y_{J''m''}|r|Y_{J'm'}\rangle^2$$ (4.70)

As can be shown, the value of the Legendre polynomial matrix element, averaged over the degenerate levels with varying m, is proportional to the larger of J' and J'', but the factor of $2J'' + 1$ is removed in the averaging, so (Herzberg, 1950, p. 127)

$$I \propto (J' + J'' + 1)e^{-J''(J''+1)\hbar^2/2IkT}$$

▶(4.71)

The intensities and positions of the allowed rotational-vibrational spectrum are illustrated schematically in Fig. 4.7. The lines are grouped about the fundamental vibrational frequency of the molecule; the separation

‡ In conventional spectroscopic notation, the doubly primed states are lower in energy than the singly primed states.

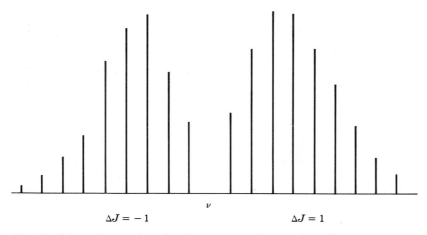

ν

$\Delta J = -1$ $\Delta J = 1$

Fig. 4.7 Schematic rotation-vibration spectrum diagram for a diatomic molecule. The intensities are approximately to scale for HCl at 300°C.

between a given pair of lines is constant and represents, in a sense, the fundamental rotational frequency.

The calculation of selection rules will be considered again in Chap. 8, when we discuss the theory of groups. We may anticipate an important result of that chapter, however, by considering the effect of a change of coordinate system on a general matrix element $\langle \Psi | \mathbf{r} | \Psi' \rangle$. Since the matrix element is a scalar quantity, it must be completely invariant to the way in which the coordinate system is chosen. Thus, for example, if the axes x, y, z are replaced by $-x$, $-y$, $-z$, an operation known as inversion, the value of the matrix element must remain the same.

Let us consider the specific case of the particle in a one-dimensional box. As noted earlier, the wavefunctions Ψ_n for this system are either symmetric or antisymmetric about the center of the box. Therefore, denoting the inversion operator as $\mathbf{P}$, we have

$$\mathbf{P}\Psi_n = (-1)^{n+1}\Psi_n \tag{4.72}$$

The symmetric functions are said to have *even parity*, whereas the antisymmetric functions have *odd parity*. The operator $\mathbf{P}$ is thus also known as the *parity operator*.

The vector $\mathbf{u} = \mathbf{r} \cdot$ constant will have odd parity if we take the center of the box as its origin. Furthermore, the parity of the matrix element $\langle \Psi_n | \mathbf{r} | \Psi_{n'} \rangle$ is clearly the product of the parities of the three quantities of which it consists. Therefore we obtain

$$\mathbf{P}\langle \Psi_n | \mathbf{r} | \Psi_{n'} \rangle = (-1)^{n+n'+1}\langle \Psi_n | \mathbf{r} | \Psi_{n'} \rangle \tag{4.73}$$

which conflicts with our previous assertion that the matrix elements must be completely independent of the choice of coordinate system. The only way that these two statements can be resolved is if

$$\langle \Psi_n | \mathbf{r} | \Psi_{n'} \rangle = 0 \qquad \text{unless } n + n' \text{ is odd} \qquad \blacktriangleright (4.74)$$

This gives us a general selection rule, not only for the particle in a box, but for many other systems as well. In its most general form, the rule states that transitions from an even-parity state to an even-parity state, and transitions from an odd-parity state to an odd-parity state, are forbidden.

We may easily apply this rule to the harmonic oscillator. As noted previously, the eigenfunctions for this problem are also alternatively symmetric and antisymmetric. Therefore we obtain at once the selection $\Delta v = \pm 1$, ± 3, . . . , which, although not as restrictive as the one previously obtained, is at least more readily derived. Similarly, selection rules for the rigid rotor may be obtained through the use of the parity operator.

One of the advantages of deriving selection rules through the use of symmetry operators is that one sees how the selection rules break down upon the loss of strict symmetry by the system. For example, real oscillators have somewhat asymmetric potential curves, and the selection rule $\Delta v = \pm 1$ no longer strictly holds since the wavefunctions will no longer be eigenfunctions of the parity operator.

PROBLEMS

4.12 What are the parities of the angular-momentum states with $l = 0$, 1, and 2?

4.13 Calculate the ratio of the zero-point energies of a CH bond vibration and a CD bond vibration.

4.14 Calculate the relative intensities of absorption in Fig. 4.6 for the HCl³⁵ molecule at 300°. How would the relative intensities change at higher temperatures?

4.15 Using (4.64), explain why a *diatomic* molecule with no permanent dipole moment has no rotation-vibration spectrum, whereas CO_2, which is linear, *does* have such a spectrum.

4.16 Under what conditions is pure rotational spectroscopy possible? Does this conflict with (4.65)? Why not?

4.6 THE HYDROGEN ATOM

The general problem of solving Schrödinger's equation for an atom cannot be solved exactly for any system other than the one-electron atoms such as H and He⁺. Nevertheless, the hydrogenic solutions do provide an introduction to atomic systems, and the form of the Hamiltonian, at least, is not so different from that for many-electron atoms.

Rather than repeat the detailed argument given in the previous section for the separability of internal and external coordinates, we shall in this case assume that the Hamiltonian for the hydrogen atom can be written as

$$\mathcal{3C} = -\frac{\hbar^2}{2\mu} \nabla^2 - \frac{Ze^2}{r} \qquad \blacktriangleright (4.75)$$

where μ is the reduced mass of the system, Z is the charge on the nucleus, and r is the separation between the electron and the nucleus. The kinetic-energy operator ∇^2 is, of course, most conveniently written out in spherical polar coordinates, and it refers only to the internal coordinates, i.e., to the relative positions of the electron and nucleus. Writing the wavefunction as $\Psi = Y_{lm}(\theta,\varphi)R(r)$, that is, as a product of radial and angular functions, we obtain the equation

$$\frac{1}{r^2}\frac{d}{dr}\left(r^2\frac{dR}{dr}\right) + \left[\frac{-l(l+1)}{r^2} + \frac{2\mu}{\hbar^2}\left(E + \frac{Ze^2}{r}\right)\right]R = 0 \qquad (4.76)$$

which is quite similar to (4.58) except for the form of the potential-energy function. Making the substitution

$$E = -\frac{Z^2}{2n^2}\frac{\mu e^4}{\hbar^2} \qquad \blacktriangleright (4.77)$$

which serves to define the parameter n, and the substitutions

$$s = \frac{2\mu e^2 rZ}{n\hbar^2} \equiv \frac{2rZ}{na_0}$$
$$R = NP(s)s^l e^{-s/2} \qquad (4.78)$$

we obtain the differential equation

$$s\frac{d^2P}{ds^2} + (2 + 2l - s)\frac{dP}{ds} + (n - l - 1)P = 0 \qquad (4.79)$$

This equation is closely related to the Laguerre differential equation. It has solutions which satisfy the boundary conditions $R = 0$ at $s = \infty$ and R finite when $s = 0$ only for n a positive integer. The boundary conditions also require that $n - l - 1$ be nonnegative, which gives us the restriction $n \geq l + 1$. Since the lowest value of l is zero, n cannot be smaller than unity.

It is possible to show that the functions

$$P = \frac{d^{2l+1}}{ds^{2l+1}} L_{n+l} \equiv L_{n+l}^{2l+1} \qquad (4.80)$$

satisfy (4.79), where

$$L_k(s) = e^s \frac{d^k}{ds^k}(s^k e^{-s}) \qquad (4.81)$$

The functions $L_k(s)$ are known as the Laguerre polynomials; the related functions $L_k{}^m(s)$ are called the associated Laguerre polynomials.

Since the functions Y_{lm} are normalized, we must also normalize R. Substituting (4.80) into (4.78) and integrating, we obtain

$$\int_0^\infty [R(r)]^2 r^2\, dr = \left(\frac{na_0}{2Z}\right)^3 N^2 \int_0^\infty s^{2l} e^{-s}[L_{n+l}^{2l+1}]^2 s^2\, ds = 1 \qquad (4.82)$$

In this way it is possible to prove that

$$R_{nl}(r) = -\sqrt{\frac{(n - l - 1)!}{2n[(n + l)!]^3}}\left(\frac{2Z}{na_0}\right)^{(2l+3)/2} r^l e^{-Zr/na_0} L_{n+l}^{2l+1}\frac{2Zr}{na_0} \qquad (4.83)$$

The functions R_{nl} with fixed l and varying n form an orthonormal set, so that

$$\int_0^\infty R_{nl}R_{n'l}r^2\, dr = \delta_{nn'} \qquad (4.84)$$

The radial functions R with different l are not orthogonal; orthogonality of the total wavefunction $\Psi_{nlm} = Y_{lm}R_{nl}$ is guaranteed by the orthogonality of the functions Y.

The general form of the R function is rather formidable, and thus it is enlightening to consider the specific radial functions with $n = 1, 2,$ and 3. These are listed in Table 4.1 and graphed in Fig. 4.8. It is conventional to label the functions with $l = 0, 1, 2, 3, \ldots$ as $s, p, d, f, \ldots$ Thus, in order of increasing energy, the wavefunctions of the hydrogen atom may be denoted

$$
\begin{aligned}
&1s \\
&2s, 2p \\
&3s, 3p, 3d \\
&4s, 4p, 4d, 4f
\end{aligned}
\qquad\blacktriangleright (4.85)
$$

where the energy, of course, depends only on the principal quantum number n. The quantum number m is usually not included in the labeling scheme since the energy is independent of m even in nonhydrogenic atoms. However, we *can* use this sort of notational scheme to specify m as well, for example, $2p_{-1}$.

The energy levels of the hydrogen atom have large degeneracies. For a given n (that is, for a given E) there are n possible values of l, each with $(2l + 1)$ values of m. Summing over l, we find that the degeneracy of the nth level is n^2. This does not include spin. For each value of n, l, and m there can be two possible values of s_z, which brings the total degeneracy of the nth level to $2n^2$. Note that spin does not appear here

---

Table 4.1 The radial functions R_{nl} for hydrogenic atoms, $n \leq 3$; $l = 0, 1, 2$

$$
A = \left(\frac{Z}{a_0}\right)^{3/2} \qquad \rho = \frac{Zr}{a_0}
$$

$$
R_{1s} = 2Ae^{-\rho}
$$

$$
R_{2s} = \frac{A}{2\sqrt{2}}\,(2 - \rho)e^{-\rho/2}
$$

$$
R_{3s} = \frac{2A}{81\sqrt{3}}\,(27 - 18\rho + 2\rho^2)e^{-\rho/3}
$$

$$
R_{2p} = \frac{A}{2\sqrt{6}}\,\rho e^{-\rho/2}
$$

$$
R_{3p} = \frac{4A}{81\sqrt{6}}\,(6\rho - \rho^2)e^{-\rho/3}
$$

$$
R_{3d} = \frac{4A}{81\sqrt{30}}\,\rho^2 e^{-\rho/3}
$$

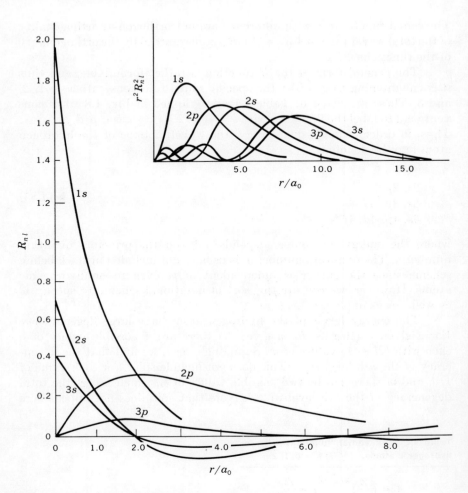

Fig. 4.8 The radial functions R_{nl} for $n < 3$, $l = 0, 1, 2$.

as a natural consequence of the form of the Hamiltonian, since our Hamiltonian is not relativistically invariant. If our treatment had followed, for example, the Dirac equation or the methods of quantum electrodynamics, several more terms would have appeared in the Hamiltonian, and some of the degeneracy would have been lifted. We shall discuss such refinements in Chap. 6 from a semi-empirical point of view.

Let us now consider the radial functions sketched in Fig. 4.8. The functions with $n = l + 1$ have an especially simple form, the radial function always having the same sign. All other radial functions pass through the zero ordinate one or more times; each such crossing is called a *node*, and the number of radial nodes is clearly equal to $n - l - 1$.

Only the ns functions are nonzero at $r = 0$. This implies that in an s state the electron has a finite probability of passing through the nucleus. Consider the function $\Psi\Psi^* \, d\tau$. According to the postulates, this function represents the probability that the electron has coordinates between r and $r + dr$, θ and $\theta + d\theta$, φ and $\varphi + d\varphi$, within a volume element of magnitude $d\tau = r^2 \sin\theta \, d\theta \, d\varphi \, dr$. Although $d\tau = 0$ at $r = 0$, $\Psi\Psi^* \, d\tau$ is at a maximum near $r = 0$ for a *fixed* constant-volume element $d\tau$.

The average value of r in a given state, that is, $\langle r \rangle$, is given by

$$\int_0^\infty R^2(r)r^3 \, dr = \langle r \rangle \tag{4.86}$$

For the $1s$ orbital of the hydrogen atom, we find $\langle r \rangle = 3a_0/2$. The so-called *most probable value* of r is that value for which the function R^2r^2 is a maximum; for the $1s$ orbital,

$$\frac{d}{dr} R^2 r^2 = -\frac{2}{a_0} r^2 e^{-2r/a_0} + 2re^{-2r/a_0} = 0 \tag{4.87}$$

so $r = a_0$ is the most probable value. The constant a_0 has another interpretation as follows: it is the radius of the first orbit in the old Bohr theory.

As noted in Chap. 3, the functions Y_{lm} are not always the most useful angular eigenfunctions for atomic theory. Table 3.1 lists the cooresponding real functions, formed from linear combinations of Y_{lm} and Y_{l-m}. These functions are quite elementary, behaving as x, y, or z for the p orbitals and as xy, xz, yz, $x^2 - y^2$, and $3z^2 - r^2$ for the d orbitals. It is of interest to plot them together with the hydrogenic functions R_{nl}, and in Fig. 4.9 we have sketched the real products Ψ_{nl}^2, which are, of course, probability distributions. We may use the real functions in place of the Y_{lm} without error since the functions Ψ_{nlm} and Ψ_{nl-m} are degenerate. The p orbitals are especially conveniently represented by the real functions since the three functions p_x, p_y, and p_z are all identical except for the choice of axis. The d orbitals have a somewhat less convenient form, with four of the functions, d_{xy}, d_{xz}, d_{yz}, and $d_{x^2-y^2}$, having the same shape except for location of axis, whereas the fifth function is completely different. The f functions, finally, are quite complex, and since they are not as important for chemical purposes, we will not graph them here.

PROBLEMS

4.17 Show that a linear combination of degenerate functions has the same energy as any one of the set.

4.18 Discuss the selection rules for the hydrogen atom. Which transitions are parity-forbidden?

4.19 Calculate the expectation value $\langle r^{-2} \rangle$ in both the $1s$ and $2s$ states. How is this value related to $R_{ns}(0)$?

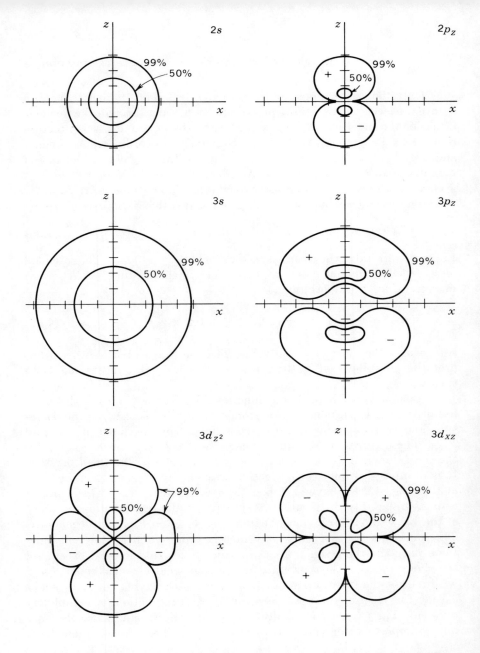

Fig. 4.9   The probability distributions for s, p, and d orbitals, Ψ^2_{nl}. Signs indicate the phase of the angular part of Ψ_{nl}; see Table 3.1. All functions are symmetric about the z axis except d_{xz}, whose spatial distribution can be generated by rotating each lobe about its own symmetry axis. Functions are approximately to scale for the hydrogen atom; grid tics are in atomic units of length. Note that the $d_{x^2-y^2}$ function has its lobes along the x and y axes. All other functions can be generated by a simple renaming of axes. [*Adapted from C. A. Coulson and E. T. Stewart, Wave Mechanics and the Alkene Bond, in S. Patai (ed.), "The Chemistry of the Alkenes," Interscience Publishers, New York, 1964.*]

4.20 Show that the operator $(1/r^2)(d/dr)[r^2(d/dr)]$ is Hermitian.

4.21 Suppose the potential $V = -Ze^2/r$ in (4.75) were replaced by the more general spherically symmetric function $V = V(r)$. What changes would take place in the function Ψ_{nlm}? To what extent would the labels n, l, and m remain valid?

4.7 SLATER-TYPE ORBITALS AND ATOMIC UNITS

In this section we will introduce two changes in nomenclature which, although they do not contribute directly to the development of the theory of atomic structure, make the writing of many equations and formulas much easier. The first change concerns the radial functions for the hydrogenic atoms. The polynomial expressions involving the associate Laguerre polynomials $L_k{}^m$ are at best awkward, and when we later discuss many-electron atoms, we will need to use radial functions with somewhat more general applicability. We therefore introduce the new set of radial functions $R_{n\zeta}^S$ defined by the expression

$$R_{n\zeta}^S = \sqrt{\frac{(2\zeta)^{2n+1}}{(2n)!}} \, a_0^{-3/2} r^{n-1} e^{-\zeta r/a_0} \tag{4.88}$$

When $n = l + 1$ and $\zeta = Z/n$, these functions are identical to the radial functions of the hydrogenic atoms, but for $n > l + 1$, they differ. In fact, the higher radial functions $R_{n\zeta}^S$ are not orthogonal to those with $n = l - 1$; the value of l is not even specified. This is, as we shall see in Chap. 6, no great disadvantage since it is always possible to construct a set of orthogonal functions from a set which is not orthogonal by a suitable transformation.

The radial functions $R_{n\zeta}^S$ are usually coupled with the real angular functions of Table 3.1 to produce *Slater-type orbitals* which we will write as $S_{nlm\zeta}$. The name originates in an observation by J. C. Slater that functions of this type can be used to provide surprisingly good approximations to the wavefunctions for many-electron atoms. It is customary to call the functions σ, π, or δ according to whether $|m|$ is 0, 1, or 2. This system of nomenclature is particularly useful in the study of diatomic molecules, and its origin and meaning will be discussed in Chap. 7.

The second change in nomenclature which we shall introduce concerns the system of units in which we express our results. For actual numerical calculations in quantum mechanics, the usual systems of units (cgs, etc.) are quite inconvenient. The quantities $\hbar$, μ, e, and a_0 occur repeatedly, causing both a certain bulkiness in the equations and a degree of difficulty in obtaining numerical results for the eigenvalues. It is thus convenient to introduce a new system of units, in which all of these quantities are defined to be unity. Formally speaking, we take

$a_0 = \hbar^2/me^2$ as the unit of length, e as the unit of charge, and twice the energy of the hydrogen atom, $2E_1 = -me^4/\hbar^2 = -e^2/a_0$, as the unit of energy. We may additionally take m, the mass of the electron, as our fundamental unit of mass, and consequently we take $\hbar$ as the unit of action. This defines the so-called atomic system of units. We must carefully note that the *correct* formula for the energy of the hydrogen $1s$ orbital is $E = -\mu e^4/2\hbar^2$, so that in atomic units the correct $1s$ energy is $E_1 = -\tfrac{1}{2}(1,836/1,837)$. The difference is of no consequence except in accurate work.

As noted in Chap. 2, the kinetic-energy operator is $-(\hbar^2/2m)\nabla^2$. In atomic units this becomes $-\tfrac{1}{2}\nabla^2$, so the Hamiltonian for a hydrogenic atom may be written

$$\mathfrak{K} = -\frac{1}{2}\nabla^2 - \frac{Z}{r} \qquad\qquad \blacktriangleright(4.89)$$

where Z is the nuclear charge. Similarly, the radial part of the Slater-type orbitals becomes

$$R_{n\zeta}^S = \sqrt{\frac{(2\zeta)^{2n+1}}{(2n)!}}\, r^{n-1}e^{-\zeta r} \qquad\qquad \blacktriangleright(4.90)$$

There is no standard method of distinguishing between different kinds of atomic units, such as mass, charge, length, and energy. The abbreviation au is conventionally used for all of them. A table of conversion factors is given in Appendix 1; note especially that one atomic unit of energy is equal to 27.20 ev, and one atomic unit of length is 0.5293 Å.

PROBLEMS

4.22 Calculate the speed of light in atomic units.

4.23 Verify that Eq. (4.90) is properly normalized.

4.24 Consider the function $\mathfrak{R} = N(R_{2,0.5}^S - SR_{1,1.0}^S)$, where N is a normalizing factor and $S = \int_0^\infty R_{2,0.5}^S(r)R_{1,1.0}^S(r)r^2\,dr$. This function is orthogonal to the $1s$ orbital of hydrogen, but it is not identical to the $2s$ orbital of hydrogen. Calculate its energy using the hydrogen Hamiltonian of (4.75).

SUGGESTIONS FOR FURTHER READING

This text is almost exclusively concerned with the *theory* of atomic and molecular electronic structure, and very little with how such structures are obtained experimentally. The interaction of radiation and matter is the subject material of spectroscopy and is of great intrinsic interest. On the one hand, it is possible to study the theory of the interaction of photons and molecules (or atoms), and on the other hand, the details of the information obtained, that is, energy levels, moments, field gradients, etc. Such information is often obtained by rather indirect spectroscopic evidence, and the relationships between the original experiments and the derived results are often fascinatingly complex.

1. Herzberg, G.: "Spectra of Diatomic Molecules" (1950), "Infrared and Raman Spectra of Polyatomic Molecules" (1945), "Electronic Spectra and Electronic Structure of Polyatomic Molecules" (1967), D. Van Nostrand Company, Inc., Princeton, N.J.

   This series of volumes presents the most comprehensive discussion available of molecular spectroscopy. The volumes contain an abundance of carefully organized and presented material.

2. Wilson, E. B., J. C. Decius, and P. C. Cross: "Molecular Spectroscopy," McGraw-Hill Book Company, New York, 1955.

   This is a more terse exposition, suitable for use as a text. However it would appear that a definitive text on spectroscopic *theory*, including the most modern advances (photoelectron spectroscopy, double resonance, lasers, etc.), has yet to be written.

5

APPROXIMATE METHODS

In the strong formulation of the
causal law "If we know exactly
the present, we can predict the
future" it is not the conclusion
but rather the premise which is
false. We *cannot* know, as a
matter of principle, the present
in all its details.
Werner Heisenberg

5.1 THE VARIATION THEOREM

Few of the interesting problems in quantum mechanics can be solved
exactly. This statement is true more or less by definition; for most of us,
a problem cannot fascinate unless it is unsolved. The direct numerical
integration of the many-dimensional Schrödinger equation has not proved
a very useful method of solving quantum problems, and so a variety of
approximate methods have been developed. Since an exact solution to
problems one ordinarily meets in treating molecular and atomic systems
is out of the question, a study of approximate methods of their solution
is of the highest importance.

Let us suppose we are given a Hamiltonian operator $\mathcal{3C}$ whose form
is of such complexity that orthonormal functions Ψ_i satisfying the eigenvalue equation

$$\mathcal{3C}\Psi_i = E_i\Psi_i \tag{5.1}$$

cannot be found. We suppose for simplicity that the eigenvalues E_i are nondegenerate and ordered, such that $E_0 < E_1 < E_2 < \cdots$. As noted in Chap. 2, the (undetermined) eigenvectors Ψ_i form a complete set, such that any observation on the system described by $\mathfrak{IC}$ can be found in the set Ψ_i. Thus, any function Φ satisfying the boundary conditions of the problem can be expressed as a linear combination of the Ψ_i as follows:

$$\Phi = \sum_i c_i \Psi_i \qquad \blacktriangleright (5.2)$$

Of course, since the Ψ_i are undetermined, the c_i are unknown; the importance of Eq. (5.2) is that *any* function Φ satisfying the boundary conditions can *in theory* be so expanded, even if the expansion is unknown.

Consider now the expectation value $\langle \Phi | \mathfrak{IC} | \Phi \rangle$. Expanding Φ on both sides, we obtain

$$\langle \Phi | \mathfrak{IC} | \Phi \rangle = \sum_i \sum_j c_i^* c_j \langle \Psi_i | \mathfrak{IC} | \Psi_j \rangle$$
$$= \sum_i c_i^2 E_i$$

Similarly,

$$\langle \Phi | E_0 | \Phi \rangle = \sum_i c_i^2 E_0$$

since E_0 is a constant. Thus,

$$\langle \Phi | \mathfrak{IC} - E_0 | \Phi \rangle = \sum_i c_i^2 (E_i - E_0)$$

and since c_i^2 and $E_i - E_0$ are both nonnegative,

$$\langle \Phi | \mathfrak{IC} - E_0 | \Phi \rangle \geq 0$$
$$\frac{\langle \Phi | \mathfrak{IC} | \Phi \rangle}{\langle \Phi | \Phi \rangle} \geq E_0 \qquad \blacktriangleright (5.3)$$

Taking Φ as a normalized function, we have

$$\langle \Phi | \mathfrak{IC} | \Phi \rangle \geq E_0 \qquad (5.4)$$

which shows that for an arbitrary function Φ, satisfying the boundary conditions, the expectation value of the energy is always greater or equal to the lowest eigenvalue of the system.

This theorem is quite powerful. Since Φ is a perfectly arbitrary function, we can write it as a function of several different parameters, say, $\Phi = \Phi(p_1, p_2, p_3, \ldots, p_n)$. We can then study the change in $\langle \Phi | \mathfrak{IC} | \Phi \rangle$ with *variation* of the parameters p_i, subject, of course, to the constraint $\langle \Phi | \Phi \rangle = 1$. Thus, writing

$$E(p_1, p_2, \ldots, p_n) = \frac{\langle \Phi | \mathcal{H} | \Phi \rangle}{\langle \Phi | \Phi \rangle} \geq E_0 \tag{5.5}$$

we can *minimize E* by solving for parameters p_i such that

$$\frac{\partial E}{\partial p_i} = 0$$

for all p_i. This gives us the *best* Φ of a given functional form, i.e., best in the sense that E is at a minimum for that particular form. Equation (5.3) is often called the *variation theorem* since, for optimal p_i, E is stable with respect to any variation in the parameters of Φ. Another widely used name is the *upper-bound variation theorem*, since E provides an upper bound to E_0.

Optimization of Φ with respect to the parameters p_i is not, in general, a trivial process. Φ may be nonlinear in the parameters, and optimization may only be possible by trial and error. Suppose, however, that Φ is linear in the p_i, so that it can be expressed as a linear sum of other functions. Then

$$\Phi = \sum_{i=1}^{n} p_i \chi_i \tag{5.6}$$

where the χ_i are not parameterized or contain only nonlinear parameters which do not concern us. Using the abbreviations $S_{ij} = \langle \chi_i | \chi_j \rangle$ and $H_{ij} = \langle \chi_i | \mathcal{H} | \chi_j \rangle$, we have

$$\frac{\partial E}{\partial p_k} = 0 = \frac{\partial}{\partial p_k} \left(\frac{\sum_i \sum_j p_i^* p_j H_{ij}}{\sum_i \sum_j p_i^* p_j S_{ij}} \right) \qquad \text{for all } k \tag{5.7}$$

Carrying out the differentiations, we obtain n equations of the form

$$\sum_i p_i H_{ik} - \langle \Phi | \mathcal{H} | \Phi \rangle \sum_i p_i S_{ik} = 0 \tag{5.8}$$

where we assume for simplicity that the parameters are all real and that $\langle \Phi | \Phi \rangle = 1$. Rewritten, the above equation becomes

$$\sum_i (H_{ik} - ES_{ik}) p_i = 0 \qquad \text{for any } k \qquad \blacktriangleright (5.9)$$

which is equivalent to the nth-order determinental equation

$$|H_{ik} - ES_{ik}| = 0 \qquad \blacktriangleright (5.10)$$

In general, this equation will have n roots which we write‡ as E_m^0; the desired root is of course the minimum value E. The coefficients p_i may also be found using (5.9), thus determining Φ according to (5.6). In this way, the problem of optimizing Φ in linear parametric form reduces to the evaluation of the *matrix elements* H_{ij} and S_{ij} and the solution of the *secular determinant* (5.10).

The variation theorem can also be used to obtain upper bounds to the excited-state eigenvalues E_1, E_2, etc. Let $\Phi_0(p_1, p_2, \ldots, p_n)$ be a function of the parameters p_k, such that

$$E_0^0 \equiv \frac{\langle \Phi_0 | \mathfrak{K} | \Phi_0 \rangle}{\langle \Phi_0 | \Phi_0 \rangle}$$

$$\frac{\partial E_0^0}{\partial p_k} = 0 \qquad \text{for all } k \tag{5.11}$$

Let us now choose an arbitrary function $\Phi_1(p_1, p_2, \ldots, p_n)$ subject to the constraint that it be orthogonal to Φ_0,

$$\langle \Phi_0 | \Phi_1 \rangle = 0$$

If we now define

$$E_1^0 = \frac{\langle \Phi_1 | \mathfrak{K} | \Phi_1 \rangle}{\langle \Phi_1 | \Phi_1 \rangle} \tag{5.12}$$

we can prove that $E_1 \leq E_1^0$. Let

$$\Phi_1 = \sum_i d_i \Psi_i \tag{5.13}$$

as in (5.2). Then, as can be shown (see Prob. 5.1), $\langle \Phi_1 | \mathfrak{K} | \Phi_0 \rangle = 0$. Defining the new function

$$\Phi_1' = c_0 \Phi_1 - d_0 \Phi_0 = \sum_i b_i \Psi_i \tag{5.14}$$

where

$$b_i = c_0 d_i - d_0 c_i$$

we obtain

$$E_1'^0 - E_1 = \frac{\displaystyle\sum_i b_i^2 (E_i - E_1)}{\displaystyle\sum_i b_i^2} \tag{5.15}$$

‡ We write the roots with a superscript to distinguish them from the eigenvalues E_i. Methods for solving the secular equation are described in Appendix 2.

where

$$E_1'^0 = \frac{\langle \Phi_1' | \mathcal{K} | \Phi_1' \rangle}{\langle \Phi_1' | \Phi_1' \rangle} \tag{5.16}$$

Since b_0 is zero by definition,

$$E_1'^0 - E_1 \geq 0 \tag{5.17}$$

Using (5.14) and (5.16), we find that

$$E_1'^0 - E_1 = \frac{c_0^2(E_1^0 - E_1) + d_0^2(E_0^0 - E_1)}{c_0^2 + d_0^2} \geq 0$$

so

$$\frac{c_0^2}{c_0^2 + d_0^2}(E_1^0 - E_1) \geq \frac{d_0^2}{c_0^2 + d_0^2}(E_1 - E_0^0) \tag{5.18}$$

Thus, either $E_0^0 > E_1^0$, which is contrary to our hypothesis, or

$$E_1^0 \geq E_1 \tag{5.19}$$

The equality holds only when $c_0 = 1$, $d_1 = 1$, and all other c_i and d_i are zero.

A somewhat stronger form of this theorem has been given by MacDonald (1933), who has shown that when Φ_0 and Φ_1 are expressed in linear parametric form, E_0^0 converges monotonically with increasing basis size. Furthermore, MacDonald has shown that if a trial function Φ_n is orthogonal to all previously optimized functions $\Phi_{n-1}, \Phi_{n-2}, \ldots, \Phi_0$, then

$$E_n^0 \geq E_n \qquad \blacktriangleright (5.20)$$

Thus, when solving the secular equation (5.10), the roots E_m^0 all provide upper bounds to the true eigenvalues E_m. This is an additional advantage of the linear parametric form since by using it, one simultaneously obtains estimates of the wavefunctions and energies of a large number of states.

The variation theorem offers the promise of enabling us to calculate functions Φ_i and energies E_i^0 with whatever accuracy is desired. Two difficulties arise. First, we have no way of determining how close a particular E_i^0 is to the true eigenvalue E_i, and secondly, we have no method of determining the accuracy of other properties calculated using the trial function Φ_i. What is needed is a method of determining *lower bounds* to the energies and a method of calculating both upper and lower estimates of expectation values $\langle \Phi_i | \mathbf{O} | \Phi_i \rangle$. Several theorems have been derived‡ concerning lower bounds for the energy, but as yet they have had little application. Very little progress has been made in theoretically

‡ See, for example, E. B. Wilson, Jr., *J. Chem. Phys.*, **43**:S172 (1965) and references contained therein.

estimating the error in other expectation values. Use of the variation theorem remains, to a great extent, a matter of treating a large number of similar problems, with errors estimated primarily by experience with a particular kind of trial function. One kind of application of the variation principle to atoms and molecules is via the well-known Hartree-Fock equations, and here so much experience has been accumulated in the literature that quite good estimates of molecular and atomic properties may be obtained.

PROBLEMS

5.1 Suppose a function $\Phi_0(p_1, p_2, \ldots, p_n)$ has been optimized with respect to the parameters p_k, so that E_0^0, as defined in (5.11), is at a minimum. Suppose further that $\Phi(p_1, p_2, \ldots, p_n)$ is orthogonal to Φ_0 and has also been optimized. Prove that $\langle \Phi_0 | \mathfrak{IC} | \Phi_1 \rangle = 0$.

Hint: Assume that $\langle \Phi_0 | \mathfrak{IC} | \Phi_1 \rangle = H_{01} \neq 0$. Construct new functions Φ_0' and Φ_1' from linear combinations of Φ_0 and Φ_1 such that the new functions are orthogonal and have off-diagonal matrix elements equal to zero. Show that these new functions have lower energy than the original functions unless $H_{01} = 0$.

5.2 PERTURBATION THEORY

There is a second approximate method of wide applicability in the general theory of quantum mechanics but of somewhat limited use in orbital theory. Perturbation theory assumes that the Hamiltonian can be expanded in the form

$$\mathfrak{IC} = \mathbf{H}_0 + \lambda \mathbf{H}' \qquad \blacktriangleright (5.21)$$

where, in a certain sense (to be seen), $\lambda \mathbf{H}'$ is "small." We assume that the eigenvalues and eigenfunctions of $\mathbf{H}_0$ are known, so that

$$\mathbf{H}_0 \Psi_n^0 = E_n^0 \Psi_n^0 \qquad \blacktriangleright (5.22)$$

where there are, in general, an infinite number of functions Ψ_n^0. We wish to solve the somewhat more complicated equation

$$\mathfrak{IC} \Psi_n = E_n \Psi_n \qquad (5.23)$$

for the eigenfunctions Ψ_n and eigenvalues E_n. Note that as $\lambda \to 0$, $\Psi_n \to \Psi_n^0$ and $E_n \to E_n^0$, which points toward an explanation of why we consider $\lambda \mathbf{H}'$ to be small. On the other hand, λ *need not* be small itself.

The fundamental assumption of perturbation theory is that both Ψ_n and E_n may be expanded as power series in λ, to be written

$$\Psi_n = \Psi_n^0 + \lambda \Psi_n^{(1)} + \lambda^2 \Psi_n^{(2)} + \cdots = \sum_{i=0} \lambda^i \Psi_n^{(i)}$$
$$E_n = E_n^0 + \lambda E_n^{(1)} + \lambda^2 E_n^{(2)} + \cdots = \sum_{i=0} \lambda^i E_n^{(i)} \qquad \blacktriangleright (5.24)$$

Substituting these expansions in (5.23), we obtain the equation

$$(\mathbf{H}_0 + \lambda\mathbf{H}') \sum_{i=0} \lambda^i \Psi_n^{(i)} = \sum_{i=0}\sum_{j=0} \lambda^{i+j} E_n^{(i)} \Psi_n^{(i)} \tag{5.25}$$

Equating powers of λ, we obtain an infinite number of equations, the first three of which are

$$\begin{aligned}
\mathbf{H}_0\Psi_n^0 &= E_n^0\Psi_n^0 \\
(\mathbf{H}_0 - E_n^0)\Psi_n^{(1)} &= (E_n^{(1)} - \mathbf{H}')\Psi_n^0 \\
(\mathbf{H}_0 - E_n^0)\Psi_n^{(2)} &= E_n^{(2)}\Psi_n^0 + (E_n^{(1)} - \mathbf{H}')\Psi_n^{(1)}
\end{aligned} \tag{5.26}$$

The solutions to the first equation are known by hypothesis. Using these, we can solve the second equations for $E_n^{(1)}$ and $\Psi_n^{(1)}$, and so forth. Ordinarily, perturbation theory is not carried beyond "the second order," and we do not attempt to solve for $\Psi_n^{(3)}$. For obvious reasons, the E_n^0 are called the zero-order energies, the $E_n^{(1)}$ are called the first-order energies, and so on, and the approximate functions $\Psi_n \sim \Psi_n^0 + \lambda\Psi_n^{(1)} + \lambda^2\Psi_n^{(2)}$ are called the second-order approximations to the Ψ_n. Equivalently, such a function is considered "correct to second order," and solving the three equations (5.26) is called second-order perturbation theory.

The most obvious way of solving the perturbation equations is to assume that all the various functions $\Psi_n^{(i)}$ can be expanded in terms of the complete set of eigenfunctions of $\mathbf{H}_0$, so that

$$\Psi_n^{(i)} = \sum_j c_j^{(i)} \Psi_j^0 \qquad \blacktriangleright (5.27)$$

Substitution in the second of Eqs. (5.26) yields

$$(\mathbf{H}_0 - E_n^0) \sum_j c_j^{(1)} \Psi_j^0 = (E_n^{(1)} - \mathbf{H}')\Psi_n^0 \tag{5.28}$$

Left-multiplying by Ψ_m^{0*} ($m \neq n$) and integrating gives the equation

$$c_m^{(1)}(E_m^0 - E_n^0) = -\langle\Psi_m^0|\mathbf{H}'|\Psi_n^0\rangle = -H'_{mn} \tag{5.29}$$

so

$$c_m^{(1)} = -\frac{H'_{mn}}{E_m^0 - E_n^0} \tag{5.30}$$

When $m = n$, we obtain the relation

$$H'_{nn} = E_n^{(1)} \qquad \blacktriangleright (5.31)$$

but no expression for $c_n^{(1)}$. The latter coefficient is found by normalizing the first-order expression for Ψ_n as follows:

$$\langle\Psi_n|\Psi_n\rangle = 1 \approx 1 + \lambda c_n^{(1)} + \lambda^2 \sum_j |c_j^{(1)}|^2$$

The terms quadratic in λ may be neglected. Thus, $c_n^{(1)} = 0$, and we obtain

$$\Psi_n \approx \Psi_n^0 - \lambda \sum_{m \neq n} \frac{H'_{mn}}{E_m^0 - E_n^0} \Psi_m^0 \qquad \blacktriangleright (5.32)$$

correct to the first order. We note, however, that to use this form of perturbation theory, the functions Ψ_n^0 must be nondegenerate, so that $E_m^0 \neq E_n^0$ for any m.

The corresponding expression for the first-order energy is simply

$$E_n \approx E_n^0 + \lambda E_n^{(1)} = E_n^0 + \lambda H'_{nn} \qquad \blacktriangleright (5.33)$$

As is so often the case in quantum mechanics, the energy is obtained with relative ease, in this instance involving only the single integral H'_{nn}. Finding the correct wavefunction, on the other hand, involves the sum of an infinite number of integrals.

Having solved the first-order perturbation problem, we may proceed to the more difficult problem of obtaining the second-order corrections.‡ Writing

$$\Psi_n^{(2)} = \sum_j c_j^{(2)} \Psi_j^0$$

and substituting in the third of Eqs. (5.26), we obtain

$$(\mathbf{H}_0 - E_n^0) \sum_j c_j^{(2)} \Psi_j^0 = E_n^{(2)} \Psi_n^0 + (E_n^{(1)} - \mathbf{H}') \sum_{j \neq n} c_j^{(1)} \Psi_j^0 \qquad (5.34)$$

Left-multiplying by Ψ_n^{0*} and integrating yields the relation

$$
\begin{aligned}
E_n^{(2)} &= \sum_{m \neq n} c_m^{(1)} \langle \Psi_n^0 | \mathbf{H}' | \Psi_m^0 \rangle \\
&= - \sum_{m \neq n} \frac{H'_{mn} H'_{nm}}{E_m^0 - E_n^0}
\end{aligned}
\qquad (5.35)
$$

Once again, $c_n^{(2)} = 0$ from the normalization condition. Left-multiplying by Ψ_m^{0*} ($m \neq n$) and integrating we find that

$$(E_m^0 - E_n^0) c_m^{(2)} = c_m^{(1)} H'_{nn} - \sum_{j \neq n} c_j^{(1)} H'_{mj}$$

$$c_m^{(2)} = - \frac{H'_{mn} H'_{nn}}{(E_m^0 - E_n^0)^2} + \sum_{j \neq n} \frac{H'_{mj} H'_{jn}}{(E_j^0 - E_n^0)(E_m^0 - E_n^0)}$$

‡ It is possible to obtain the energy correct to the order $2n + 1$ with a wavefunction correct only to order n.

Therefore, the second-order perturbation expressions are

$$E_n \approx E_n^0 + \lambda H_{nn}' - \lambda^2 \sum_{m \neq n} \frac{H_{nm}' H_{mn}'}{E_m^0 - E_n^0}$$

and

$$\Psi_n \approx \Psi_n^0 - \lambda \sum_{m \neq n} \frac{H_{mn}'}{E_m^0 - E_n^0} \Psi_m^0$$

$$- \lambda^2 \sum_{m \neq n} \left[\frac{H_{mn}' H_{nn}'}{(E_m^0 - E_n^0)^2} - \sum_{j \neq n} \frac{H_{mj}' H_{jn}'}{(E_j^0 - E_n^0)(E_m^0 - E_n^0)} \right] \Psi_m^0 \quad \blacktriangleright (5.36)$$

There is a very close and important relationship between perturbation theory and the variation principle, which we will use both to obtain approximate solutions of the secular equation and to find the correct perturbation expressions when there is degeneracy in the zero-order functions Ψ_n^0. We again write

$$\mathcal{3C} = \mathbf{H}_0 + \lambda \mathbf{H}'$$

and seek this time to solve for the best possible Ψ_n in the form

$$\Psi_n = \sum_i c_{ni} \Psi_i^0 \tag{5.37}$$

This may be considered either as an expansion of Ψ_n in terms of the eigenfunctions of $\mathbf{H}_0$ or, equivalently, as a way of writing Ψ_n in (infinite) linear parametric form. According to Sec. 5.1, the secular equation becomes

$$\begin{aligned} |H_{ij} - E\delta_{ij}| &= 0 \\ H_{ij} &= \langle \Psi_i^0 | \mathcal{3C} | \Psi_j^0 \rangle = E_i^0 \delta_{ij} + \lambda H_{ij}' \\ H_{ij}' &= \langle \Psi_i^0 | \mathbf{H}' | \Psi_j^0 \rangle \end{aligned} \tag{5.38}$$

with Ψ_n determined by the solutions to

$$\sum_j (H_{kj} - E_n \delta_{kj}) c_{nj} = 0 \qquad \text{for all } k,n \tag{5.39}$$

Of course, the secular equation is of infinite order.

Suppose there exists a nondegenerate state Ψ_n^0 of $\mathbf{H}_0$ for which the relation

$$|H_{mn}| \ll |E_n^0 - E_m^0| \qquad\qquad \blacktriangleright (5.40)$$

holds for all m. We write (5.39) in the form

$$\lambda \sum_{j \neq n} H_{nj}' c_{nj} = (E_n - E_n^0 - \lambda H_{nn}') c_{nn} \tag{5.41}$$

for $k = n$ and as

$$\lambda \sum_{j \neq n \neq k} H_{kj}' c_{nj} + \lambda H_{kn}' c_{nn} = (E_n - E_k^0 - \lambda H_{kk}') c_{nk} \tag{5.42}$$

for $k \neq n$. Defining $b_{nj} = c_{nj}/c_{nn}$, we obtain the relation

$$E_n = E_n^0 + \lambda H_{nn}' + \lambda \sum_{j \neq n} b_{nj} H_{nj}' \qquad (5.43)$$

from (5.41). The coefficients b_{nj} may be found by iteration from the second equation, assuming that $c_{nn} \approx 1$, $c_{nj} \approx 0$. Transforming (5.42), we may write

$$b_{nk} = \frac{\lambda H_{kn}'}{E_n - E_k^0 - \lambda H_{kk}'} + \lambda \sum_{j \neq n \neq k} \frac{H_{kj}' b_{nj}}{E_n - E_k^0 - \lambda H_{kk}'} \qquad (5.44)$$

where the sum is presumed to be small compared to the first term. Thus, substituting the first term of the right-hand side of (5.44) into (5.43), we get

$$E_n \approx E_n^0 + \lambda H_{nn}' + \lambda^2 \sum_{j \neq n} \frac{H_{nj}' H_{jn}'}{E_n - E_j^0 - \lambda H_{jj}'} \qquad \blacktriangleright (5.45)$$

which is identical to (5.36) if (5.40) is true for all m. [It is possible to show that the wavefunctions Ψ_n derived from the secular equation are also identical to those of (5.36).] Thus perturbation theory can be viewed as an approximate method of solving an infinite-order secular equation.

The variational form of the perturbation equations have the following decided advantage over the formalism developed in the first part of this section: it is not necessary to know $\mathbf{H}_0$. In (5.45), we may set $\lambda = 1$, replace H_{nj}' by H_{nj}, and replace $E_n^0 + H_{nn}'$ by H_{nn}. Thus

$$E_n \approx H_{nn} - \sum_{j \neq n} \frac{H_{nj} H_{jn}}{H_{jj} - H_{nn}} \qquad \blacktriangleright (5.46)$$

which provides a simple approximate expansion of any secular determinant.

The case of degenerate zero-order eigenfunctions is also most easily handled within the framework of the variational approach. The essence of the procedure is to mix in excited states using perturbation theory, omitting those excited states which are degenerate or near-degenerate with the state being determined. The degenerate states are then diagonalized. Consider, for example, the case in which E_p^0 and E_q^0 are equal or nearly equal. Then we define

$$E_{p1} = E_p^0 + \lambda H_{pp}' - \lambda^2 \sum_{j \neq p \neq q} \frac{H_{pj}' H_{jp}'}{E_j^0 - E_p^0}$$

$$E_{q1} = E_q^0 + \lambda H_{qq}' - \lambda^2 \sum_{j \neq p \neq q} \frac{H_{qj}' H_{jq}'}{E_j^0 - E_q^0} \qquad \blacktriangleright (5.47)$$

The final values E_r and E_q are to be found by diagonalizing the 2×2 secular equation

$$\begin{vmatrix} E_{p1} - E & E_{pq} \\ E_{qp} & E_{q1} - E \end{vmatrix} = 0 \qquad\qquad \blacktriangleright (5.48)$$

where $E_{pq} = \lambda H'_{pq} - \lambda^2 \sum_{j \neq p \neq q} [H'_{qj}H'_{jp}/(E_j^0 - E_q^0)]$. The generalization to the case in which several states are degenerate is obvious.

It should be mentioned in closing this section that the perturbation formalism has one advantage over the variational approach as follows: the "convergence" of the perturbation expression allows at least a qualitative estimate of the error in terminating the treatment at the first or second order. If, for example, the second-order energy $E_n^{(2)}$ is very much smaller than the first-order correction $E_n^{(1)}$, we may well expect that the error in E_n is smaller, perhaps much smaller, than $E_n^{(2)}$. Occasionally, a careful analysis of the error can be given. However, the problem of having to evaluate and sum an infinite number of integrals remains, and it is only in fairly special cases that this type of error analysis can be done.

5.3 AN APPLICATION OF PERTURBATION THEORY

In this text, we shall use perturbation theory primarily as a method of obtaining approximate solutions to the variation equations. In order not to leave the reader with the impression that perturbation theory is of no other value, we present a simple example which illustrates at least some of the power of the method.

In Sec. 4.4 we considered the rigid rotor, with Hamiltonian

$$\mathfrak{K} = -\frac{\hbar^2}{2\mu} \left[\frac{1}{r^2 \sin\theta} \frac{\partial}{\partial\theta} \left(\sin\theta \frac{\partial}{\partial\theta} \right) + \frac{1}{r^2 \sin^2\theta} \frac{\partial^2}{\partial\varphi^2} \right]$$

and eigenfunctions $Y_{lm} = \Theta_{lm}(\theta)\Phi_m(\varphi)$, where $\Phi_m = (1/\sqrt{2\pi})e^{im\varphi}$. The functions Θ_{lm} are given by Eq. (3.32). Let us now simplify this problem somewhat by assuming that the rotor is restricted to move in a plane, with $\theta = \pi/2$. The Hamiltonian then becomes

$$\mathbf{H}_0 = -\frac{\hbar^2}{2\mu} \left[\frac{1}{r^2} \frac{\partial^2}{\partial\varphi^2} \right] \qquad\qquad (5.49)$$

with eigenvectors

$$\Psi_n^0 = \frac{1}{\sqrt{2\pi}} e^{in\varphi} \qquad\qquad (5.50)$$

and eigenvalues

$$E_n^0 = \frac{n^2\hbar^2}{2I} \tag{5.51}$$

We now suppose that the rotor is placed in an electric field **F**. The Hamiltonian may now be written

$$\mathcal{K} = \mathbf{H}_0 + \mu F \cos\varphi \tag{5.52}$$

where we have assumed that the dipole moment of the rotor is directed along the internuclear axis. Then, with $\lambda = F$ and $\mathbf{H'} = \mu\cos\varphi$, we obtain

$$H'_{mn} = \frac{\mu}{2\pi}\int_0^{2\pi} e^{i(m-n)\varphi}\cos\varphi\, d\varphi = \begin{cases} \dfrac{\mu}{2} & \text{for } n = m \pm 1 \\[2mm] 0 & \text{otherwise} \end{cases} \tag{5.53}$$

The first-order perturbation term is thus zero, and in the second order treatment, only two terms in the infinite sum can contribute. Thus

$$\begin{aligned} E_n &\approx E_n^0 - F^2\left(\frac{\mu^2/4}{E_{n+1}^0 - E_n^0} + \frac{\mu^2/4}{E_{n-1}^0 - E_n^0}\right) \\[2mm] &\approx \frac{n^2\hbar^2}{2I} - \frac{F^2\mu^2 I}{2\hbar^2}\left(\frac{1}{1+2n} - \frac{1}{1-2n}\right) \\[2mm] &\approx \frac{n^2\hbar^2}{2I} + \frac{F^2\mu^2 I}{\hbar^2(4n^2-1)} \end{aligned} \tag{5.54}$$

From this relation one can calculate the polarizability α_n of the rigid planar rotor since

$$\alpha_n \equiv -\frac{1}{F}\frac{\partial E_n}{\partial F} = -\frac{2\mu^2 I}{\hbar^2(4n^2-1)} \tag{5.55}$$

Note that the induced moment is in the opposite direction of the field F when $n \neq 0$, in agreement with the classical result. (Classically, the induced moment is a consequence of the rotor speeding up when the dipole is attracted to the field, and slowing down when the dipole points against the field.)

There are a number of similar cases in which the perturbation treatment is extremely useful. Most of these involve the disturbance of a system by an electric or magnetic field. Other applications include the effects of an anharmonic perturbation on the harmonic oscillator and the coupling between rotation and vibration in a diatomic molecule.

PROBLEMS

5.2 Use the first-order wavefunction (5.32) to calculate $\langle \Psi_n | \mathcal{K} | \Psi_n \rangle$. Show that to the second order in λ, the energy expression is identical to (5.36).

5.3 The energy correct to order $2n + 1$ can always be obtained from a wavefunction correct to order n. Find the third-order energy expression using (5.32).

5.4 Show that the first-order wavefunction (5.32) cannot be normalized unless Ψ_n^0 is replaced by $N\Psi_n^0$. Find N. Show that

$$\Psi_n \approx \Psi_n^0 - \lambda \sum_{m \neq n} k_m(\Psi_m^0 + k_m\Psi_n^0)$$

where $k_m = H'_{mn}/(E_m^0 - E_n^0)$. What difference does this correction make to the previous two problems?

5.5 The potential energy for a real oscillator can be expanded in the power series

$$V = \tfrac{1}{2}k_1q^2 + \tfrac{1}{6}k_2q^3 + \tfrac{1}{24}k_3q^4 + \cdots$$

where

$$k_{n-1} = \frac{\partial^n V}{\partial q^n}\bigg|_{q=0}$$

Let $\mathbf{H}_0 = \tfrac{1}{2}k_1q^2$ and $\mathbf{H}' = \tfrac{1}{6}k_2q^3 + \cdots$. Prove the following:

    (a) The first-order energy due to the perturbation involving q^3 is zero.

    (b) The first-order energy due to the perturbation involving q^4 is $\langle \Psi_v|\tfrac{1}{24}k_3q^4|\Psi_v\rangle = (a^4k_3/24)(6v^2 + 6v + 3)$, where $a = (\hbar^2/4k_1\mu)^{1/4}$.

    (c) The second-order energy due to the perturbation involving q^3 consists of only four nonvanishing terms of the form $\langle \Psi_{v+3}|q^3|\Psi_v\rangle = a^3[(v + 1)(v + 2)(v + 3)]^{1/2}$ or $\langle \Psi_{v+1}|q^3|\Psi_v\rangle = a^3[(v + 1)^3]^{1/2}$. [**Note:** Use (4.22) and (4.25)].

5.6 The Hamiltonian for a nonrigid rotor can be expanded in the power series (Pilar, p. 264)

$$\mathcal{H} = \frac{\mathbf{J}^2}{2\mu r_0{}^2} - \frac{\mathbf{J}^4}{2\mu^2 k_1 r_0{}^6} + \cdots$$

Show that the first-order energy correction is given by

$$\frac{J^2(J + 1)^2\hbar^4}{2\mu^2 k_1 r_0{}^6}$$

Also prove that the second-order perturbation vanishes. Compare these results with Eq. (4.59).

5.4 THE HELIUM ATOM

As an introduction to the next chapter, we present an extended treatment of the helium atom. This discussion will provide both an illustration of the power of the variation method as well as several examples of different approaches to the general many-electron problem in atoms.

    The Hamiltonian for the helium atom, written in atomic units, is

$$\mathcal{H} = -\frac{1}{2}\nabla_1{}^2 - \frac{1}{2}\nabla_2{}^2 - \frac{2}{R_1} - \frac{2}{R_2} + \frac{1}{r_{12}} \qquad \blacktriangleright(5.56)$$

where R_i is the distance between the nucleus and electron i, and r_{12} is the distance between the two electrons. We seek a trial function for our variation. One such function, suggested on a variety of physical and mathematical grounds, is

$$\Psi_0 = \alpha\Psi_{1s}(1)\Psi_{1s}(2) = \alpha\varphi_{1s}(1)\varphi_{1s}(2) \qquad \blacktriangleright (5.57)$$

where $\Psi_{1s} = \varphi_{1s}$ is‡ the $1s$ eigenfunction of the hydrogenic He$^+$ and α is the antisymmetrizer, inserted in order to make our trial function satisfy the Pauli exclusion principle. The function Ψ_0 would be the exact solution to the Hamiltonian (5.56) if the r_{12} term were omitted. A perturbation treatment can be carried out using this relation, but the convergence is rather slow.

The function Ψ_0 contains as yet no parameters, so it is of interest to see what type of energy we can obtain using it. Writing out the antisymmetric function Ψ_0 in determinantal form, we find§

$$\begin{aligned}
\Psi_0 &= \sqrt{\tfrac{1}{2}}(\varphi_{1s}^+(1)\varphi_{1s}^-(2) - \varphi_{1s}^-(1)\varphi_{1s}^+(2)) \\
&= \sqrt{\tfrac{1}{2}}[\varphi_{1s}(1)\varphi_{1s}(2)(\alpha_1\beta_2 - \beta_1\alpha_2)]
\end{aligned} \qquad (5.58)$$

where in the second equation we have factored out the spin functions. Evaluating the expectation energy of Ψ_0, we obtain

$$\begin{aligned}
E = \langle\Psi_0|\mathcal{K}|\Psi_0\rangle = \ &\tfrac{1}{2}\int\varphi_{1s}^{*+}(1)\varphi_{1s}^{*-}(2)\mathcal{K}\varphi_{1s}^+(1)\varphi_{1s}^-(2)\,d\tau_1\,d\tau_2 \\
&+ \tfrac{1}{2}\int\varphi_{1s}^{*-}(1)\varphi_{1s}^{*+}(2)\mathcal{K}\varphi_{1s}^-(1)\varphi_{1s}^+(2)\,d\tau_1\,d\tau_2 \\
&- \tfrac{1}{2}\int\varphi_{1s}^{*+}(1)\varphi_{1s}^{*-}(2)\mathcal{K}\varphi_{1s}^-(1)\varphi_{1s}^+(2)\,d\tau_1\,d\tau_2 \\
&- \tfrac{1}{2}\int\varphi_{1s}^{*-}(1)\varphi_{1s}^{*+}(2)\mathcal{K}\varphi_{1s}^+(1)\varphi_{1s}^-(2)\,d\tau_1\,d\tau_2
\end{aligned} \qquad (5.59)$$

The last two integrals are zero since, if we factor out the spin coordinates, the integration over the spin space of electron 1 yields integrals of the form

$$\int\alpha_1\mathcal{K}\beta_1\,d\xi_1$$

which are zero by the orthogonality of α and β ($\mathcal{K}$ does not contain the spin). Furthermore, the first two integrals are equal since $\mathcal{K}$ is symmetric in the coordinates of the two electrons. Thus,

$$E = \int\varphi_{1s}^*(1)\varphi_{1s}^*(2)\mathcal{K}\varphi_{1s}(1)\varphi_{1s}(2)\,d\tau_1\,d\tau_2 \qquad \blacktriangleright (5.60)$$

‡ In the future, we reserve the use of Ψ for the total wavefunction, and we will use φ and ψ to represent one-electron functions, i.e., orbitals.
§ Note that we are using the determinantal form in Eq. (3.57) via the notation scheme specified by Eq. (3.60).

where we have carried out the integration over the spin coordinates. Writing out $\mathcal{H}$ explicitly, we obtain

$$
\begin{aligned}
E &= 2 \int \varphi_{1s}^*(i) \left(-\frac{\nabla^2}{2} - \frac{2}{R_i} \right) \varphi_{1s}(i)\, d\tau_i \\
&\quad + \int \varphi_{1s}^*(1) \varphi_{1s}^*(2) \frac{1}{r_{12}} \varphi_{1s}(1) \varphi_{1s}(2)\, d\tau_1\, d\tau_2 \\
&= 2 \langle \varphi_{1s} | -\frac{\nabla^2}{2} - \frac{2}{R} | \varphi_{1s} \rangle + \langle \varphi_{1s}\varphi_{1s} | \frac{1}{r_{12}} | \varphi_{1s}\varphi_{1s} \rangle
\end{aligned} \tag{5.61}
$$

The first term is clearly equal to twice the energy of a single electron bound to an He$^+$ ion and is therefore equal to -4 atomic units (au). The second integral is more difficult to evaluate; as shown on the following page, it is equal to 1.25 au. Thus, $E = -2.750$ au, which compares (not too badly) with the experimental $E = -2.904$.

We now seek a better trial function containing variable parameters. For a beginning, we will choose only functions which can be written as Slater determinants, in the form of (5.57) and (5.58). The easiest way of doing this is to let $\varphi_{1s} = Ne^{-\zeta r}$, where ζ is a variable parameter. This orbital is a special case of the general Slater-type orbital discussed in Chap. 4. Evaluating the energy of this new function, we may at once write

$$
E = 2 \langle \varphi_{1s} | -\frac{\nabla^2}{2} - \frac{2}{R} | \varphi_{1s} \rangle + \langle \varphi_{1s}\varphi_{1s} | \frac{1}{r_{12}} | \varphi_{1s}\varphi_{1s} \rangle \qquad \blacktriangleright (5.62)
$$

but in this case φ_{1s} is not an eigenfunction of the operator $-\nabla^2/2 - 2/R$. It is, however, an eigenfunction of the operator $-\nabla^2/2 - \zeta/R$, with eigenvalue $-\frac{1}{2}\zeta^2$, so we find

$$
E = -\zeta^2 + 2 \langle \varphi_{1s} | \frac{\zeta - 2}{R} | \varphi_{1s} \rangle + \langle \varphi_{1s}\varphi_{1s} | \frac{1}{r_{12}} | \varphi_{1s}\varphi_{1s} \rangle \tag{5.63}
$$

The first integral, a nuclear attraction, may be readily evaluated to give

$$
\langle \varphi_{1s} | \frac{1}{R} | \varphi_{1s} \rangle = N^2 \int_0^\infty \frac{r^2 e^{-2\zeta r}}{r}\, dr = N^2 \frac{1}{4\zeta^2} = \zeta \tag{5.64}
$$

since

$$
\frac{1}{N^2} = \int_0^\infty r^2 e^{-2\zeta r}\, dr = \frac{1}{4\zeta^3}
$$

The second integral is more difficult because of the $1/r_{12}$ term. To evaluate it, we make use of the well-known Legendre polynomial expansion of $1/r_{12}$

in the form‡

$$\frac{1}{r_{12}} = \sum_{l=0}^{\infty} \sum_{m=-l}^{l} \frac{(l-|m|)!}{(l+|m|)!} \frac{r_<^l}{r_>^{l+1}} P_l^m(\cos\theta_1) P_l^m(\cos\theta_2) e^{im(\varphi_1-\varphi_2)}$$

▶(5.65)

where $r_<(r_>)$ is the less (greater) of R_1 and R_2, and θ_i and φ_i are the polar angles the R_i make with a fixed arbitrary coordinate system whose origin is also the origin of R_1 and R_2 (see Fig. 5.1). Since the P_l^m form an orthogonal set, the infinite sum in (5.65) reduces in most applications to just a few terms. In this example, since the φ_{1s} Slater-type orbitals involve only P_0, the second integral in (5.63) reduces to

$$16\zeta^6 \int_0^{\infty} \int_0^{\infty} \frac{r_1^2 r_2^2}{r_>} e^{-2\zeta r_1} e^{-2\zeta r_2} \, dr_1 \, dr_2$$

$$= 16\zeta^6 \int_0^{\infty} r_1^2 e^{-2\zeta r_1} \left(\frac{1}{r_1} \int_0^{r_1} r_2^2 e^{-2\zeta r_2} \, dr_2 + \int_{r_1}^{\infty} r_2 e^{-2\zeta r_2} \, dr_2 \right) dr_1 = \frac{5}{8}\zeta$$

(5.66)

Thus

$$E = -\zeta^2 + 2\zeta^2 - 4\zeta + \tfrac{5}{8}\zeta = \zeta^2 - 4\zeta + \tfrac{5}{8}\zeta \qquad (5.67)$$

where the energy is again expressed in atomic units. Solving for ζ, we obtain

$$\frac{\partial E}{\partial \zeta} = 0 = 2\zeta - 4 + \tfrac{5}{8}$$

so

$$\zeta = 2 - \tfrac{5}{16} = {}^{27}\!/_{16}$$

‡ For a proof, see, for example, Eyring, Walter, and Kimball, p. 369.

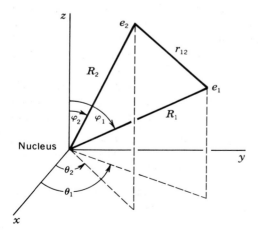

Fig. 5.1 Coordinate system for the helium atom.

and

$$E = (2 - \tfrac{5}{16})^2 = 2.848 \qquad (5.68)$$

The energy shows a pleasing reduction and is in considerably better agreement with experiment. The constant ζ has an interesting interpretation; it may be considered an "effective" nuclear charge, with the constant $\tfrac{5}{16}$ regarded as a "screening factor." In this interpretation, each electron moves in the nuclear field partially screened by the other electron; the repulsion between the two electrons is averaged out and treated as an effective nuclear field.

Keeping to the functional form (5.57), we can make one more improvement. We write φ_{1s} as a perfectly arbitrary function of r and minimize the energy of Ψ_0 of this form. One way of writing φ_{1s} as a general function of r is to expand it as a linear combination of $1s$ Slater-type orbitals as follows:

$$\varphi_{1s} = \sum_i c_i \chi_{1s}(\zeta_i)$$
$$\chi_{1s}(\zeta_i) = N e^{-\zeta_i r} \qquad \blacktriangleright (5.69)$$

Using some two-to-five functions χ_i, (Clementi, 1965) one can obtain this particular‡ minimum E with an accuracy of from four to six significant figures. However, the actual result, $E = -2.862$, is not much of an improvement over the energy of our previous functions, and a discrepancy of 0.042 au (which is equal to 1.1 ev or 26 kcal/mole) between experimental and calculated E remains.

The functional form (5.57) assumes that the two electrons move independently of each other, with the electronic repulsion term averaged out as an effective nuclear shielding. Actually, of course, the interelectronic repulsion is instantaneous, and the motion of the two electrons is "correlated," definitely not independent. Considering this, the agreement between calculated and experimental energy using (5.57) (that is, within 1.5 percent) is both surprising and somewhat disturbing; we suspect that a wavefunction which neglects electron correlation may not be very suitable for calculating other properties of helium.

It is not especially difficult to improve on the functional form (5.57) in the case of helium, and as early as 1930 Hylleraas§ was able to obtain an energy within 0.001 au of the experimental, using very general func-

---

‡ The method by which this energy is obtained, the method of Hartree and Fock, will be investigated in the next chapter. In the Hartree-Fock method, the total wavefunction Ψ_0 is limited to an antisymmetrized product of orbitals, and we solve for "best possible" orbitals which minimize E.

§ For an excellent summary of Hylleraas' work, see L. Pauling and E. B. Wilson, Jr., "Introduction to Quantum Mechanics," p. 224, McGraw-Hill Book Company, New York, 1935.

tions optimized by the variation method. Unfortunately, the methods used by Hylleraas have not and probably cannot be generalized to many-electron systems. Other techniques, such as the method of configuration interaction‡ to be discussed in the next chapter, are applicable in principle to all systems but are found to be slowly converging and quite tedious. The search for a reasonably easy and accurate method of treating the correlation problem remains an area of great research interest, but the hope of finding such a method is not great. For most purposes, we shall be satisfied with orbital functions of the form (5.57); they are usually capable of producing an energy within 1 or 2 percent of the experimental. When using the orbital functions to calculate other properties, it will be necessary to take the results with a large grain of salt, together with a strong dose of experience.

The reader may possibly have expected better results from quantum mechanics. To this one can only answer that the variation principle guarantees that one can come as close as one likes to the correct answers, the only limitation being the amount of time, effort, and ingenuity one is willing to spend. In quantum chemistry we deal with systems of many interacting particles, and one should be properly amazed that in systems of such enormous complexity one can calculate any property at all. Furthermore, as we shall see, there are a number of atomic and molecular properties which can be obtained quite rigorously, through the use of symmetry-related arguments. Finally, of course, quantum-mechanical arguments can often rationalize and qualitatively explain behavior which is inexplicable in the framework of classical mechanics.

SUGGESTIONS FOR FURTHER READING

1. Hameka, Hendrik F.: "Advanced Quantum Chemistry," Addison-Wesley Publishing Company, Inc., Reading, Mass. 1965.

   The best text-book discussion of perturbation theory of which I am aware is to be found in Chap. 4 of this book. Chapter 5 also contains an excellent discussion of time-dependent perturbation theory, a subject completely omitted from the present text. The student more interested in applications than in theory may find the references footnoted in the present chapter of value.

‡ R. K. Nesbet and R. E. Watson [*Phys. Rev.*, **110**:1073 (1958)] have carried out a very accurate configuration-interaction treatment of helium.

6

THE ELECTRONIC STRUCTURE OF ATOMS

According to convention there is
a sweet and a bitter, a hot and a
cold, and according to convention there is order. In truth there
are atoms and a void.
Democritus

6.1 ELECTRON CONFIGURATIONS

The concept of an electron configuration is a natural outgrowth of our earlier discussions of the hydrogen and helium atoms. In the latter case, we were able to put two electrons in hydrogenic (Slater-type) $1s$ orbitals; we are prevented by the Pauli exclusion principle, however, from putting more than two electrons in the $1s$ orbital of other n-electron ($n > 2$) systems. Keeping to the Slater determinant form, that is, writing the total wavefunction as an antisymmetrized product of mutually ortho-gonal one-electron orbitals, we see that a third electron cannot be added to the $1s$ shell and must be placed instead in the $2s$ (or higher) orbital. Similarly, the $2s$ shell ($l = 0$) can hold two electrons (one with spin α, one with spin β), and the $2p$ shell ($l = 1$) can hold six electrons (two each, one with α spin and one with β spin, for $m_l = 1$, 0, and -1). By contrast with the hydrogen atom, the $2s$ and $2p$ orbitals do not have the same energy, and the $2s$ shell is filled before the $2p$ shell. According to the well-known aufbau principle, the orbitals are filled in the order $1s$, $2s$, $2p$, $3s$, $3p$, $3d$, $4s$, . . . , although in the neutral and singly ionized transition metal ions, and in the alkali metals and alkaline earths, the order may break down somewhat, with the $4s$ shell filled or half-filled while the $3d$ shell is still only partly filled or empty. A *configuration* is just a specification (n and l only) of which orbitals are occupied, and we say that the carbon atom has the configuration $1s^2 2s^2 2p^2$ and the neutral chromium atom the configuration $1s^2 2s^2 2p^6 3s^2 3p^6 3d^5 4s^1$. A shell having its full compliment of electrons is called *closed*; both the carbon and chromium atoms have *open-shell configurations*, the latter having two open, half-filled shells.

Of course, the concept of a configuration is a theoretical construct, arising from the approximate solution of Schrödinger's equation using Slater determinants as trial wavefunctions. However, there is also a great deal of experimental evidence for the "existence" of configurations. Beginning with Mendelyeev and continuing through the modern synthesis of transuranic elements, chemists have noted that many of the properties of atoms are determined by position in the periodic table. The shape and form of the modern table is, of course, determined entirely by the aufbau principle. The "rule of eight," as formulated by Lewis in 1915, may be considered the first expression of quantum chemistry and is again determined by configuration; only eight electrons can be placed in the $n = 2$ shell (or the $n = 3$ shell when limited to s and p electrons).‡ There is also excellent spectroscopic evidence for configurations, the best being offered by the absorption of x-rays by inner-shell electrons at fixed wavelengths, and by x-ray scattering experiments which confirm the shell structure of the atom.

Despite this evidence, electron configurations are an approximation, arising entirely from the one-electron treatment of many-electron atoms. In the case of helium considered in the previous section, we were able to interpret our wavefunction in terms of an independent electron model, with the electron repulsion averaged out as an effective nuclear shielding. Although this is an obviously crude model, it gave about 98.5 percent of the total energy of helium when the orbitals were optimized using the variation method. The model can be generalized to many-electron systems by using the *central-field approximation*, which assumes directly that the correct Hamiltonian can be replaced by an effective Hamiltonian in which the electron repulsions are averaged out and replaced by an effective spherically symmetric central field. As we will show, this approxi-

‡ Older usage would restrict the term "shell" to a specification of the principal quantum number and would employ the word "subshell" for a specification of l. In quantum chemistry, we use the word "shell" both ways and say, for example, that the configuration $1s^2 2s 2p^3$ contains two open shells.

mation is useful for the labeling and classification of experimental atomic states, and for the approximate calculation of atomic energies and properties.

The nonrelativistic Hamiltonian for an n-electron atom is

$$\mathcal{3C} = -\frac{\hbar^2}{2m} \sum_{i=1}^{n} \nabla_i^2 - \sum_{i=1}^{n} \frac{Ze^2}{R_i} + \sum_{i=2}^{n} \sum_{\substack{j=1 \\ j<i}}^{i-1} \frac{e^2}{r_{ij}} \qquad \blacktriangleright (6.1)$$

where we have, for the moment, omitted certain potential terms occurring in the Dirac relativistic treatment.‡ In the central-field approximation, we replace the Hamiltonian (6.1) by an effective Hamiltonian of the form

$$\mathcal{3C}_0 = -\frac{\hbar^2}{2m} \sum_{i=1}^{n} \nabla_i^2 + \sum_{i=1}^{n} \mathbf{V}_i(r_i) \qquad \blacktriangleright (6.2)$$

where we shall leave the specific nature of the $\mathbf{V}_i$ otherwise undetermined except to say that they are spherically symmetric around the nucleus. When we later discuss the Hartree-Fock method, we shall introduce a specific form for the $\mathbf{V}_i$, but this is not essential to the central-field approximation.

Since the central-field Hamiltonian is written as a sum of one-electron operators, its eigenfunctions are products (which we antisymmetrize) of one-electron functions (orbitals). Referring to Prob. 4.21 and to the discussion of the helium atom in the previous section, we see that these orbitals are a product of a radial function, an angular part, and a spin function

$$\varphi_{nlms} = R_{nl}(r) Y_{lm}(\theta,\varphi) \xi \qquad \blacktriangleright (6.3)$$

The Y_{lm} are the spherical harmonics already discussed, but the form of the R_{nl} cannot be determined unless the $\mathbf{V}_i$ are known. By analogy with the hydrogenic functions, we label the R_{nl} as $1s$, $2s$, $2p$, . . . , although the radial functions need no longer be Laguerre polynomials. The total eigenfunction of the approximate Hamiltonian (6.2) is thus

$$\Psi_0 = \mathcal{C} | \varphi_{n_1 l_1 m_{l_1} m_{s_1}}(1) \varphi_{n_2 l_2 m_{l_2} m_{s_2}}(2) \cdots \varphi_{n_n l_n m_{l_n} m_{s_n}}(n) \rangle \qquad \blacktriangleright (6.4)$$

which is conveniently written as a Slater determinant.§ A specification of an electronic configuration gives the nl values of all the orbitals, but, in general, it does not specify Ψ_0 completely, since there are m_{l_i} and m_{s_i} quantum numbers as well. For example, the configuration $1s^2 2s^2 2p^2$

---

‡ We will introduce these terms later as perturbations. Note that we have also assumed that the nuclear and electronic motions are separable, which can be rigorously demonstrated. To be precise, the electronic mass m should be replaced by the reduced mass $\mu = mM/(m + M)$.

§ See Eq. (3.57).

implies the existence of 15 independent Slater determinants, one for each (ordered) arrangement of the two electrons in the open shell. For a closed-shell configuration, there is but one possible determinant.

The approximate wavefunction of (6.4) implies that the eigenvalues l_i and m_i are all observables, since the constituent orbitals contain the Y_{lm} as factors. Furthermore, the discussion in Chap. 3 on the addition of angular-momentum vectors suggests that the total angular momentum of the atom, expressed as the values of L and M_L, may also be observable. This conclusion, obviously valid for the approximate Hamiltonian (6.2), will also be valid for the nonrelativistic Hamiltonian (6.1), provided it commutes with $\mathbf{L}^2$ and $\mathbf{L}_z$. We have already shown in Chap. 3 that these angular-momentum operators commute with the kinetic-energy operator ∇^2, and we need only prove that they commute with the potential-energy term.

It is easy enough to show that the individual $\mathbf{l}_{z_i}$ do *not* commute with the Hamiltonian (6.1) since they do not commute with the potential terms $1/r_{ij}$; thus

$$[\mathbf{l}_{z_i}, 1/r_{ij}] = -i\hbar \frac{x_i y_j - y_i x_j}{r_{ij}} \tag{6.5}$$

However, the sum $\mathbf{l}_{z_i} + \mathbf{l}_{z_j}$ commutes with $1/r_{ij}$ since

$$[\mathbf{l}_{z_i}, 1/r_{ij}] + [\mathbf{l}_{z_j}, 1/r_{ij}] = 0 \tag{6.6}$$

Thus $\mathbf{L}_z$ commutes with $1/r_{ij}$. More trivially, the $\mathbf{l}_{z_i}$ commute with the potential terms Z/R_i, so $\mathbf{L}_z$ commutes with $\mathcal{3C}$ and is an observable. By symmetry, $\mathbf{L}_x$ and $\mathbf{L}_y$ commute with $\mathcal{3C}$ and so $\mathcal{3C}$, $\mathbf{L}^2$, and $\mathbf{L}_z$ form a mutually commuting set. It is also possible to show that the individual $\mathbf{l}_i^2$ do not commute with $\mathcal{3C}$, so the set is apparently complete as far as ordinary angular momentum is concerned. Since $\mathcal{3C}$ does not contain any spin coordinates, $\mathbf{S}^2$ and $\mathbf{S}_z$ also commute with $\mathcal{3C}$, and we finally obtain $\mathcal{3C}$, $\mathbf{L}^2$, $\mathbf{S}^2$, $\mathbf{L}_z$, and $\mathbf{S}_z$ as a set of five mutually commuting observables. Note that although the central-field approximation allows a different choice of commuting observables, we take the set appropriate to the *correct* Hamiltonian.

PROBLEMS

6.1 Prove that $\mathbf{L}_z$ commutes with $\mathcal{3C}$ by verifying Eqs. (6.5) and (6.6).

6.2 TERMS AND COUPLING SCHEMES

The set $\mathcal{3C}$, $\mathbf{S}^2$, $\mathbf{L}^2$, $\mathbf{S}_z$, and $\mathbf{L}_z$ is not a unique set of commuting observables, in the sense that there exists an equivalent set $\mathcal{3C}$, $\mathbf{S}^2$, $\mathbf{L}^2$, $\mathbf{J}^2$, and $\mathbf{J}_z$ which is also mutually commuting and which spans the same space. The vector **J** was defined in Sec. 3.5 as the *total* angular momentum, that is,

$$\mathbf{J} = \mathbf{L} + \mathbf{S}.$$

It is obvious that $\mathbf{J}_z$ commutes with $\mathfrak{JC}$, and thus that $\mathfrak{JC}$ commutes with $\mathbf{J}^2$. The other commutation relations are also easily proved, and, using the discussion on the addition of angular momenta in Sec. 3.4, we find that the two sets of operators indeed span the same operator space.

Since we have sets of five commuting observables, the states of atomic systems can be classified according to either the values of E, S, L, M_S, and M_L or the values of E, S, L, J, and M_J. Either coupling scheme will do since they are equivalent. (This equivalence breaks down when we alter the Hamiltonian, for example, by placing the atom in a field or by including the spin-orbit interaction in the Hamiltonian.) Both schemes are known as Russell-Saunders coupling, after the originators, in contradistinction to the more complex j-j coupling schemes, in which the individual $\mathbf{s}_i$ and $\mathbf{l}_i$ are coupled together to form $\mathbf{j}_i$, after which the $\mathbf{j}_i$ are coupled to form $\mathbf{J}$.

For the present, we choose the SLM_SM_L scheme because it is somewhat simpler. We label the eigenkets $|ESLM_SM_L\rangle$, where E specifies the energy of the system. Often we do not know the value of E, and so we omit it or give it arbitrary values ($E_1, E_2, \ldots$) to distinguish the ket $|ESLM_SM_L\rangle$ from others with the same S, L, M_S, and M_L, but with differing E. The energy does not [with the present Hamiltonian (6.2)] depend on M_L or M_S since the Hamiltonian does not distinguish the z direction from any other. The energy does depend quite strongly on S and L, however, and so it is customary to label the energies of states according to the values of S and L, using the notation ^{2S+1}L. This symbol is called a *term*, and the collection of $(2S + 1)(2L + 1)$ states,‡ one for each pair of values M_S and M_L, all have the same term symbol (and the same energy).

The values of L are not denoted numerically but are given letter symbols in the same way that orbitals are labeled. For $L = 0, 1, 2, 3, 4, 5, \ldots$ we write S, P, D, F, G, H, $\ldots$, and a typical term may be written ^3P. The superscript $2S + 1$ is called the *multiplicity* of the term, and we call the term a *singlet, doublet, triplet, quartet, quintet,* or *sextet,* according to whether $2S + 1 = 1, 2, 3, 4, 5,$ or 6. Thus the term ^3P is called a *triplet P*. Later on we will wish to specify J as well, and we do so by adding a subscript to the term symbol, $^{2S+1}L_J$. Such a symbol defines what is called a *level*. The values of J range from $S + L$ to $|S - L|$ as discussed in Chap. 3. The states of a given level (and for the present, the levels of a given term) all have the same energy, and we say that the term ^{2S+1}L is

‡ See the discussion following Eq. (3.25).

$(2S + 1)(2L + 1)$-fold degenerate, whereas the level $^{2S+1}L_J$ is $(2J + 1)$-fold degenerate.

As mentioned above, specification of an atomic electronic configuration is not sufficient to specify a state. Within a given configuration (with one or more open shells) there are many Slater determinants, each determinant corresponding to different possible values for the l_i, m_{l_i}, and m_{s_i}. We can form linear combinations of these determinants which are eigenfunctions of $\mathbf{L}^2$, $\mathbf{S}^2$, $\mathbf{L}_z$, and $\mathbf{S}_z$ and which will therefore serve as approximate eigenfunctions of the exact $\mathcal{3C}$. In order to do this, we must first consider which terms may occur in a given configuration.

If a shell is closed, it is clear that both the m_{l_i} and m_{s_i} add to give zero, so the only possible term is a ^1S. Furthermore, if an atom has one or more shells outside a closed core, the closed shells may be neglected in the coupling since they contribute no net angular momentum of any kind. A single electron outside a closed core therefore has $L = l$, $S = \frac{1}{2}$, and the appropriate term symbol for the boron atom is ^2P. For two electrons, the situation is considerably more complicated. At first we might suppose that L can range from $l_1 + l_2$ to $|l_1 - l_2|$ and that S can be either 0 or 1, but the Pauli exclusion principle ordinarily restricts the number of different terms which can occur. Let us begin, therefore, with an example, and try to derive the allowed terms of the lowest configuration of neutral carbon, $1s^2 2s^2 2p^2$.

As noted in the previous section, the $1s^2 2s^2 2p^2$ configuration gives rise to 15 different Slater determinants, one for each possible set of values m_{l_1}, m_{l_2}, m_{s_1}, and m_{s_2}. (Note that certain kets, with $m_{l_1} = m_{l_2}$ and $m_{s_1} = m_{s_2}$, are forbidden by the Pauli exclusion principle.) The determinants can be

Table 6.1 The 15 Slater determinants for the $2p^2$ configuration

			M_L	M_S
$\lvert D_1 \rangle$	$= \lvert 1\ 1\ \frac{1}{2}\ -\frac{1}{2} \rangle$	$= \lvert 1^+ 1^- \rangle$	2	0
$\lvert D_2 \rangle$	$= \lvert 1\ 0\ \frac{1}{2}\ \frac{1}{2} \rangle$	$= \lvert 1^+ 0^+ \rangle$	1	1
$\lvert D_3 \rangle$	$= \lvert 1\ 0\ \frac{1}{2}\ -\frac{1}{2} \rangle$	$= \lvert 1^+ 0^- \rangle$	1	0
$\lvert D_4 \rangle$	$= \lvert 1\ 0\ -\frac{1}{2}\ \frac{1}{2} \rangle$	$= \lvert 1^- 0^+ \rangle$	1	0
$\lvert D_5 \rangle$	$= \lvert 1\ 0\ -\frac{1}{2}\ -\frac{1}{2} \rangle$	$= \lvert 1^- 0^- \rangle$	1	-1
$\lvert D_6 \rangle$	$= \lvert 1\ -1\ \frac{1}{2}\ \frac{1}{2} \rangle$	$= \lvert 1^+ -1^+ \rangle$	0	1
$\lvert D_7 \rangle$	$= \lvert 1\ -1\ \frac{1}{2}\ -\frac{1}{2} \rangle$	$= \lvert 1^+ -1^- \rangle$	0	0
$\lvert D_8 \rangle$	$= \lvert 1\ -1\ -\frac{1}{2}\ \frac{1}{2} \rangle$	$= \lvert 1^- -1^+ \rangle$	0	0
$\lvert D_9 \rangle$	$= \lvert 1\ -1\ -\frac{1}{2}\ -\frac{1}{2} \rangle$	$= \lvert 1^- -1^- \rangle$	0	-1
$\lvert D_{10} \rangle$	$= \lvert 0\ 0\ \frac{1}{2}\ -\frac{1}{2} \rangle$	$= \lvert 0^+ 0^- \rangle$	0	0
$\lvert D_{11} \rangle$	$= \lvert 0\ -1\ \frac{1}{2}\ \frac{1}{2} \rangle$	$= \lvert 0^+ -1^+ \rangle$	-1	1
$\lvert D_{12} \rangle$	$= \lvert 0\ -1\ \frac{1}{2}\ -\frac{1}{2} \rangle$	$= \lvert 0^+ -1^- \rangle$	-1	0
$\lvert D_{13} \rangle$	$= \lvert 0\ -1\ -\frac{1}{2}\ \frac{1}{2} \rangle$	$= \lvert 0^- -1^+ \rangle$	-1	0
$\lvert D_{14} \rangle$	$= \lvert 0\ -1\ -\frac{1}{2}\ -\frac{1}{2} \rangle$	$= \lvert 0^- -1^- \rangle$	-1	-1
$\lvert D_{15} \rangle$	$= \lvert -1\ -1\ \frac{1}{2}\ -\frac{1}{2} \rangle$	$= \lvert -1^+ -1^- \rangle$	-2	0

specified in the $|l_1 l_2 m_{l_1} m_{l_2} m_{s_1} m_{s_2}\rangle$ scheme, which is conveniently abbreviated to $|m_{l_1} m_{l_2} m_{s_1} m_{s_2}\rangle$, since $l_1 = l_2 = 1$ and $s_1 = s_2 = \frac{1}{2}$. The determinants are given in Table 6.1. Values of $M_L = m_{l_1} + m_{l_2}$ and $M_S = m_{s_1} + m_{s_2}$ are also listed. An alternative system of nomenclature, in which the spin values are specified by superscript plus and minus signs, is also given. This system has the virtue of extreme compactness, and, for example, an entire determinant is specified‡ by the ket $|1^+ 1^-\rangle$.

There is only one ket with $M_L = 2$, and it has $M_S = 0$. Since this is the maximum value of M_L, the maximum value of L is 2, and the maximum value of S for this L is 0. Thus we have a ^1D term for the $2p^2$ configuration. The remaining terms are most easily derived by means of Table 6.2. Here are listed the *numbers* of determinants with particular values of M_L and M_S. For example, there are three different kets with $M_L = 0$ and $M_S = 0$, and so the number 3 appears in the appropriate box of the topmost implied terms table of Table 6.2.

‡ The ket written in this way is implicitly normalized and antisymmetric unless indicated otherwise.

Table 6.2 Implied-terms tables for the $2p^2$ configuration

			M_S	
		1	0	−1
	2	0	1	0
	1	1	2	1
M_L	0	1	3	1
	−1	1	2	1
	−2	0	1	0

$2p^2$ Table

		1	0	−1
	2	0	0	0
	1	1	1	1
M_L	0	1	2	1
	−1	1	1	1
	−2	0	0	0

$2p^2$ Table with the ^1D implication removed

		1	0	−1
	2	0	0	0
	1	0	0	0
M_L	0	0	1	0
	−1	0	0	0
	−2	0	0	0

$2p^2$ Table with the ^1D and ^3P implications removed

A ^1D term implies the existence of five functions with $M_S = 0$ and $M_L = 2, 1, 0, -1$ and -2. These functions will be linear combinations of the Slater determinants with $M_S = 0$ and $m_{l_1} + m_{l_2} = M_L$. In order to find what *other* terms are implied by the table, however, it is *not* necessary to know the precise combinations. We only need to know *how many* determinants with particular M_L and M_S are implied by the ^1D term. *We then subtract these numbers from the implied terms table to produce a new table with the implications of the ^1D term removed.* In this new table, the highest M_L value is 1, and the highest M_S with this M_L is also 1. Therefore there exists a ^3P term, with nine constituent states having the appropriate M_S and M_L values. Subtracting *these* implications from the table, we obtain a single state, with $M_L = 0$ and $M_S = 0$, and this clearly implies the existence of a ^1S term. Thus, for the ground configuration of carbon, there are three terms, ^1D, ^3P, and ^1S.

This method is easily generalized to systems with more than two electrons in open shells. It is only necessary to consider the total number of independent kets with fixed values of M_L and M_S (it is not actually necessary to write them out as in Table 6.1) and to obtain the implied terms table. In Table 6.3 the allowed terms for the p^n and d^n configurations are listed; the reader should verify a few of the entries. The allowed terms for various excited configurations (for example, $2p^23p$ and $2p^23d$) are not given, but they are readily derived using the same method. The reader should note that it is not in fact necessary to construct the entire implied terms table, but only that portion with $M_L, M_S \geq 0$.

The terms, and their constituent states, are expressed in the $|SLM_SM_L\rangle$ Russell-Saunders coupling scheme. The individual Slater determinants are given in the $|l_1l_2 \cdots l_n m_{l_1} m_{l_2} \cdots m_{l_n} m_{s_1} m_{s_2} \cdots m_{s_n}\rangle$ scheme. In order to construct eigenfunctions of $\mathbf{L}^2$ and $\mathbf{L}_z$ and $\mathbf{S}^2$ and $\mathbf{S}_z$, we must linearly combine the individual determinants. In general, this task is far more difficult than finding the allowed terms. We must couple together all $\mathbf{l}_i$ and $\mathbf{s}_i$ and build up appropriate states. Fortunately, we

---

Table 6.3   Allowed terms for p^n and d^n configurations

p, p^5		^2P			
p^2, p^5	$^1S\ ^1D$		^3P		
p^3		$^2P\ ^2D$		^4S	
d^2, d^8	$^1S\ ^1D\ ^1G$		$^3P\ ^3F$		
d^3, d^7		$^2P\ ^2D_a\ ^2D_b$		$^4P\ ^4F$	
		$^2F\ ^2G\ ^2H$			
d^4, d^6	$^1S_a\ ^1S_b\ ^1D_a\ ^1D_b$		$^3P_a\ ^3P_b\ ^3D$		^5D
	$^1F\ ^1G_a\ ^1G_b\ ^1I$		$^3F_a\ ^3F_b\ ^3G$		
d^5		$^2S\ ^2P\ ^2D_a\ ^2D_b\ ^2D_c$		$^4P\ ^4D\ ^4F\ ^4G$	^6S
		$^2F_a\ ^2F_b\ ^2G_a\ ^2G_b\ ^2H\ ^2I$			

usually only need to know the expansion for one state in any given term, and so we do not have to solve the problem in its full generality. We illustrate this with the p^2 configuration. The $|^1D\ 0\ 2\rangle$ ($|^{2S+1}LM_SM_L\rangle$) state‡ is clearly equal to the ket $|1^+\ 1^-\rangle$, the first determinant of Table 6.1. Using the shift operator $\mathbf{L}^-$, we then obtain

$$\mathbf{L}^-|^1D\ 0\ 2\rangle = 2\,\hbar|^1D\ 0\ 1\rangle = \mathbf{l}_1^-|1^+\ 1^-\rangle + \mathbf{l}_2^-|1^+\ 1^-\rangle$$
$$= \sqrt{2}\,\hbar|0^+\ 1^-\rangle + \sqrt{2}\,\hbar|1^+\ 0^-\rangle$$

so

$$|^1D\ 0\ 1\rangle = \frac{1}{\sqrt{2}}\,(|1^+\ 0^-\rangle - |1^-\ 0^+\rangle) \tag{6.7}$$

Similarly,

$$|^1D\ 0\ 0\rangle = \frac{1}{\sqrt{6}}\,(|1^+\ -1^-\rangle + 2|0^+\ 0^-\rangle - |1^-\ -1^+\rangle) \tag{6.8}$$

We do not ordinarily require kets with M_S or M_L negative, and so we turn our attention to the ^3P term. Since $|^3P\ 1\ 1\rangle = |1^+\ 0^+\rangle$, we have

$$\mathbf{L}^-|^3P\ 1\ 1\rangle = \sqrt{2}\,\hbar|^3P\ 1\ 0\rangle = \mathbf{l}_1^-|1^+\ 0^+\rangle + \mathbf{l}_2^-|1^+\ 0^+\rangle$$

so

$$|^3P\ 1\ 0\rangle = |1^+\ -1^+\rangle \tag{6.9}$$

Furthermore, by acting on this last ket with the shift operator $\mathbf{S}^-$, we can show that

$$|^3P\ 0\ 0\rangle = \frac{1}{\sqrt{2}}\,(|1^+\ -1^-\rangle + |1^-\ -1^+\rangle) \tag{6.10}$$

The eigenfunction for $|^1S\ 0\ 0\rangle$ can be obtained by orthogonality with the two kets $|^1D\ 0\ 0\rangle$ and $|^3P\ 0\ 0\rangle$. It is not difficult to show that

$$|^1S\ 0\ 0\rangle = \frac{1}{\sqrt{3}}\,(|1^+\ -1^-\rangle - |1^-\ -1^+\rangle - |0^+\ 0^-\rangle) \tag{6.11}$$

where the phase of the ket is determined by the convention that when $M_S = S$ and $M_L = L$, the Wigner coefficient [see Eq. (3.43)] has the sign $(-1)^{l_1-m_{l_1}}(-1)^{1/2-m_{s_1}}$.

The kets for configurations with more than two electrons can be derived in a similar way. For example, in the p^3 configuration with ^2D, ^2P, and ^4S terms we may write, by inspection, $|^2D\ \frac{1}{2}\ 2\rangle = |1^+\ 1^-\ 0^+\rangle$ and $|^4S\ \frac{3}{2}\ 0\rangle = |1^+\ 0^+\ -1^+\rangle$. Using shift operators $\mathbf{L}^- = \mathbf{l}_1^- + \mathbf{l}_2^- + \mathbf{l}_3^-$ and $\mathbf{S}^- = \mathbf{s}_1^- + \mathbf{s}_2^- + \mathbf{s}_3^-$, we may construct the other constituent ^2D and ^4S states and obtain ^2P states by orthogonality.

‡ Other authors may write $|^{2S+1}LM_LM_S\rangle$. Beware!

The terms for the configurations p^{6-n} and p^n are identical, as are the terms for d^{10-n} and d^n. This is a consequence of the following hole-particle relation: a ket may be considered either as an n-electron or an n-hole description. For example, the ket $|1^+ 0^+\rangle$ can describe either a two-particle state or a two-hole state. It is thus clear that there is a one-to-one correspondence between kets in the particle and hole schemes, and that the terms of p^n are identical with those for p^{6-n}. The relationship between individual states is somewhat more complex, and the phase relations between the two sets must be considered carefully. For further information, see the references at the end of this chapter.

PROBLEMS

6.2 Show that the number of independent Slater determinants in the l^n configuration is the binomial coefficient $\binom{q}{n} = q!/[(q-n)!n!]$ where $q = 2(2l+1)$.

6.3 Verify the expressions derived for the states $|^{2S+1}LM_SM_L\rangle$ obtained for the p^2 configuration in this section.

◆ **6.4** Derive expressions for *all* states with $M_S + M_L \geq 0$ for any *one* of the three configurations p^2, p^3, or d^2.

6.3 GENERAL EVALUATION OF THE MATRIX ELEMENTS OF $\mathcal{3C}$

Since $\mathbf{L}^2$ and $\mathbf{S}^2$ commute with $\mathcal{3C}$, the Hamiltonian has no matrix elements between states belonging to different terms, and in fact

$$\langle {}^{2S+1}LM_SM_L|\mathcal{3C}|{}^{2S'+1}L'M'_SM'_L\rangle = E(SL)\delta_{SS'}\delta_{LL'}\delta_{M_SM_S'}\delta_{M_LM_L'} \qquad \blacktriangleright (6.12)$$

This relation greatly simplifies the evaluation of the energies since we need only evaluate diagonal elements, using only one state in any given term or at most a few extra off-diagonal elements if the configuration gives rise to more than one term of a given S and L, for example, the two ^2D terms of the d^3 configuration.

We begin by considering the elements $\mathcal{3C}_{ij}$, defined by the relation

$$\mathcal{3C}_{ij} = \langle \Psi_i|\mathcal{3C}|\Psi_j\rangle \qquad \blacktriangleright (6.13)$$

where $|\Psi_i\rangle$ and $|\Psi_j\rangle$ are Slater determinants, not necessarily eigenfunctions of $\mathbf{L}^2$ or $\mathbf{S}^2$. Such matrix elements, with i and j perhaps different, will occur in the expansion of (6.12) in terms of the constituent determinants of each state. Using the *exact* Hamiltonian of (6.1), we may regroup its parts and write

$$\mathcal{3C} = \sum_{k=1}^{n} \mathbf{H}(k) + \sum_{\substack{k=2\\k>l}}^{n} \sum_{l=1}^{k-1} \mathbf{G}(k,l) \qquad \blacktriangleright (6.14)$$

where $\mathbf{H}(k) = -(\hbar^2/2m)\nabla_k^2 - Ze^2/R_k$ and $\mathbf{G}(k,l) = e^2/r_{kl}$. $\mathbf{H}(k)$ is a *one-electron operator*, whereas $\mathbf{G}(k,l)$ is a *two-electron operator*. Separating their matrix elements, we see clearly that

$$\mathcal{3C}_{ij} = H_{ij} + G_{ij} \qquad \blacktriangleright (6.15)$$

where, for example,

$$H_{ij} \equiv \langle \Psi_i | \sum_{k=1}^{n} \mathbf{H}(k) | \Psi_j \rangle \qquad \blacktriangleright (6.16)$$

We treat the one-electron interaction first. Let us write $|\Psi_i\rangle$ and $|\Psi_j\rangle$ as antisymmetrized products of one-electron orbitals;‡ thus we have

$$|\Psi_j\rangle = \frac{1}{n!} \sum_{P} (-1)^P P |\varphi_1(1)\varphi_2(2) \cdots \varphi_n(n)\rangle \qquad (6.17)$$

Thus,

$$H_{ij} = \frac{1}{n!} \sum_{P} \sum_{P'} (-1)^{P+P'} PP' \langle \varphi_1'(1)\varphi_2'(2)$$

$$\cdots \varphi_n'(n) | \sum_{k=1}^{n} \mathbf{H}(k) | \varphi_1(1)\varphi_2(2) \cdots \varphi_n(n)\rangle \quad (6.18)$$

where the orbitals φ_p' are not necessarily identical§ to the φ_p. Because $\mathbf{H}(k)$ is a one-electron operator and because the φ_p are *orthonormal*, each term in the sum vanishes unless the permutation of the arguments in the bra is the same as the permutation of the arguments in the ket. Thus, integrals of the form

$$\langle \varphi_1'(1)\varphi_2'(2) \cdots \varphi_n'(n) | \mathbf{H}(k) | \varphi_1(1)\varphi_2(2) \cdots \varphi_n(n)\rangle$$

reduce to a product of overlaps and a single one-electron integral,

$$\langle \varphi_1'(1)|\varphi_1(1)\rangle\langle \varphi_2'(2)|\varphi_2(2)\rangle$$
$$\cdots \langle \varphi_k'(k)|\mathbf{H}(k)|\varphi_k(k)\rangle \cdots \langle \varphi_n'(n)|\varphi_n(n)\rangle$$

Since the φ_p are orthogonal, each such term is zero unless all $\varphi_p' = \varphi_p$ ($p \neq k$). Therefore, $H_{ij} = 0$ if $|\Psi_i'\rangle$ and $|\Psi_j'\rangle$ differ by *more than one orbital*. If $|\Psi_i\rangle$ and $|\Psi_j\rangle$ are identical, we have at once

$$H_{ii} = \sum_{k=1}^{n} \langle \varphi_k(k)|\mathbf{H}(k)|\varphi_k(k)\rangle \qquad (6.19)$$

‡ In this section the orbital φ implicitly includes a specification of the *spin*, and φ is thus a *spin orbital*. It is therefore identical to the u of Eq. (3.55). See also the discussion following Eq. (3.68).

§ However, the orbitals φ_p' and φ_p are obtained from the same set, so a *given* orbital φ_p' is orthogonal to all φ_p except at most one.

or simply

$$H_{ii} = \sum_{k=1}^{n} \langle \varphi_k | \mathbf{H} | \varphi_k \rangle \qquad \blacktriangleright (6.20)$$

where the normalizing factor $1/n!$ is canceled by the existence of $n!$ terms in the sum over all permutations. If $|\Psi_i\rangle$ and $|\Psi_j\rangle$ differ by a single orbital, so that $\varphi'_p = \varphi_p$ except for a single $\varphi'_k \neq \varphi_k$, then

$$H_{ij} = \langle \varphi'_k | \mathbf{H} | \varphi_k \rangle \qquad i \neq j \qquad \blacktriangleright (6.21)$$

Let us illustrate the above argument with the two-electron configuration with determinants $|\Psi_j\rangle = |\varphi_1(1)\varphi_2(2)\rangle$, $|\Psi_i\rangle = |\varphi'_1(1)\varphi'_2(2)\rangle$. Expanding the antisymmetrizer and the matrix element, we have

$$\begin{aligned}
H_{ij} = \langle \Psi_i | \mathbf{H}(1) + \mathbf{H}(2) | \Psi_j \rangle = \tfrac{1}{2} [\langle \varphi'_1(1)\varphi'_2(2) | \mathbf{H}(1) \\
+ \mathbf{H}(2) | \varphi_1(1)\varphi_2(2) \rangle + \langle \varphi'_1(1)\varphi'_2(2) | \mathbf{H}(1) + \mathbf{H}(2) | \varphi_2(1)\varphi_1(2) \rangle \\
+ \langle \varphi'_2(1)\varphi'_1(2) | \mathbf{H}(1) + \mathbf{H}(2) | \varphi_1(1)\varphi_2(2) \rangle + \langle \varphi'_2(1)\varphi'_1(2) | \mathbf{H}(1) \\
+ \mathbf{H}(2) | \varphi_2(1)\varphi_1(2) \rangle] \quad (6.22)
\end{aligned}$$

The second and third integrals are zero since, from the way in which the orbitals are ordered,‡ $\langle \varphi'_1 | \varphi_2 \rangle = \langle \varphi'_2 | \varphi_1 \rangle = 0$. The first integral equals $\langle \varphi'_1 | \mathbf{H} | \varphi_1 \rangle \langle \varphi'_2 | \varphi_2 \rangle + \langle \varphi'_2 | \mathbf{H} | \varphi_2 \rangle \langle \varphi'_1 | \varphi_1 \rangle$, and similarly for the fourth integral. Thus,

$$H_{ij} = \begin{cases} \langle \varphi_1 | \mathbf{H} | \varphi_1 \rangle + \langle \varphi_2 | \mathbf{H} | \varphi_2 \rangle & \text{if } i = j, \ \varphi'_1 = \varphi_1, \ \varphi'_2 = \varphi_2 \\ \langle \varphi'_1 | \mathbf{H} | \varphi_1 \rangle & \text{if } \varphi'_1 \neq \varphi_1, \ \varphi'_2 = \varphi_2 \\ \langle \varphi'_2 | \mathbf{H} | \varphi_2 \rangle & \text{if } \varphi'_1 = \varphi_1, \ \varphi'_2 \neq \varphi_2 \\ 0 & \text{if } \varphi_1 \neq \varphi_1, \ \varphi'_2 \neq \varphi_2 \end{cases} \qquad (6.23)$$

which confirms the general rule. Of course, this rule holds for *any* one-electron operator, and not just $\mathbf{H}(k)$; in particular, it holds for the unit operator, so the Slater determinants form a mutually orthogonal set.

Let us now consider the two-electron operator $\mathbf{G}(k,l)$. As in Eq. (6.18), we have

$$G_{ij} = \frac{1}{n!} \sum_P \sum_{P'} (-1)^{P+P'} P P' \langle \varphi'_1(1) \varphi'_2(2)$$

$$\cdots \varphi'_n(n) | \sum_{\substack{k=2 \\ k>l}}^{n} \sum_{l=1}^{k-1} \mathbf{G}(k,l) | \varphi_1(1)\varphi_2(2) \cdots \varphi_n(n) \rangle \quad (6.24)$$

It should be clear that G_{ij} is zero if $|\Psi_i\rangle$ differs from $|\Psi_j\rangle$ in *more than* two orbitals. We may therefore distinguish three special cases, (I) $|\Psi_i\rangle = |\Psi_j\rangle$; (II) $|\Psi_i\rangle \neq |\Psi_j\rangle$ but all $\varphi'_p = \varphi_p$ except for a single $\varphi'_k \neq \varphi_k$; and (III)

‡ We do not consider perverse definitions such as $\varphi'_1 = \varphi_2$ and $\varphi'_2 = \varphi_1$. The determinants must be ordered for *maximum coincidence*.

$|\Psi_i\rangle \neq |\Psi_j\rangle$ but all $\varphi_p' = \varphi_p$ except for a pair $\varphi_k' \neq \varphi_k$, $\varphi_l' \neq \varphi_l$. In the first case, integrals of the form

$$\langle \varphi_1(1)\varphi_2(2) \cdots \varphi_k'(k) \cdots \varphi_l'(l) \cdots \varphi_n(n)|\mathbf{G}(k,l)|\varphi_1(1)\varphi_2(2) \\ \cdots \varphi_k(k) \cdots \varphi_l(l) \cdots \varphi_n(n)\rangle$$

are nonzero provided that *either* $\varphi_k' = \varphi_k$ and $\varphi_l' = \varphi_l$ or $\varphi_k' = \varphi_l$ and $\varphi_l' = \varphi_k$. When this is true, the integral reduces to $\langle \varphi_k(k)\varphi_l(l)|\mathbf{G}(k,l)|\varphi_k(k)\varphi_l(l)\rangle$ or $\langle \varphi_k(k)\varphi_l(l)|\mathbf{G}(k,l)|\varphi_l(k)\varphi_k(l)\rangle$. Since the permutation operators guarantee that if one occurs, the other will as well, but with opposite sign, we obtain the expression

$$G_{ii} = \sum_{\substack{k=2 \\ k>l}}^{n} \sum_{l=1}^{k-1} \langle \varphi_k(k)\varphi_l(l)|\mathbf{G}(k,l)|\varphi_k(k)\varphi_l(l)\rangle$$

$$- \langle \varphi_k(k)\varphi_l(l)|\mathbf{G}(k,l)|\varphi_l(k)\varphi_k(l)\rangle \quad (6.25)$$

The *arguments* k and l are only dummy indices, since $\mathbf{G}(k,l)$ is symmetric in the electron arguments, and we may equally write

$$G_{ii} = \sum_{\substack{k=2 \\ k>l}}^{n} \sum_{l=1}^{k-1} \langle \varphi_k(1)\varphi_l(2)|\mathbf{G}(1,2)|\varphi_k(1)\varphi_l(2)\rangle$$

$$- \langle \varphi_k(1)\varphi_l(2)|\mathbf{G}(1,2)|\varphi_l(1)\varphi_k(2)\rangle \quad (6.26)$$

or, more tersely,‡

$$G_{ii} = \sum_{\substack{k=2 \\ k>l}}^{n} \sum_{l=1}^{k-1} \langle \varphi_k\varphi_l|\frac{1}{r_{12}}|\varphi_k\varphi_l\rangle - \langle \varphi_k\varphi_l|\frac{1}{r_{12}}|\varphi_l\varphi_k\rangle \qquad \blacktriangleright(6.27)$$

The first term is known as a *coulomb integral* since it represents the classical interaction of a charge located in the region φ_k with a charge in the region φ_l. It is often abbreviated J_{kl}. Other common abbreviations include $(\varphi_k\varphi_k|\varphi_l\varphi_l)$, $(\varphi_k\varphi_l|\varphi_k\varphi_l)$, $\langle \varphi_k\varphi_l\|\varphi_k\varphi_l\rangle$ and so forth, with no agreement whatever on how the arguments are to be ordered. Confusion in the literature is avoided only by the fact that in every paper the author is usually careful to define his convention. In *this* book we define

$$J_{kl} = \langle \varphi_k(1)\varphi_l(2)|\frac{1}{r_{12}}|\varphi_k(1)\varphi_l(2)\rangle = \langle \varphi_k\varphi_l|\frac{1}{r_{12}}|\varphi_k\varphi_l\rangle$$

$$= \langle \varphi_k\varphi_l\|\varphi_k\varphi_l\rangle \qquad \blacktriangleright(6.28)$$

The second integral, $\langle \varphi_k\varphi_l|1/r_{12}|\varphi_l\varphi_k\rangle$, is called the *exchange integral* since it arises from the coulomb integral through the exchange of electrons 1 and 2. It is therefore a consequence of the indistinguishability of elec-

‡ For brevity the factor e^2 is often dropped, since it is equal to unity in the atomic system of units.

trons and has no strict classical analog. Nevertheless, it can be considered the self-energy of a charge distribution located simultaneously on φ_k and φ_l, that is, the interaction between two electrons having the same spatial distribution $\varphi_k \varphi_l$. The exchange integral is customarily abbreviated K_{kl}, and we specifically define

$$K_{kl} = \langle \varphi_k(1) \varphi_l(2) | \frac{1}{r_{12}} | \varphi_l(1) \varphi_k(2) \rangle = \langle \varphi_k \varphi_l | \frac{1}{r_{12}} | \varphi_l \varphi_k \rangle$$

$$= \langle \varphi_k \varphi_l \| \varphi_l \varphi_k \rangle = \langle \varphi_l \varphi_k \| \varphi_k \varphi_l \rangle \qquad \blacktriangleright (6.29)$$

Since φ_k and φ_l include a specification of the spin, K_{kl} will be zero when the spins of φ_k and φ_l are different. Thus there is no "exchange force" between electrons of unlike spin.

In this notation,

$$G_{ii} = \sum_{k=2}^{n} \sum_{l=1}^{k-1} (J_{kl} - K_{kl}) \qquad \blacktriangleright (6.30)$$

or, since $J_{kk} = K_{kk}$,

$$G_{ii} = \frac{1}{2} \sum_{k=1}^{n} \sum_{l=1}^{n} (J_{kl} - K_{kl}) \qquad \blacktriangleright (6.31)$$

The next matrix element to be considered is the expression for G_{ij} when $i \neq j$ but for which $|\Psi_i\rangle$ differs from $|\Psi_j\rangle$ in just one orbital, $\varphi_k' \neq \varphi_k$ (case II). Equation (6.24) then becomes

$$G_{ij} = \sum_{l=1}^{n} [\langle \varphi_1(1) \varphi_2(2) \cdots \varphi_k'(k) \cdots \varphi_l(l)$$
$$\cdots \varphi_n(n) | \mathbf{G}(k,l) | \varphi_1(1) \varphi_2(2) \cdots \varphi_k(k) \cdots \varphi_l(l) \cdots \varphi_n(n) \rangle$$
$$- \langle \varphi_1(1) \varphi_2(2) \cdots \varphi_k'(k) \cdots \varphi_l(l)$$
$$\cdots \varphi_n(n) | \mathbf{G}(k,l) | \varphi_1(1) \varphi_2(2) \cdots \varphi_l(k) \cdots \varphi_k(l) \cdots \varphi_n(n) \rangle]$$

$$= \sum_{l=1}^{n} \left[\langle \varphi_k' \varphi_l | \frac{1}{r_{12}} | \varphi_k \varphi_l \rangle - \langle \varphi_k' \varphi_l | \frac{1}{r_{12}} | \varphi_l \varphi_k \rangle \right] \qquad \blacktriangleright (6.32)$$

Finally, for $|\Psi_i\rangle$ and $|\Psi_j\rangle$ differing in two orbitals, $\varphi_k' \neq \varphi_k$, $\varphi_l' \neq \varphi_l$ (case III), only two terms in (6.24) survive, and we obtain

$$G_{ij} = \langle \varphi_k' \varphi_l' | \frac{1}{r_{12}} | \varphi_k \varphi_l \rangle - \langle \varphi_k' \varphi_l' | \frac{1}{r_{12}} | \varphi_l \varphi_k \rangle \qquad \blacktriangleright (6.33)$$

Although Eqs. (6.20), (6.21), (6.31), (6.32), and (6.33) are valid for any one- and two-electron operators acting between determinantal functions composed of orthonormal orbitals, certain simplifications can be introduced which apply only to atoms. First of all, we consider the one-electron operator $\mathbf{H}(k)$. The orbitals are products of a spin part, an angular part, and a radial part, but since $\mathbf{H}(k)$ commutes with $\mathbf{l}$ and $\mathbf{s}$, we can integrate over angular and spin coordinates in the expression for H_{ij}. Doing so, we find

$$H_{ij} = \langle R_{n'l'}|\mathbf{H}|R_{nl}\rangle \delta_{ll'}\delta_{m_l m_{l'}}\delta_{m_s m_{s'}} \tag{6.34}$$

The integral is zero unless $l = l'$, $m_l = m'_l$, $m_s = m'_s$, which is impossible by the Pauli principle for $i \neq j$, that is, for different terms. Therefore $H_{ij} = 0$. For the diagonal term,

$$H_{ii} = \sum_k \langle R_{nl}(k)|\mathbf{H}(k)|R_{nl}(k)\rangle \tag{6.35}$$

Thus H_{ii} is independent of m_{l_i} and m_{s_i} and is therefore identical‡ for all determinants of a given configuration. It therefore cannot be responsible for the dependence of the state energies on L and S. Such dependence must arise entirely from the two-electron interactions.

A similar analysis can be carried out for the matrix element G_{ij} in which $|\Psi_i\rangle$ and $|\Psi_j\rangle$ differ by one orbital $\varphi'_k \neq \varphi_k$. However, since $1/r_{12}$ does not commute with $\mathbf{l}_{z_1}$ or $\mathbf{l}_{z_2}$, we cannot factor out the Y_{lm} and integrate over the angular coordinates (the integration over spin coordinates proceeds as before). Nevertheless, $1/r_{12}$ commutes with $\mathbf{l}_{z_1} + \mathbf{l}_{z_2}$ [see Prob. 6.1 and Eq. (6.6)]. Therefore an integral such as

$$\langle \varphi_{nlm_l m_s}\varphi_{nlm_{l'} m_{s'}}|\frac{1}{r_{12}}|\varphi_{nlm_{l''} m_{s''}}\varphi_{nlm_{l'''} m_{s'''}}\rangle$$

will be zero unless $m_l + m'_l = m''_l + m'''_l$, $m_s = m''_s$, and $m'_s = m'''_s$. In the first integral of (6.32), $m_l = m''_l$, so this integral vanishes unless $m'_l = m'''_l$, which is contrary to the assumption that φ' and φ''' are different. In the second integral, $m_l = m'''_l$, so it too vanishes. Thus G_{ij} in case II is zero for atoms. In general, G_{ij} for $|\Psi_i\rangle$ and $|\Psi_j\rangle$ differing by two orbitals is nonzero, however, and this term is in part responsible for the differing energies of the terms of a given configuration. Finally, G_{ii} is not the same for all determinants, and so it also contributes to the term separations. G_{ii} is thus unlike H_{ii}, which *is* the same for all determinants of a given configuration.

Interactions with the filled shells do not contribute to the separations among terms of an open shell. We prove this statement by considering the H_{ii}, G_{ii}, and G_{ij} separately. We have already noted that H_{ii} is the same for all terms of a configuration and so cannot contribute to

‡ Note the hidden assumption that the radial functions are independent of m_l and m_s. This is a consequence of assuming spherical symmetry in the central-field Hamiltonian (6.2). It is not a direct consequence of the orbital approximation, although, if our trial functions $|\Psi_i\rangle$ are to be eigenfunctions of $\mathbf{L}^2$ and $\mathbf{S}^2$, it is a necessary assumption.

the separation energies. Expanding G_{ii}, we obtain

$$G_{ii} = \frac{1}{2} \sum_{k=1}^{n_c} \sum_{l=1}^{n_c} (J_{kl} - K_{kl}) + \sum_{k=1}^{n_c} \sum_{l=n_c+1}^{n_c+n_o} (J_{kl} - K_{kl})$$
$$+ \frac{1}{2} \sum_{k=n_c+1}^{n_c+n_o} \sum_{l=n_c+1}^{n_c+n_o} (J_{kl} - K_{kl}) \quad (6.36)$$

where n_c is the number of electrons in closed shells and n_o is the number of electrons in open shells. The first sum represents the energy of electrons in closed shells and contributes the same‡ energy to all terms. The second sum, representing the interaction between closed and open shells, contributes the same energy to all terms because the closed shells are spherically symmetric. Therefore,

$$\mathcal{H}_{ii} = U + \frac{1}{2} \sum_{k=n_c+1}^{n_c+n_o} \sum_{l=n_c+1}^{n_c+n_o} (J_{kl} - K_{kl}) \qquad \blacktriangleright (6.37)$$

where the summation extends only over the electrons in the open shells. The constant U is common to all determinants of a given configuration. As for the off-diagonal elements $\mathcal{H}_{ij} = G_{ij}$, it should be clear that the matrix element (6.33) cannot contain orbitals belonging to the closed shells. Therefore the closed shells may be neglected in any discussion of the relative energies of the terms of a given configuration. Of course, should we be interested in total atomic energies or in more than one configuration, the extra terms involving the closed shells must be considered explicitly.

Let us now apply the formulas obtained thus far to the determination of the *relative* energies of the three terms of the p^2 configuration. Since $|{}^1D\ 0\ 2\rangle = |1^+\ 1^-\rangle$, we have at once§

$$\begin{aligned} E({}^1D) &= \langle 1^+\ 1^- || 1^+\ 1^- \rangle - \langle 1^+\ 1^- || 1^-\ 1^+ \rangle \\ &= J_{11} \end{aligned} \qquad (6.38)$$

since the integration over spin coordinates contributes a factor of unity to the first integral and eliminates the second (exchange) integral. For the ${}^3P$ term, taking the $|{}^3P\ 1\ 1\rangle = |1^+\ 0^+\rangle$ component, we obtain

$$\begin{aligned} E({}^3P) &= \langle 1^+\ 0^+ || 1^+\ 0^+ \rangle - \langle 1^+\ 0^+ || 0^+\ 1^+ \rangle \\ &= J_{10} - K_{10} \end{aligned} \qquad (6.39)$$

Finally, for the $|{}^1S\ 0\ 0\rangle = 1/\sqrt{3}\{|1^+\ -1^-\rangle - |0^+\ 0^-\rangle - |1^-\ -1^+\rangle\}$ term, we find $E({}^1S)$ by using the well-known *diagonal-sum rule*. This rule states

‡ Tacitly we are assuming that when we form a new determinant from $|\Psi_i\rangle$ by replacing φ'_k and/or φ'_l by φ_k and/or φ_l, all *other* orbitals remain the same. This assumption is not necessary, and its removal is discussed in Prob. 6.10.
§ We omit the "core potential" U throughout. Note that we can evaluate the energy of any component state of the ${}^1D$ term since all such components are degenerate.

that the sum of the elements $\mathcal{H}_{ii}$ is equal to the sum of the roots of the secular equation. Thus, since the ^1D, ^3P, and ^1S terms all contain $M_S = 0$ and $M_L = 0$ components and since each state has only one such component, we obtain

$$E(^1D) + E(^3P) + E(^1S) = E(|1^+ \ -1^-\rangle)$$
$$+ E(|0^+ \ 0^-\rangle) + E(|1^- \ -1^+\rangle)$$
$$= 2J_{1-1} + J_{00} \qquad (6.40)$$

so

$$E(^1S) = 2J_{1-1} + J_{00} + K_{10} - J_{11} - J_{10} \qquad (6.41)$$

We are primarily interested in the relative energies of the three terms, yet we have obtained the energies in terms of five different integrals, J_{1-1}, J_{00}, K_{10}, J_{11}, and J_{10}. In the next section we show how these integrals are related to each other, and we shall reduce the relative energy expression to a single parameter.

The formulas and methods we have derived in this section are of the greatest importance throughout all of orbital theory, including molecular orbital theory. It is therefore desirable to collect them all in one place, and this we do in Table 6.4. If possible, the reader should try to commit the contents of this table to memory.

Table 6.4 Matrix elements of one- and two-electron operators between determinantal wavefunctions. $|\Psi_i\rangle = |\varphi_1'(1)\varphi_2'(2) \cdots \varphi_n'(n)\rangle$ and $|\Psi_j\rangle = |\varphi_1(1)\varphi_2(2) \cdots \varphi_n(n)\rangle$†

	Formula	Specialization to atoms with a single open shell
Case I	$\|\Psi_i\rangle = \|\Psi_j\rangle$, $\varphi_p' = \varphi_p$ for all p	
	$\langle\Psi_i\|\mathbf{H}\|\Psi_i\rangle = H_{ii} = \sum_{k=1}^{n} \langle\varphi_k\|\mathbf{H}\|\varphi_k\rangle$	$= \sum_{k=1}^{n} \langle R_{nl}(k)\|\mathbf{H}(k)\|R_{nl}(k)\rangle$
	$\langle\Psi_i\|\mathbf{G}\|\Psi_i\rangle = G_{ii} = \frac{1}{2}\sum_{k=1}^{n}\sum_{l=1}^{n} (J_{kl} - K_{kl})$	
Case II	$\|\Psi_i\rangle \neq \|\Psi_j\rangle$, but $\varphi_p' = \varphi_p$ for all p except $\varphi_k' \neq \varphi_k$	
	$H_{ij} = \langle\varphi_k'\|\mathbf{H}\|\varphi_k\rangle$	$= 0$
	$G_{ij} = \sum_{l=1}^{n} \left[\langle\varphi_k'\varphi_l\|\dfrac{1}{r_{12}}\|\varphi_k\varphi_l\rangle - \langle\varphi_k'\varphi_l\|\dfrac{1}{r_{12}}\|\varphi_l\varphi_k\rangle \right]$	$= 0$
Case III	$\|\Psi_i\rangle \neq \|\Psi_j\rangle$, but $\varphi_p = \varphi_p$ for all p except $\varphi_k' \neq \varphi_k$, $\varphi_l' \neq \varphi_l$	
	$H_{ij} = 0$	—
	$G_{ij} = \langle\varphi_k'\varphi_l'\|\dfrac{1}{r_{12}}\|\varphi_k\varphi_l\rangle - \langle\varphi_k'\varphi_l'\|\dfrac{1}{r_{12}}\|\varphi_l\varphi_k\rangle$	—

† The operators are defined in Eq. (6.14) and are followed by derivations. The φ_i are spin orbitals, and the sums extend over all n electrons in the system.

PROBLEMS

6.5 Consider a pair of determinants $|\Psi_i\rangle = |\varphi_1\varphi_2 \cdots \varphi_n\rangle$ and $|\Psi_j\rangle = |\varphi_1'\varphi_2' \cdots \varphi_n'\rangle$ in which the sets of orbitals φ_p' and φ_p are not orthogonal to each other. Obtain a general expression for $\langle\Psi_i|\Psi_j\rangle$.

6.6 Verify (6.38), (6.39), (6.40), and (6.41).

6.7 Consider the secular equation

$$|H_{ij} - E\delta_{ij}| = 0$$

Show that the sum of the roots E is equal to the sum of the diagonal elements H_{ii} (this is the diagonal-sum rule).

6.8 Carry out the problem above for the general one-electron matrix element $\langle\Psi_i|\mathbf{H}|\Psi_j\rangle$.

6.9 Do the relations $H_{ij} = 0$ and $G_{ij} = 0$ ($|\Psi_i\rangle$, $|\Psi_j\rangle$ differing by a single orbital) apply to a configuration with more than one open shell?

◆ **6.10** Carry out Prob. 6.5 for the two-electron matrix element $\langle\Psi_i|\mathbf{G}|\Psi_j\rangle$.

6.4 SLATER–CONDON PARAMETERS

We wish to solve, insofar as possible, for the two-electron matrix elements of the general form

$$\langle\varphi_p\varphi_q|\frac{e^z}{r_{12}}|\varphi_r\varphi_s\rangle = \langle pq\|rs\rangle \qquad\qquad \blacktriangleright(6.42)$$

The φ are products of a radial function, a spin part, and an angular part (the Y_{lm}), so we may try at least to integrate over the spin and angular functions since they are known. As noted in the previous section, the angular parts of the φ cannot be factored out directly since 1_{z_i} does not commute with $1/r_{12}$. On the other hand, the integration over spin coordinates is trivial, with $\langle pq\|rs\rangle = 0$ unless $m_{s_p} = m_{s_r}$ and $m_{s_q} = m_{s_s}$. We may therefore restrict the rest of the discussion to spatial orbitals without regard to spin.

The operator $1/r_{12}$ may be expanded as in Sec. 5.4 in the slightly different form

$$\frac{1}{r_{12}} = \sum_{k=0}^{\infty} \sum_{m_k=-k}^{k} \frac{4\pi}{2k+1} \frac{r_<^k}{r_>^{k+1}} Y_{km_k}(\theta_1,\varphi_1) Y_{km_k}^*(\theta_2,\varphi_2) \qquad \blacktriangleright(6.43)$$

We can now factor out *all* the angular factors in (6.42). Integrating over them, we obtain

$$\langle pq\|rs\rangle = \sum_{k=0}^{\infty} \langle R_{n_pl_p}R_{n_ql_q}|\frac{e^2r_<^k}{r_>^{k+1}}|R_{n_rl_r}R_{n_sl_s}\rangle \frac{4\pi}{2k+1} \sum_{m_k=-k}^{k}$$
$$\langle Y_{l_pm_p}|Y_{km_k}|Y_{l_rm_r}\rangle\langle Y_{l_qm_q}|Y_{km_k}^*|Y_{l_sm_s}\rangle \qquad (6.44)$$

The first integral over the Y_{lm} is zero unless $m_p = m_k + m_r$ since the product $Y_{km_k} Y_{l_r m_r}$ is an eigenfunction of $\mathbf{1}_z$ with eigenvalue $m_k + m_r$. Similarly, the second integral is zero unless $m_q = -m_k + m_s$. Therefore the entire angular part is zero unless $m_p + m_q = m_s + m_r$, and the sum over m_k reduces to a single term. Writing, as is conventional,

$$c^k(l_p m_p; l_r m_r) \equiv \left(\frac{4\pi}{2k+1} \right)^{1/2} \langle Y_{l_p m_p} Y_{k, m_p - m_r} Y_{l_r m_r} \rangle \qquad \blacktriangleright (6.45)$$

we obtain‡

$$\langle pq \| rs \rangle = \sum_{k=0}^{\infty} \langle R_{n_p l_p} R_{n_q l_q} | \frac{e^2 r_<^k}{r_>^{k+1}}$$
$$| R_{n_r l_r} R_{n_s l_s} \rangle c^k(l_p m_p; l_r m_r) c^k(l_q m_q; l_s m_s) \delta_{m_p + m_q, m_r + m_s} \quad (6.46)$$

As can be seen from the discussion above, the c^k are zero unless

$$|l_p - l_r| \leq k \leq l_p + l_r \qquad (6.47)$$

so the range of k is in fact not infinite but restricted to just a few terms. Furthermore, $k + l_p + l_r$ must be even.

Since the sum over k is finite, only a few different radial integrals (integrals over the R_{nl}) need be considered. Corresponding to the coulomb and exchange integrals, we define the Slater-Condon parameters

$$F^k(n_p l_p; n_q l_q) = \langle R_{n_p l_p} R_{n_q l_q} | \frac{e^2 r_<^k}{r_>^{k+1}} | R_{n_p l_p} R_{n_q l_q} \rangle$$
$$\qquad \blacktriangleright (6.48)$$
$$G^k(n_p l_p; n_q l_q) = \langle R_{n_p l_p} R_{n_q l_q} | \frac{e^2 r_<^k}{r_>^{k+1}} | R_{n_q l_q} R_{n_p l_p} \rangle$$

We also define

$$a^k(l_p m_p; l_q m_q) = c^k(l_p m_p; l_p m_p) c^k(l_q m_q; l_q m_q)$$
$$b^k(l_p m_p; l_q m_q) = [c^k(l_p m_p; l_q m_q)]^2 \qquad \blacktriangleright (6.49)$$

Therefore, for the coulomb integrals,

$$J_{pq} = \sum_{k=0}^{\infty} a^k(l_p m_p; l_q m_q) F^k(n_p l_p; n_q l_q) \qquad \blacktriangleright (6.50)$$

and for the exchange integrals,

$$K_{pq} = \sum_{k=0}^{\infty} b^k(l_p m_p; l_q m_q) G^k(n_p l_p; n_q l_q) \delta_{m_{sp} m_{sq}} \qquad \blacktriangleright (6.51)$$

For a single-open-shell configuration of the type l^n, $n_p = n_q$, and $l_p = l_q$, so $G^k = F^k$, and in fact the radial integrals for any set of four functions

‡ The superscript values k on c^k, F^k, G^k, a^k, and b^k are not to be confused with exponential powers.

Table 6.5 **Values of** $c^k(l_p m_p; l_q m_q)$ **for** l^n **configurations**

l_p	l_q	m_p	m_q	c^k		
				$k=0$	2	4
s	s	0	0	1		
p	p	± 1	± 1	1	$-\sqrt{1/25}$	
		± 1	0	0	$\sqrt{3/25}$	
		0	0	1	$\sqrt{4/25}$	
		± 1	∓ 1	0	$-\sqrt{6/25}$	
d	d	± 2	± 2	1	$-\sqrt{4/49}$	$\sqrt{1/441}$
		± 2	± 1	0	$\sqrt{6/49}$	$-\sqrt{5/441}$
		± 2	0	0	$-\sqrt{4/49}$	$\sqrt{15/441}$
		± 1	± 1	1	$\sqrt{1/49}$	$-\sqrt{16/441}$
		± 1	0	0	$\sqrt{1/49}$	$\sqrt{30/441}$
		0	0	1	$\sqrt{4/49}$	$\sqrt{36/441}$
		± 2	∓ 2	0	0	$\sqrt{70/441}$
		± 2	∓ 1	0	0	$-\sqrt{35/441}$
		± 1	∓ 1	0	$-\sqrt{6/49}$	$-\sqrt{40/441}$

φ_p, φ_q, φ_r, φ_s will be identical to the corresponding F^k. For such a configuration, we may write

$$J_{pq} = \sum_{k=0}^{2l} a^k F^k$$

$$K_{pq} = \delta_{m_{sp} m_{sq}} \sum_{k=0}^{2l} b^k F^k$$

▶(6.52)

where the arguments of a^k, b^k, and F^k are understood and where the sum extends only over even values of k.

The integrals c^k may be calculated for given values of the arguments; a limited list‡ is given in Table 6.5. The radial integrals F^k and G^k may be calculated using a suitable wavefunction (see Sec. 6.8) or they may be determined empirically by comparison with spectroscopic data. It is possible to show theoretically that both parameters are positive. In general, $F^k \geq G^k$.

Equations (6.52), for an l^n configuration, are of the greatest importance. Using them, in the next two sections we shall derive the term separation energies for the p^n and d^n configurations.

PROBLEMS

6.11 Consider the p^n configuration. We have derived the coulomb and exchange integrals between p orbitals which are eigenfunctions of 1^2 and 1_z and which

‡ A far more complete list will be found in Condon and Shortley, p. 178–180.

therefore contain the complex functions Y_{lm} as factors. Using Table 6.5, derive the coulomb and exchange integrals between *real* orbitals as defined in Table 3.1.

6.5 TERM SEPARATION ENERGIES FOR THE p^n CONFIGURATIONS

It is easy to determine the relative energies of the three terms of the p^2 configuration by combining the results of the previous two sections. From (6.38) and (6.39), we have $E(^1D) = J_{11}$ and $E(^3P) = J_{10} - K_{10}$. Using (6.52) and Table 6.5, we obtain $J_{11} = F^0 + \frac{1}{25}F^2$, $J_{10} = F^0 - \frac{2}{25}F^2$, and $K_{10} = \frac{3}{25}F^2$, so we have the definite prediction that the ^3P lies below the ^1D. The energy of the ^1S may be derived by using the diagonal-sum rule, and we obtain

$$
\begin{aligned}
E(^1S) &= 2J_{1-1} + J_{00} - J_{11} - J_{10} + K_{10} \\
&= 2F^0 + \frac{2}{25}F^2 + F^0 + \frac{4}{25}F^2 - F^0 - \frac{1}{25}F^2 \\
&\qquad\qquad\qquad\qquad - F^0 + \frac{2}{25}F^2 + \frac{3}{25}F^2 \\
&= F^0 + \frac{10}{25}F^2
\end{aligned}
\tag{6.53}
$$

It is convenient, *for p^n configurations*, to define $F_0 = F^0$ and $F_2 = \frac{1}{25}F^2$, so that

$$
\begin{aligned}
E(^3P) &= F_0 - 5F_2 \\
E(^1D) &= F_0 + F_2 \\
E(^1S) &= F_0 + 10F_2
\end{aligned}
\qquad\blacktriangleright(6.54)
$$

Two predictions are immediate. First, the theory predicts that the ^3P term lies lowest. This is also a consequence of *Hund's rule;* which states that in a given configuration, the term with largest S lies lowest, and if several terms have the same maximal S, the one with largest L is the ground term. A variant of Hund's rule is that terms are ordered first according to S and then according to L, and we see that the p^2 configuration obeys this rule as well. However, although Hund's rule invariably predicts the ground term correctly, it usually fails to order the excited terms correctly.

Secondly, the theory predicts that the difference between the first two terms is two-thirds the difference between the second and third terms, that is,

$$
\frac{E(^1D) - E(^3P)}{E(^1S) - E(^1D)} = \frac{2}{3}
\qquad\blacktriangleright(6.55)
$$

Experimentally, for C, Si, Ge, and Sn, the ratios are 0.88, 0.68, 0.67, and 0.72, so the agreement is only qualitative.

For the p^3 configuration, we again have three terms. For the ^4S, we may take the $M_S = \frac{3}{2}$ component, $|^4S\ \frac{3}{2}\ 0\rangle = |1^+\ 0^+\ -1^+\rangle$, so

$$
\begin{aligned}
E(^4S) &= J_{10} + J_{1-1} + J_{0-1} - K_{10} - K_{1-1} - K_{0-1} \\
&= F_0 - 2F_2 + F_0 + F_2 + F_0 - 2F_2 - 3F_2 - 6F_2 - 3F_2 \\
&= 3F_0 - 15F_2
\end{aligned}
\tag{6.56}
$$

We suspect, according to Hund's rule, that this term will be the lowest. For the ^2D, we may write $|^2D\ \frac{1}{2}\ 2\rangle = |1^+\ 1^-\ 0^+\rangle$, so

$$
\begin{aligned}
E(^2D) &= J_{11} + 2J_{10} - K_{10} \\
&= F_0 + F_2 + 2F_0 - 4F_2 - 3F_2 \\
&= 3F_0 - 6F_2
\end{aligned}
\tag{6.57}
$$

Finally, by the diagonal-sum rule,

$$
\begin{aligned}
E(^2P) &+ E(^2D) + E(^4S) \\
&= E(|1^+\ 0^+\ -1^-\rangle) + E(|1^+\ 0^-\ -1^+\rangle) + E(|1^-\ 0^+\ -1^+\rangle) \\
&= 3(J_{10} + J_{0-1} + J_{1-1}) - K_{10} - K_{1-1} - K_{0-1} \\
&= 9F_0 - 21F_2
\end{aligned}
$$

so

$$
E(^2P) = 3F_0
\tag{6.58}
$$

The terms are again ordered by S and L. A ratio rule for the p^3 configuration can be derived, and its agreement with experiment is much as in the p^2 case.

The terms for the p^4 configuration are identical to those for p^2, and the energies are the same except that each term has $6F_0$ instead of $3F_0$ in the energy expression. The reader may wish to verify this in detail or to refer to the references on the hole-particle relation.

In each configuration, it is possible to define an *average configuration energy*. We average the energies of all states or, equivalently, average over terms weighted by their degeneracy. For example, the p^2 configuration has a total of 15 states, so

$$
\begin{aligned}
\bar{E}(p^2) &= \tfrac{1}{15}[9(F_0 - 5F_2) + 5(F_0 + F_2) + (F_0 + 10F_2)] \\
&= F_0 - 2F_2
\end{aligned}
\tag{6.59}
$$

Similarly,

$$
\bar{E}(p^3) = 3F_0 - 6F_2
\tag{6.60}
$$

These definitions are quite useful in discussing semiempirical theories of molecules. The value of F_2 may be obtained by comparison of the predicted atomic spectrum with the observed. F_0 may be obtained from ionization potential data. For example, ionizing an atom from the ^4S ground term of p^3 to the ^3P ground term of p^2 requires energy $3F_0 - 15F_2 - F_0 + 5F_2 = 2F_0 - 10F_2$ which may be set equal to the ionization potential. In doing so, however, we tacitly assume that the

Slater-Condon parameters F_0 and F_2 are identical in the atom and the ion, which is certainly incorrect since a contraction of the radial function is expected for the ion. Nevertheless, this is the way "experimental" values of F_0 are obtained, and the procedure can be partially justified depending on the use to which F_0 is put.

Often, when doing molecular calculations including first-row atoms, one is not so much interested in the standard configurations p^n as one is in certain excited configurations, such as sp^3 for carbon. It is possible to obtain Slater-Condon parameters for these configurations as well, and so to obtain the average energy of a variety of excited configurations, including a number of "hybrid" configurations.‡

6.6 TERM SEPARATION ENERGIES FOR THE d^n CONFIGURATIONS

Working out the relative term energies for the d^n configurations is a considerably more complicated task than for the p^n configurations, simply because there are so many more Slater determinants and terms. For example, the d^5 configuration includes 252 states. We will therefore restrict the present discussion to a summary of the results and to a brief mention of some peculiarities which do not arise in the p^n cases.

It is first convenient to introduce a change in nomenclature. As in the p^n case, we define new parameters

$$
\begin{aligned}
F_0 &= F^0 \\
F_2 &= \tfrac{1}{49}F^2 \\
F_4 &= \tfrac{1}{441}F^4
\end{aligned}
$$
▶(6.61)

Although these new parameters are widely used, an even more convenient set, defined by

$$
\begin{aligned}
A &= F_0 - 49F_4 \\
B &= F_2 - 5F_4 \\
C &= 35F_4
\end{aligned}
$$
▶(6.62)

is also found in the literature. A, B, and C are known as the Racah parameters. Table 6.6 gives the energies of all relevant coulomb and exchange integrals in terms of them. This table will come in handy in Chap. 9 as well, where we discuss ligand field theory.

The term energies of d^2 may be worked out in a straightforward manner, exactly as in the case of p^2. The reader will find it a valuable exercise. The d^3 configuration is the first case in which two terms of the same S and L occur, and it is therefore of more interest. The two ^2D terms represent states with small S and L, however, so their state func-

‡ For a detailed series of calculations of configurational energies for the first-row atoms, see G. Prichard and H. A. Skinner, *J. Inorg. Nucl. Chem.*, **24**:937 (1962).

Table 6.6 Coulomb and exchange integrals for d orbitals in terms of Racah parameters

m_l		J	K
± 2	± 2	$A + 4B + 2C$	$A + 4B + 2C$
± 2	∓ 2	$A + 4B + 2C$	$2C$
± 2	± 1	$A - 2B + C$	$6B + C$
± 2	∓ 1	$A - 2B + C$	C
± 2	0	$A - 4B + C$	$4B + C$
± 1	± 1	$A + B + 2C$	$A + B + 2C$
± 1	∓ 1	$A + B + 2C$	$6B + 2C$
± 1	0	$A + 2B + C$	$B + C$
0	0	$A + 4B + 3C$	$A + 4B + 3C$

tions are quite difficult to work out. The reader will find it a valuable but exhausting exercise. The end result is that there are off-diagonal matrix elements between the two ^2D states, and hence one must diagonalize a 2×2 secular determinant.

Table 6.7 lists the energies of the terms for all d^n configurations. The relative energies for the configurations d^{10-n} are, of course, the same as for d^n. In every case, Hund's rule is followed for the ground term, although the terms above the ground state are not necessarily ordered by S and L. The physical basis behind Hund's rule is of some interest. In the state of maximum S, the electrons are placed in orbitals so that they are as far apart as possible, thereby minimizing the energy. This effect is often referred to as *spin correlation* since, to an extent, the motion of the electrons is correlated spatially by the Pauli exclusion principle. For example, in the ^6S ground state of d^5, the five electrons are in five different spatial orbitals.

The theory for the d^n configurations is in many ways more satisfying than the theory for p^n. For example, in the d^5 configuration, the theory predicts the relative energies of 16 terms with only two parameters. Furthermore, the agreement with experiment is quite reasonable. In addition, one finds experimentally that B is on the order of four or five times C, effectively reducing the number of independent parameters (this is often expressed as F_2 being 12 to 14 times F_4).

The energies of all terms for d^2 through d^5 are graphed in Figs. 6.1 through 6.4 in units of C plotted against B/C. These diagrams can be used to fit experimental terms to the theory. Just how the fit is made is a matter of choice; one may require the ground term to have energy zero, ignore the more highly excited states, or simply fit all terms by least squares. Even the method of least squares leaves many choices as to weighting. One must therefore beware, in using experimental values of B and C, of how they were obtained. For many purposes, direct graphical

Table 6.7 **Energies of the terms of the** d^n **configurations in Racah parameters. For the energies of the terms of** d^{10-n}**, add** $(45 - 9n)A + (-70 + 14n)B + (35 - 7n)C$ **to the energies of** d^n **(from Racah, 1942)**

	d^2			d^3	
^3F	$A - 8B$		^4F		$3A - 15B$
^3P	$A + 7B$		^4P		$3A$
^1G	$A + 4B + 2C$		$^2H,$	^2P	$3A - 6B + 3C$
^1D	$A - 3B + 2C$		^2G		$3A - 11B + 3C$
^1S	$A + 14B + 7C$		^2F		$3A + 9B + 3C$
$\bar{E}$	$\dfrac{45A - 70B + 35C}{45}$		^2D		$3A + 5B + 5C \pm (193B^2$ $+ 8BC + 4C^2)^{1/2}$
			$\bar{E}$		$\dfrac{120A - 560B + 280C}{120}$

	d^4		d^5	
^5D	$6A - 21B$	^6S	$10A - 35B$
^3H	$6A - 17B + 4C$	^4G	$10A - 25B + 5C$
^3G	$6A - 12B + 4C$	^4F	$10A - 13B + 7C$
^3F	$6A - 5B + 5\frac{1}{2}C \pm \frac{3}{2}(68B^2$ $+ 4BC + C^2)^{1/2}$	^4D	$10A - 18B + 5C$
		^4P	$10A - 28B + 7C$
^3D	$6A - 5B + 4C$	^2I	$10A - 24B + 8C$
^3P	$6A - 5B + 5\frac{1}{2}C \pm \frac{1}{2}(912B^2$ $- 24BC + 9C^2)^{1/2}$	^2H	$10A - 22B + 10C$
		^2G	$10A - 13B + 8C$
^1I	$6A - 15B + 6C$	$^2G'$	$10A + 3B + 10C$
^1G	$6A - 5B + 7\frac{1}{2}C \pm \frac{1}{2}(708B^2$ $- 12BC + 9C^2)^{1/2}$	^2F	$10A - 9B + 8C$
		$^2F'$	$10A - 25B + 10C$
^1F	$6A + 6C$	^2D	$10A - 4B + 10C$
^1D	$6A + 9B + 7\frac{1}{2}C \pm \frac{3}{2}(144B^2$ $+ 8BC + C^2)^{1/2}$	$^2D'$	$10A - 3B + 11C \pm 3(57B^2$ $+ 2BC + C^2)^{1/2}$
^1S	$6A + 10B + 10C \pm 2(193B^2$ $+ 8BC + 4C^2)^{1/2}$	^2P	$10A + 20B + 10C$
		^2S	$10A - 3B + 8C$
$\bar{E}$	$\dfrac{210A - 1960B + 980C}{210}$	$\bar{E}$	$\dfrac{252A - 3920B + 1960C}{252}$

comparison is sufficient, and in Fig. 6.2, such a comparison of the electronic structure of V^{+2} (d^3) is shown. The agreement is really quite good except for the ^2P.

As in the p^n configurations, one may define an average d^n configurational energy, and these are given in Table 6.7. In practice, average d^n energies are difficult to obtain experimentally since one must observe all the states of a given configuration. This has rarely been done. This difficulty has led to a large number of different ways of calculating experimental average d^n energies in the literature. Some of these methods

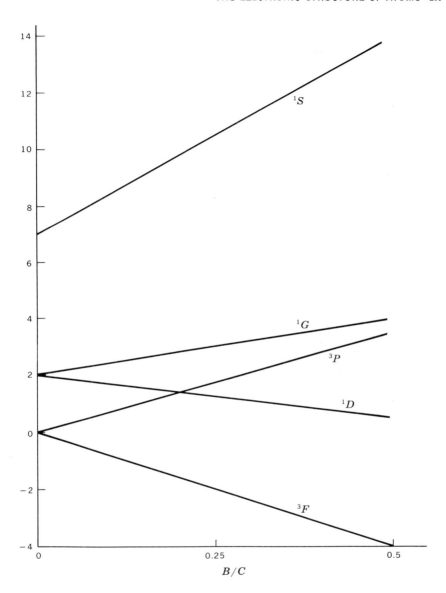

Fig. 6.1 Term energies for d^2. Energies are given in units of C, plotted against B/C. (*Figs. 6.1 through 6.4 are from J. S. Griffith, "The Theory of Transition Metal Ions," pp. 87–90, Cambridge University Press, London, 1961. Reproduced by permission of the publisher.*)

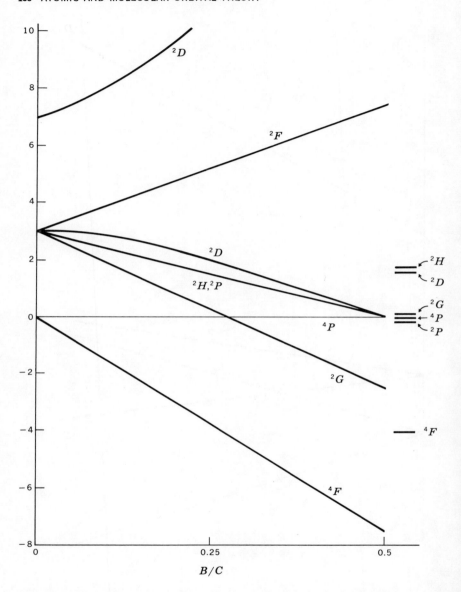

Fig. 6.2 Term energies of d^3. See Fig. 6.1. The experimental electronic structure of V^{2+} is shown on the right, assuming $B/C = 0.2683$. A fit of the ^4F term has been forced.

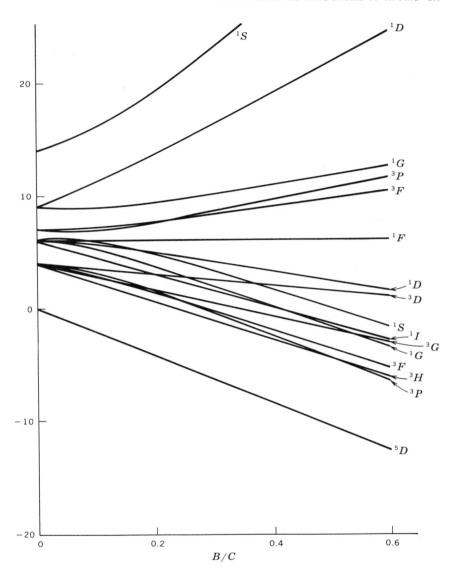

Fig. 6.3 Term energies for d^4. See Fig. 6.1.

involve fitting the average energies to smooth functions of the atomic number or charge.‡

‡ See for example, Ballhausen and Gray, p. 120. For a careful choice of B and C parameters fitted to individual atoms, see Griffith, 1961, p. 437. See also Slater, 1960, p. 339 *ff.*, p. 387.

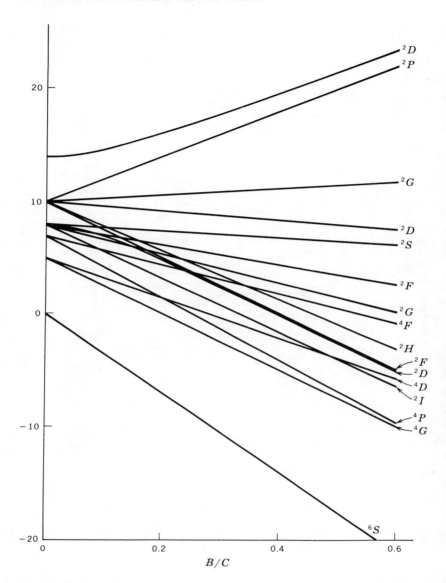

Fig. 6.4 Term energies for d^5. See Fig. 6.1.

6.7 MAGNETIC EFFECTS: SPIN–ORBIT COUPLING
AND ZEEMAN SPLITTINGS

The theory outlined thus far is based entirely on the Hamiltonians (6.1) and (6.2). These Hamiltonians omit certain relativistic terms which arise in the Dirac treatment. The most important of these is the spin-orbit interaction; in a semiclassical sense, it arises from the interaction between two magnetic fields due to spin and angular momentum. When the angular momentum of an electron is zero, the two spin states have equal energy, but for nonzero angular momentum, the spin and angular-momentum vectors can couple to separate the energies of the two spin components. According to the Dirac equation, this coupling is of the form

$$\mathbf{H}_{so} = -\frac{e}{2m^2c^2r}\frac{dV}{dr}\mathbf{l}\cdot\mathbf{s} = \xi(r)\mathbf{l}\cdot\mathbf{s} \qquad \blacktriangleright(6.63)$$

where V is the electrostatic potential of the central field. For a system of n electrons,

$$\mathbf{H}_{so} = \sum_{i=1}^{n} \xi(r_i)\mathbf{l}_i\cdot\mathbf{s}_i \qquad \blacktriangleright(6.64)$$

and we must add this interaction to our Hamiltonians.

It is not difficult to show (see Prob. 6.12) that $\mathbf{L}^2$ and $\mathbf{S}^2$ do not commute with $\mathbf{H}_{so}$, although $\mathbf{J}^2$ and $\mathbf{J}_z$ do. Therefore the eigenstates of the full Hamiltonian are no longer eigenstates of $\mathbf{L}^2$ and $\mathbf{S}^2$, and we cannot strictly classify states as before. Nevertheless, the classification remains "approximately correct" since $\mathbf{H}_{so}$ is usually quite small. Spin-orbit coupling destroys the $(2S + 1)(2L + 1)$ degeneracy of the terms, however, so it becomes convenient to reclassify the states according to the second Russell-Saunders scheme. We will seek functions of the form $|^{2S+1}LJM_J\rangle$. It is clear that matrix elements between states of different J or M_J will be zero, and that the effect of the spin-orbit coupling will be to cause an interaction between levels with the same value of J. Furthermore, it can be shown (Griffith, 1961, p. 108) that only states with L differing by zero or unity and S differing by zero or unity will have nonzero off-diagonal matrix elements; $\langle^{2S+1}L_JM|\mathbf{H}_{so}|^{2S'+1}L'_{J'}M'\rangle$ is zero unless $J = J'$, $M = M'$, $S = S'$, or $S = S' \pm 1$, $L = L'$ or $L = L' \pm 1$. This provides a set of "selection rules" for spin-orbit coupling.

The most direct effect of spin-orbit coupling is to lift the degeneracy of the levels of a given term. It is possible to show‡ that within a term,

$$\langle^{2S+1}L_JM|\mathbf{H}_{so}|^{2S+1}L_JM\rangle = \lambda\langle^{2S+1}L_JM|\mathbf{L}\cdot\mathbf{S}|^{2S+1}L_JM\rangle \qquad \blacktriangleright(6.65)$$

and we therefore obtain

$$\langle^{2S+1}L_JM|\mathbf{H}_{so}|^{2S+1}L_JM\rangle = \tfrac{1}{2}\lambda[J(J + 1) - L(L + 1) - S(S + 1)] \qquad (6.66)$$

since $2\mathbf{L}\cdot\mathbf{S} = \mathbf{J}^2 - \mathbf{L}^2 - \mathbf{S}^2$. The parameter λ is difficult to relate to the integrals over the functions $\xi(r_i)$, and we shall do so only on a case-to-

‡ The proof is not as obvious as might first appear, since $\mathbf{L}\cdot\mathbf{S} \neq \Sigma\mathbf{l}_i\cdot\mathbf{s}_i$. See Griffith, 1961, p. 108, or Condon and Shortley, p. 196.

case basis. For the present, we point out the fact that the splitting of the levels of a given term is proportional to J; that is,

$$\langle {}^{2S+1}L_J M | \mathbf{H}_{so} | {}^{2S+1}L_J M \rangle - \langle {}^{2S+1}L_{J-1} M | \mathbf{H}_{so} | {}^{2S+1}L_{J-1} M \rangle = \lambda J \quad (6.67)$$

This is the Landé interval rule, which states that the separation between levels of a given term (known as the fine structure) is proportional to J.

We illustrate this discussion with the p^2 configuration. The levels are ${}^3P_2$, ${}^3P_1$, ${}^3P_0$, ${}^1D_2$, and ${}^1S_0$. The splitting of the ${}^3P$ term is

$$\langle {}^3P_J | \mathbf{H}_{so} | {}^3P_J \rangle = \tfrac{1}{2}\lambda[J(J+1) - 4]$$
$$= \lambda, -\lambda, -2\lambda \quad \text{for } J = 2, 1, 0 \quad (6.68)$$

In order to evaluate λ, we must find the spin-orbit coupling matrix element between one of the ${}^3P$ states. The $|{}^3P_2\, 2\rangle = |1^+\, 0^+\rangle$ is the natural choice. Since $\mathbf{H}_{so} = \xi(r)(\mathbf{l}_1 \cdot \mathbf{s}_1 + \mathbf{l}_2 \cdot \mathbf{s}_2)$, we obtain

$$\langle {}^3P_2 | \mathbf{H}_{so} | {}^3P_2 \rangle = \langle 1^+\, 0^+ | \xi(r)(\mathbf{l}_1 \cdot \mathbf{s}_1 + \mathbf{l}_2 \cdot \mathbf{s}_2) | 1^+\, 0^+ \rangle \quad (6.69)$$

The integral over $\xi(r)$ involves the radial functions. It is customary to abbreviate this radial integral as a parameter,

$$\int R_{nl} \xi(r) R_{nl}\, r^2\, dr = \frac{\zeta_{nl}}{\hbar^2} \qquad\qquad \blacktriangleright (6.70)$$

This separation is possible because $\xi(r)$ is a function only of r, whereas l is a function only of the angular coordinates. In addition, we have used the fact that the spin-orbit coupling operator is a one-electron operator, so the discussion of Sec. 6.3 applies. Expanding $\mathbf{l} \cdot \mathbf{s}$ (see problem 6.13),

$$\mathbf{l} \cdot \mathbf{s} = l_z s_z + \tfrac{1}{2}(l^+ s^- + l^- s^+) \qquad\qquad \blacktriangleright (6.71)$$

we find

$$\langle {}^3P_2 | \mathbf{H}_{so} | {}^3P_2 \rangle = \langle 1^+\, 0^+ | l_{z_1} s_{z_1} + l_{z_2} s_{z_2} | 1^+\, 0^+ \rangle \frac{\zeta_{2p}}{\hbar^2} = \tfrac{1}{2}\zeta_{2p} \quad (6.72)$$

Therefore,

$$\lambda = \tfrac{1}{2}\zeta_{2p} \quad (6.73)$$

for the ${}^3P$ term‡ of p^2.

The mixing between ${}^3P_2$ and ${}^1D_2$ may be handled in a similar way. Taking the components with largest M_S and M_L, that is, the components

‡ There can be, in a given configuration, a different value of λ for each term.

with $M_J = J$, we have

$$\langle {}^3P_2|\mathbf{H}_{so}|{}^1D_2\rangle = \langle 1^+0^+|\mathbf{H}_{so}|1^+1^-\rangle$$

$$= \langle 1^+0^+|\tfrac{1}{2}1_2^- s_2^+|1^+1^-\rangle \frac{\zeta_{2p}}{\hbar^2}$$

$$= \frac{\sqrt{2}}{2}\,\zeta_{2p} \tag{6.74}$$

Finally, we examine the mixing of the ${}^3P_0$ and the ${}^1S$. Starting with $|{}^3P_2\,2\rangle = |{}^3P\,1\,1\rangle$ (in the $|{}^{2S+1}LM_SM_L\rangle$ scheme) and applying $\mathbf{J}^- = \mathbf{L}^- + \mathbf{S}^-$, we obtain

$$\mathbf{J}^-|{}^3P_2\,2\rangle = 2\hbar|{}^3P_2\,1\rangle = \sqrt{2}\hbar|{}^3P\,0\,1\rangle + \sqrt{2}\hbar|{}^3P\,1\,0\rangle$$

so

$$|{}^3P_2\,1\rangle = \frac{1}{\sqrt{2}}\,(|{}^3P\,0\,1\rangle + |{}^3P\,1\,0\rangle)$$

Applying $\mathbf{J}^-$ once again, we find

$$|{}^3P_2\,0\rangle = \frac{1}{\sqrt{6}}\,(2|{}^3P\,0\,0\rangle + |{}^3P\,-1\,1\rangle + |{}^3P\,1\,-1\rangle) \tag{6.75}$$

By orthogonality with the $|{}^3P_2\,1\rangle$,

$$|{}^3P_1\,1\rangle = \frac{1}{\sqrt{2}}\,(|{}^3P\,1\,0\rangle - |{}^3P\,0\,1\rangle)$$

so, using $\mathbf{J}^-$, we have

$$|{}^3P_1\,0\rangle = \frac{1}{\sqrt{2}}\,(|{}^3P\,1\,-1\rangle - |{}^3P\,-1\,1\rangle) \tag{6.76}$$

Finally, the $|{}^3P_0\,0\rangle$ state may be obtained by orthogonality with the $|{}^3P_2\,0\rangle$ and $|{}^3P_1\,0\rangle$ as follows:

$$|{}^3P_0\,0\rangle = \frac{1}{\sqrt{3}}\,(|{}^3P\,-1\,1\rangle + |{}^3P\,1\,-1\rangle - |{}^3P\,0\,0\rangle) \tag{6.77}$$

Each one of the states in the $|{}^{2S+1}LM_SM_L\rangle$ scheme can be written in terms of the basic Slater determinants. In this way we find

$$|{}^3P_0\,0\rangle = \frac{1}{\sqrt{6}}\,(\sqrt{2}|1^-0^-\rangle + \sqrt{2}|0^+-1^+\rangle - |1^+-1^-\rangle - |1^--1^+\rangle) \tag{6.78}$$

Combining this with the relation

$$|{}^1S\rangle = \frac{1}{\sqrt{3}}\,(|1^+-1^-\rangle - |0^+0^-\rangle - |1^--1^+\rangle) \tag{6.79}$$

from Sec. 6.2, we derive

$$\langle {}^1S|\mathbf{H}_{so}|{}^3P_0\, 0\rangle = -\sqrt{2}\,\zeta_{2p} \tag{6.80}$$

Spin-orbit coupling is also responsible for small shifts in the separation between levels of different terms due to diagonal elements. As long as the spin-orbit coupling parameter is small compared to F_2, these shifts are entirely undetectable. In any case, for the p^2 configuration,‡ both $\langle {}^1D_2|\mathbf{H}_{so}|{}^1D_2\rangle$ and $\langle {}^1S|\mathbf{H}_{so}|{}^1S\rangle$ are zero.

For the upper left-hand side of the periodic table, the ζ_{nl} values are quite small. We may therefore treat the mixing of these levels as perturbations. To end our example, then, we find the following energies for the levels of p^2:

$$
\begin{aligned}
E({}^3P_0) &= U + F_0 - 5F_2 - \zeta_{2p} - \frac{2\zeta_{2p}^2}{15F_2}\\[6pt]
E({}^3P_1) &= U + F_0 - 5F_2 - \tfrac{1}{2}\zeta_{2p}\\[6pt]
E({}^3P_2) &= U + F_0 - 5F_2 + \tfrac{1}{2}\zeta_{2p} - \frac{\zeta_{2p}^2}{6F_2}\\[6pt]
E({}^1D_2) &= U + F_0 + F_2 + \frac{\zeta_{2p}^2}{6F_2}\\[6pt]
E({}^1S_0) &= U + F_0 + 10F_2 + \frac{2\zeta_{2p}^2}{15F_2}
\end{aligned}
\qquad\blacktriangleright(6.81)
$$

These energies are sketched qualitatively in Fig. 6.5. For atoms on the lower right-hand side of the periodic table, the ζ_{nl} values may be so large as to make it impossible to treat the spin-orbit interaction as a perturbation. One must then diagonalize the entire matrix of spin-orbit interaction together with the two-electron terms.

The second magnetic effect we consider in this section is the influence§ of an external magnetic field on the energy levels of an atom. The field will interact with both the orbital angular momentum and the spin angular momentum according to the perturbation

$$\mathbf{H}_m = \frac{\beta \mathbf{H}\cdot(\mathbf{L} + 2\mathbf{S})}{\hbar} \qquad\blacktriangleright(6.82)$$

where $\beta = e\hbar/2mc$ is the Bohr magneton. Taking the field parallel to the z axis and writing $\mathbf{J}_z = \mathbf{L}_z + \mathbf{S}_z$, we have

$$\mathbf{H}_m = \frac{\beta H(\mathbf{J}_z + \mathbf{S}_z)}{\hbar} \qquad\blacktriangleright(6.83)$$

‡ Matrices for the spin-orbit coupling interaction for a number of configurations may be found on p. 268 of Condon and Shortley. The first-order splitting of a term always leaves the "center of gravity" of the term unchanged.
§ This influence is known as the Zeeman effect.

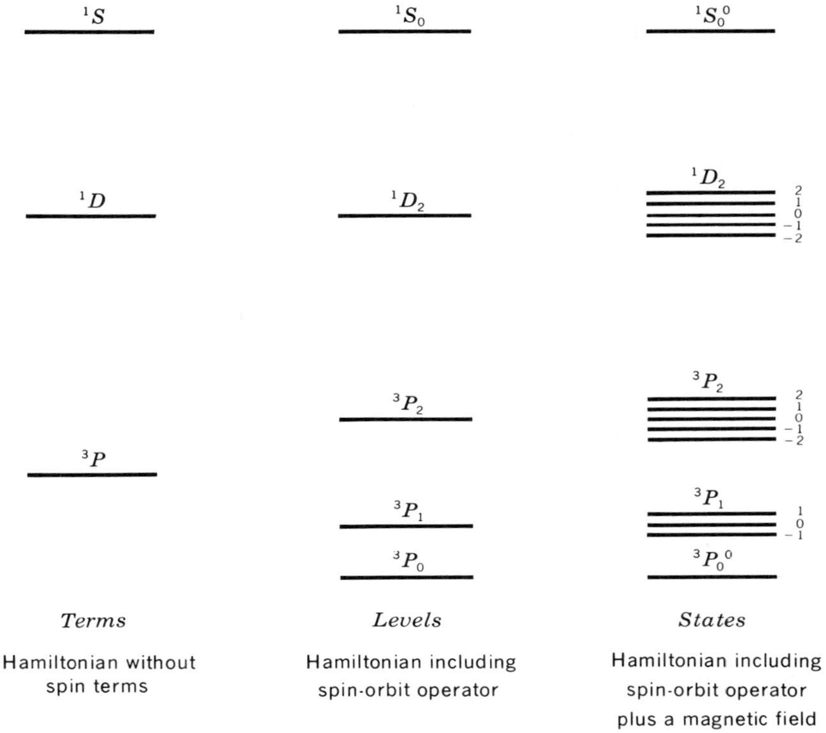

| | ^1S | $^1S_0$ | $^1S_0^0$ |

^1D $^1D_2$ $^1D_2$

^3P $^3P_2$ $^3P_2$

$^3P_1$ $^3P_1$

$^3P_0$ $^3P_0^0$

Terms	*Levels*	*States*
Hamiltonian without spin terms	Hamiltonian including spin-orbit operator	Hamiltonian including spin-orbit operator plus a magnetic field

Fig. 6.5 Energy separations for the p^2 configuration. Energy separations are emphatically *not* to scale. The term separations are on the order of 10,000 cm^{-1}, whereas the levels of a given term are separated by 10 to 100 cm^{-1}. The magnetic separations are on the order of 1 cm^{-1}.

The states of a given level are then split according to

$$E_m = \langle ^{2S+1}L_JM|\mathbf{H}_m|^{2S+1}L_JM\rangle \tag{6.84}$$

Since $\mathbf{S}_z$ does not commute with $\mathbf{J}^2$, $\mathbf{H}_m$ will mix terms of different J together. However, the magnetic interaction is in practice so small that it is negligible compared even to the spin-orbit interaction. Except in special situations, this mixing can be neglected altogether. We thus need only consider the lifting of the M-degeneracy. We have at once

$$E_m = \beta H[M + \langle ^{2S+1}L_JM|\mathbf{S}_z|^{2S+1}L_JM\rangle] \tag{6.85}$$

The matrix element of $\mathbf{S}_z$ can be evaluated with some difficulty (Condon and Shortley, p. 64; Griffith, 1961, p. 130). In the end we find

$$E_m = g\beta HM \qquad \blacktriangleright (6.86)$$

where

$$g = 1 + \frac{J(J + 1) + S(S + 1) - L(L + 1)}{2J(J + 1)} \qquad \blacktriangleright (6.87)$$

The constant g is the Landé factor. Each level will have its own well-defined Landé g factor, and experimentally this is a useful means of distinguishing one level from another. Note that in a magnetic field, all degeneracy is removed, except when $g = 0$.

We may conclude this section with a discussion of the selection rules which determine the intensities observed in atomic spectroscopy. The dipole moment operator $\mathbf{\mu} = \sum_i e\mathbf{r}_i$ is a one-electron operator, commuting with $\mathbf{S}^2$ and $\mathbf{S}_z$, so we can immediately derive the selection rule $\Delta S = 0$. This selection rule is valid even outside the framework of the orbital approximation. It can be broken through spin-orbit coupling, since this interaction mixes levels with differing S. However, in the lighter atoms, the coupling is so small that the $\Delta S = 0$ selection rule remains quite good. For example, if one prepares the ^1D excited state of the carbon atom in an arc, the atom will, on the average, take several seconds before phosphorescing and returning to the ^3P ground state. In fact, spin-forbidden transitions are responsible for all known phosphorescence.

A second selection rule of importance is the parity rule. As noted in Chap. 4, only transitions from an even-parity state to an odd-parity state, or vice versa, are allowed. It is easily shown that the parity operator commutes with the Hamiltonian (excepting the $\mathbf{H}_m$ part), and so this rule is also independent of any approximations. This is because the atom has a center of symmetry. Since the parity of an orbital is independent of m_l and m_s, all states of a given configuration have the same parity, and so all transitions within a configuration are forbidden.

A final selection rule forbids so-called two-electron jumps. Since the dipole moment operator is a one-electron operator, it has zero matrix elements between determinants differing by more than one spin orbital. Thus, excited states which differ from the ground state by two spin orbitals cannot be reached by a one-photon absorption. Unlike the other selection rules, this one is not rigorous, having meaning only within the assumption of definite configurations.

PROBLEMS

6.12 Show that $\mathbf{L}^2$ and $\mathbf{S}^2$ do not commute with $\mathbf{H}_{so}$, but that $\mathbf{J}^2$ does.

6.13 Prove that $\mathbf{l} \cdot \mathbf{s} = \mathbf{l}_z\mathbf{s}_z + \frac{1}{2}(\mathbf{l}^+\mathbf{s}^- + \mathbf{l}^-\mathbf{s}^+)$, Eq. (6.71).

6.14 Verify Eq. (6.80).

6.15 Using perturbation theory, verify the second-order spin-orbit corrections to the energies of the terms of the p^2 configuration, Eq. (6.81). Derive the corresponding wavefunctions, and show that the ground state is no longer a pure triplet.

6.16 Derive a set of selection rules for the off-diagonal matrix elements of the operator $\mathbf{H}_m$.

6.17 Calculate all off-diagonal matrix elements of $\mathbf{H}_m$ within the states of the p^2 configuration.
Hint: Expand each $|^{2S+1}L_J M\rangle$ state in terms of kets in the $|^{2S+1}L M_S M_L\rangle$ scheme in order to evaluate the matrix elements of $\mathbf{S}_z$.

6.8   THE METHOD OF THE SELF–CONSISTENT FIELD

We still have no knowledge of the radial wavefunctions for many-electron atoms, except for the empirically determined Slater-Condon factors. For many purposes it is sufficient to use Slater-type orbitals (STO's; see Sec. 4.7) with the parameters ζ_i chosen according to the well-known Slater rules. According to these rules, the (nodeless) radial functions are

$$R_{nl}(r) = N r^{n-1} e^{-\zeta r/a_0} \qquad\qquad \blacktriangleright (6.88)$$

The parameter ζ can be defined in terms of a shielding factor s according to the relation

$$\zeta = \frac{Z - s}{n} \qquad\qquad (6.89)$$

For each radial function, the shielding factor is determined as follows. For every electron in an orbital with principal quantum number less than n by 2 or more, we include a contribution of unity to s—complete shielding. For each electron in an orbital with principal quantum number less than n by 1, we include a contribution of 0.85, except in the case of the $3d$ valence shell, where we include a contribution of 1 for each $2s$, $2p$, $3s$, and $3p$ electron. For each electron in an orbital with the same principal quantum number, we include a contribution of 0.35 except when dealing with the $1s$ shell, where each $1s$ electron shields the other with $s = 0.30$. Note that we have effectively divided the electrons into four groups, $1s$; $2s$, $2p$; $3s$, $3p$; $3d$; . . . . Electrons in the outer groups do not shield those in the inner groups; the $3d$ valence electrons do not shield the $3s$ and $3p$ electrons.

An additional set of rules, governing the $4s$, $4p$, $4d$, . . . shells, includes the replacement of the actual principal quantum number by nonintegral values in the STO's, but orbitals constructed in this way have proved neither accurate nor practical.

For carbon, with nuclear charge $Z = 6$, the $1s$ Slater orbital is thus

$$\varphi_{1s} = N_{1s}e^{-5.70r/a_0} \tag{6.90}$$

The radial part of the $2p$ Slater orbital is

$$R_{2p} = N_{2p}re^{-1.625r/a_0} \tag{6.91}$$

The $2s$ Slater orbital has the same radial part as the $2p$ and is therefore *not* properly orthogonal to the $1s$. This difficulty is easily overcome by orthogonalizing the $2s$ to the $1s$, so that

$$\varphi_{2s} = N_{2s}(re^{-1.625r/a_0} - SN_{1s}e^{-5.70r/a_0}) \tag{6.92}$$

where

$$S = \int_0^\infty r^3 e^{-1.625r/a_0}e^{-5.70r/a_0}\, dr \tag{6.93}$$

For $1s$, $2s$, $2p$, $3s$, and $3p$ functions, the Slater orbitals do not provide too bad a description of atoms, although they appear to be rather better for inner shells than for valence shells. For negative ions and for the $3d$ valence shell of transition metal ions, they seem to be decidedly inadequate. Clementi and Raimondi (1963) have recently derived a set of ζ values which are somewhat better than the Slater values, but they provide no major improvement.

We turn now to the method of Hartree and Fock, which poses the following question: in the framework of a configurational wavefunction, i.e., using wavefunctions composed of one-electron orbitals, what is the *best-possible* set of radial functions for the orbitals? We use the phrase "best possible" in the sense of the variation theorem and seek that set of radial functions which minimizes the energy of the wavefunction for the configuration. This can strictly be done only if each function is expressed in essentially infinite parametric form, but it has been found that a linear combination of just a few Slater-type orbitals is quite adequate to produce energies within 0.001 percent of the true "Hartree-Fock minimum." We therefore seek radial functions of the form

$$R_{nl}(r) = \sum_i c_i \, \mathrm{STO}(n_i, \zeta_i) \qquad \blacktriangleright (6.94)$$

The problem of finding the best-possible functions is thus reduced to finding the linear parameters c_i for each orbital.

The exponents ζ_i and "principal quantum numbers" n_i are nonlinear parameters and are usually not varied, or are varied only after the minimum energy for a particular set has been determined, in order to see if a better set can be found. The technique of solving for the best c_i for a fixed-basis set is due to Roothaan (1951a). (See also Chap. 10.)

Before the Roothaan formulation of the Hartree-Fock equations, it was customary to find the radial functions by numerical methods and to tabulate the R_{nl} point by point.

Let us begin by considering the general problem of evaluating the energy of a Slater determinant containing *only closed shells*. The wavefunction is thus a ^1S, and there are no unpaired spins. Since the number of electrons is even, we call it $2n$. We then may write

$$\Psi = \mathcal{Q}|\varphi_1(1)\alpha_1\varphi_1(2)\beta_2\varphi_2(3)\alpha_3\varphi_2(4)\beta_4$$
$$\cdots \varphi_n(2n-1)\alpha_{2n-1}\varphi_n(2n)\beta_{2n}\rangle \qquad \blacktriangleright(6.95)$$

Each spatial‡ orbital φ_i contains two electrons, one with $m_s = \frac{1}{2}$ and one with $m_s = -\frac{1}{2}$. Each orbital φ_i is a linear combination of basis functions§ χ_p, such that

$$\varphi_i = \sum_{p=1}^{m} c_{ip}\chi_p \qquad \blacktriangleright(6.96)$$

We assume the φ_i form an orthonormal set, $\langle\varphi_i|\varphi_j\rangle = \delta_{ij}$. The χ_p are not necessarily orthogonal, although it is often convenient to assume that they are normalized, such that $S_{pp} = \langle\chi_p|\chi_p\rangle = 1$.

The Hamiltonian for the system is

$$\mathfrak{IC} = \sum_{i=1}^{2n} \mathbf{H}(i) + \sum_{i=2}^{2n}\sum_{j=1}^{i-1} \frac{e^2}{r_{ij}} \qquad \blacktriangleright(6.97)$$

where $\mathbf{H}(i)$ is the one-electron operator (kinetic plus nuclear attraction energies), and where the sums extend over all $2n$ electrons. We now define the matrix elements‖

$$H_i = \langle\varphi_i(1)|\mathbf{H}(1)|\varphi_i(1)\rangle$$
$$J_{ij} = \langle\varphi_i(1)\varphi_j(2)|\frac{1}{r_{12}}|\varphi_i(1)\varphi_j(2)\rangle \qquad (6.98)$$
$$K_{ij} = \langle\varphi_i(1)\varphi_j(2)|\frac{1}{r_{12}}|\varphi_j(1)\varphi_i(2)\rangle$$

Each matrix element can be expanded in terms of integrals over the constituent basis functions. We abbreviate these integrals as

$$h_{pq} = \langle\chi_p(1)|\mathbf{H}(1)|\chi_q(1)\rangle$$
$$\langle pq\|rs\rangle = \langle\chi_p(1)\chi_q(2)|\frac{1}{r_{12}}|\chi_r(1)\chi_s(2)\rangle \qquad (6.99)$$

‡ We now switch notational schemes and let the φ_i be spatial orbitals instead of spin orbitals.

§ The χ_p are usually Slater-type orbitals. If φ_i is a $1s$ orbital, the c_{ip} for χ_p a $2p$ orbital will be zero, so we can include the entire basis set of STO's without having to write separate equations for each shell.

‖ Since these matrix elements are now defined over *spatial* orbitals, spin is irrelevant, and K_{ij} is ordinarily nonzero.

and it is clear that

$$
\begin{aligned}
H_i &= \sum_{p=1}^{m} \sum_{q=1}^{m} c_{ip}^* c_{iq} h_{pq} \\
J_{ij} &= \sum_{p=1}^{m} \sum_{q=1}^{m} \sum_{r=1}^{m} \sum_{s=1}^{m} c_{ip}^* c_{jq}^* c_{ir} c_{js} \langle pq \| rs \rangle \\
K_{ij} &= \sum_{p=1}^{m} \sum_{q=1}^{m} \sum_{r=1}^{m} \sum_{s=1}^{m} c_{ip}^* c_{jq}^* c_{jr} c_{is} \langle pq \| rs \rangle
\end{aligned}
\tag{6.100}
$$

The integrals h_{pq} and $\langle pq \| rs \rangle$ may be calculated in a straightforward manner since both the Hamiltonian and the form of the χ_p are presumed to be known. In practice, with the χ_p Slater-type orbitals, their evaluation is reasonably elementary. However, the integrals H_i, J_{ij}, and K_{ij} cannot be determined without knowledge of the coefficient c_{ip}, and these in turn are actually the *solutions* we seek! The process is therefore circular—we cannot solve for the c_{ip} unless we know them initially. Fortunately, this poses no difficulty in practice, as shall be seen. One can determine the c_{ip} through a rather elegant procedure involving the concept of a self-consistent field.

The energy of the configuration, according to the discussion of Sec. 6.3, is equal to

$$
E = 2 \sum_{i=1}^{n} H_i + \sum_{i=1}^{n} \sum_{j=1}^{n} (2J_{ij} - K_{ij})
\qquad \blacktriangleright (6.101)
$$

The sums extend over *spatial* functions, i.e., over *doubly occupied* orbitals. This accounts for the factor of 2 multiplying the sum over one-electron matrix elements. The sum over coulomb and exchange integrals has a more complicated origin—we have used the relations $J_{ii} = K_{ii}$, $J_{ij} = J_{ji}$, $K_{ij} = K_{ji}$, and the fact that the exchange integral over *spin orbitals*,

$$
\langle \varphi_i(1) \xi_i \varphi_j(2) \xi_j | \frac{1}{r_{12}} | \varphi_j(1) \xi_j \varphi_i(2) \xi_i \rangle
$$

is zero unless $\xi_i = \xi_j$ ($\xi = \alpha$ or β). The reader should verify (6.101) as an exercise (Prob. 6.18).

Rather than expanding the energy expression (6.101) directly in terms of the coefficients, we now write

$$
E = 2 \sum_{i=1}^{n} \langle \varphi_i | \mathbf{H} | \varphi_i \rangle + \sum_{i=1}^{n} \sum_{j=1}^{n} \langle \varphi_i | 2\mathbf{J}_j - \mathbf{K}_j | \varphi_i \rangle
\tag{6.102}
$$

where we explicitly define

$$\langle\varphi_k|\mathbf{J}_j|\varphi_l\rangle = \langle\varphi_k(1)\varphi_j(2)|\frac{1}{r_{12}}|\varphi_l(1)\varphi_j(2)\rangle$$

▶(6.103)

$$\langle\varphi_k|\mathbf{K}_j|\varphi_l\rangle = \langle\varphi_k(1)\varphi_j(2)|\frac{1}{r_{12}}|\varphi_j(1)\varphi_l(2)\rangle$$

so that

$$\langle\varphi_i|\mathbf{J}_j|\varphi_i\rangle = J_{ij} \qquad (6.104)$$
$$\langle\varphi_i|\mathbf{K}_j|\varphi_i\rangle = K_{ij} \qquad (6.105)$$

Note that we have, in effect, replaced the Hamiltonian $\mathcal{3C}$ by a new operator. We now seek to minimize the energy expression

$$E = \sum_{i=1}^{n} \{\langle\varphi_i|\mathbf{H}|\varphi_i\rangle + \langle\varphi_i|\mathbf{F}|\varphi_i\rangle\}$$

▶(6.106)

where $\mathbf{F} = \mathbf{H} + \sum_{j=1}^{n} (2\mathbf{J}_j - \mathbf{K}_j)$.

As has been shown by Roothaan, it is possible to find $\mathbf{F}$ such that

$$\mathbf{F}|\varphi_i\rangle = \epsilon_i|\varphi_i\rangle$$

▶(6.107)

for all φ_i. This has the form of an eigenvalue equation, except that the operator $\mathbf{F}$ is determined by the φ_i, effectively coupling all n equations. The physical interpretation, however, is straightforward. The shape of each orbital, i.e., the probability distribution of each electron, is determined by a single operator, which includes the kinetic energy, nuclear attraction, and *average* electron repulsion of all other electrons. The coupling of the equations is simply a consequence of the fact that the average electron-repulsion term contained in one equation depends on the shape of all the other orbitals, and the shape of the other orbitals depends in turn on the shape (and consequent electron-repulsion term) of the first orbital.

One feature of Eq. (6.107) which is of the highest importance is the fact that since each φ_i is expressed in linear parametric form, the original problem of finding the lowest energy of the system is reduced to the solution of n (coupled) eigenvalue equations in which all wavefunctions are in this linear form. We can therefore apply Eq. (5.10). The problem thus reduces to solving the determinantal equation

$$|F_{pq} - S_{pq}\epsilon_i| = 0 \qquad (6.108)$$

for all i.

Roothaan's self-consistent-field (SCF) method is an efficient and accurate means of solving the Hartree-Fock equations, although, of course, for an *exact* solution to these equations, an infinitely large basis

χ must be used. In practice, one uses a relatively small basis and solves for the *best energy within that basis;* this is an SCF method, but it is not truly a solution to the Hartree-Fock equations. Nevertheless, this limited-basis set method is often loosely referred to as the Hartree-Fock method. A better name is the Roothaan SCF method. The *self-consistent* matrix F_{pq} is generally known as the Fock matrix, whether or not it represents a basis large enough so that the energy reaches the Hartree-Fock limit. In the intermediate stages of a given calculation, the matrix F_{pq} is *not* called the Fock matrix. The term is specifically restricted to the self-consistent operator which has a pseudo-one-electron form and which satisfies the pseudo-eigenvalue equations (6.107). Since the dimensions of the matrices F_{pq} and S_{pq} in (6.108) are equal to m, the operator $\mathbf{F}$ also determines ϵ_i and φ_i for $i > n$ as well as for $i \leq n$. These represent so-called *virtual* orbitals and have no direct physical meaning, although, as we shall see, they are extremely important to any discussion of the excited states of an atom.

Since the matrix elements of $\mathbf{F}$ between *orbitals* are diagonal, we may use (6.106) to show that

$$E = \sum_{i=1}^{n} (H_i + \epsilon_i)$$
▶(6.109)

Note that the total energy of the system is *not* a sum of orbital energies. Using Eq. (6.102), we see at once that

$$\epsilon_i = H_i + \sum_{j=1}^{n} (2J_{ij} - K_{ij})$$
▶(6.110)

This confirms the interpretation of the ϵ's as orbital energies; each contains the usual kinetic and nuclear attraction terms, together with a pair of electron repulsions from each doubly occupied orbital and an exchange contribution from each electron with the same spin, all as expected.

A number of problems arise in applying the Roothaan formalism to *open* shells. In an intuitive sense, the origin of these problems is quite clear—the potential seen by an up-spin electron is, in general, different from that seen by a down-spin electron. This implies that there should be different orbitals for the two sets of electrons.‡ These difficulties are avoided by averaging the exchange terms and by requiring a single radial function for each shell. One can define an *average* orbital energy for each shell. The actual procedure is complicated, however, and

‡ In fact, the potential seen by an electron can also depend on m_l, as may be seen by calculating ϵ_{1+} and ϵ_{0+} for the ^4S term of the p^3 configuration.

the proof that such a procedure can be rigorously derived from the variation theorem is quite difficult. ‡

There are other "extended" or "unrestricted" Hartree-Fock methods which do not insist on a single radial function per shell. However, in such treatments, the resultant atomic wavefunctions are not eigenfunctions of S^2 or L^2, and so we do not consider them here.

We shall not go further into the theory or practice of Roothaan-Hartree-Fock calculations in this chapter. A thorough discussion is contained in Chap. 10. Instead, we shall discuss some of the results which have been obtained for atoms, including atoms with open shells. The student should therefore concentrate on understanding and interpreting these results, without too much regard for how they are obtained.

For helium, Clementi (1965) has shown that the simple function

$$\varphi = 0.83415 \, STO(1.44608 \colon 1s) + 0.19060 \, STO(2.86222 \colon 1s) \quad (6.111)$$

provides an excellent approximation to the exact Hartree-Fock function. The calculated energy using (6.111) is -2.8616700 au, compared to the exact -2.8616799 au. In general, a linear combination of two Slater-type orbitals per atomic orbital, known as a double-ζ (double-zeta) basis, is adequate to reproduce all but the most subtle details of the atomic radial wavefunctions. The function of (6.111) is sketched in Fig. 6.6 and is compared graphically with the simple single-ζ Slater-type orbital with $\zeta = 27/16$. Small but important differences are apparent.

For systems with open shells, one can define a set of radial functions for each possible state (term). For example, the radial function φ_{2r} of neutral carbon is plotted for the ^3P, ^1D, and ^1S terms in Fig. 6.7. The reader who believes these differ but negligibly should be aware of the sensitivity of p-p overlap between a pair of neighboring carbon atoms to choice of radial function φ_{2p}. For example, at 2.6 au, the sigma overlap integral between a pair of carbon atoms having the $2p$ radial function characteristic of the ^3P term is 0.23474, whereas the corresponding overlaps are 0.21237 and 0.17471 for the functions arising from the ^1D and ^1S terms, respectively (the pi overlaps are 0.32972, 0.34891, and 0.37849). The actual overlap between a pair of carbon atoms in a compound may therefore be sensitive to the type of compound and the location and hybridization of the carbon orbitals, and cannot be approximated accurately by the overlap of two Slater-orbitals with exponents chosen according to Slater's rules.

In transition metal ions, the single-ζ functions are even less adequate. In Fig. 6.8 the Hartree-Fock d orbitals of $Ni^{2+}(^3F, 3d^8)$ are compared to the single-ζ function, and the discrepancies are seen to be

‡ For a description of some of the computer programs used in the Roothaan SCF treatment of atoms, with both closed and open shells, see Roothaan and Bagus (1963).

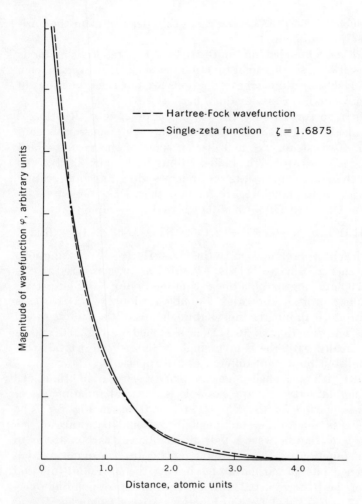

Fig. 6.6 Radial wavefunctions for the $1s$ orbital of helium.

enormous. On the other hand, a double-ζ function, due to Richardson et al. (1962), fits the exact Hartree-Fock radial function quite well. It is of interest to note that the d function for ^3F $3d^8$ Ni^{2+} is absolutely indistinguishable from the d function for ^3F $3d^8$ $4s^2$ neutral nickel. On the other hand, both functions are greatly different from the d function for $3d^{10}$ neutral nickel. Occupancy of the $4s$ orbitals seems to have absolutely no effect on the shape of the $3d$ functions, although there is considerable change and distortion in the $3d$ functions when the occupancy of the $3d$ level is changed.

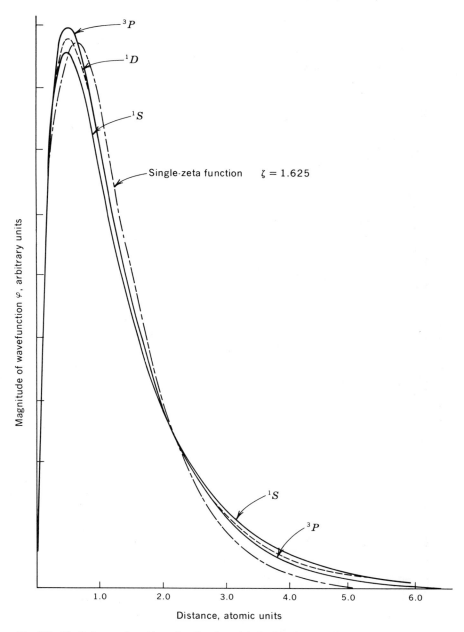

^3P

^1D

^1S

Single-zeta function $\zeta = 1.625$

^1S

^3P

Magnitude of wavefunction φ, arbitrary units

1.0   2.0   3.0   4.0   5.0   6.0

Distance, atomic units

Fig. 6.7   Radial wavefunctions for the $2p$ orbital of carbon, $1s^2 2s^2 2p^2$.

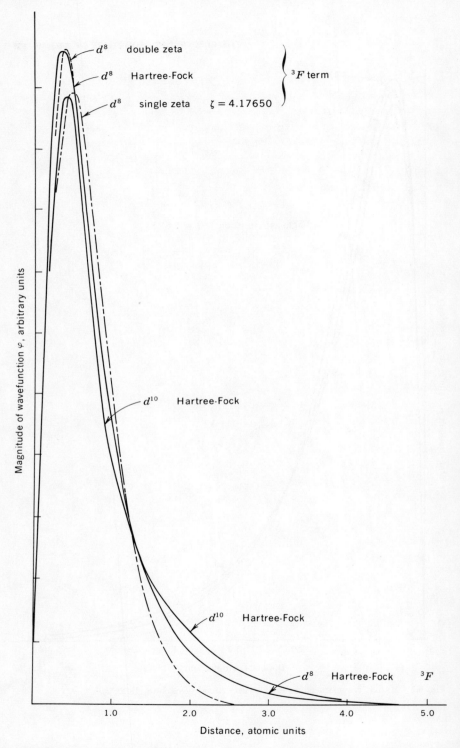

Fig. 6.8 Radial wavefunctions for the nickel atom. The configuration d^8 refers to Ni^{+2}, and d^{10} refers to the neutral atom.

It is also of interest to compare the quantities F_0, F_2, and F_4 (or the Racah parameters A, B, and C) calculated using Hartree-Fock functions with those obtained empirically. This is done for F^2 ($= 49F_2$) in Fig. 6.9. The results can be summarized by noting that the calculated values are over twice as large as the empirical ones for the neutral atoms, with the percent error decreasing steadily with increasing charge on the

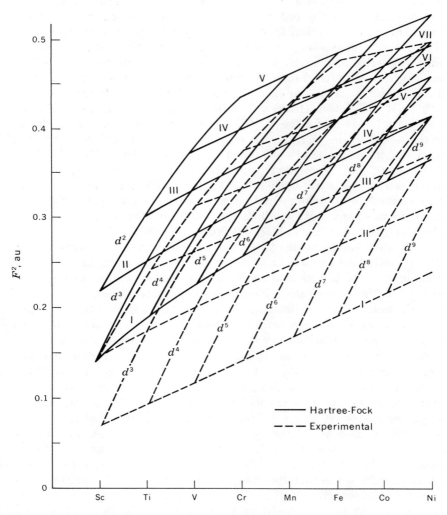

Fig. 6.9 A comparison of experimental and calculated (Hartree-Fock) values of F^2 ($= 49F_2$) for $3d^n$ configurations. The state I represents the neutral atom, II the singly positive ion, and so on. [*After Slater, vol. I, p. 387, 1960. Used by permission of the publisher.*]

atom. Studies on the parameter F_4 show precisely the same trend, with the calculated ratio F_2/F_4 remaining close to that found empirically. Since F_2 and F_4 (or B and C) determine the separations among the excited states (see Table 6.7), we see that for neutral atoms, Hartree-Fock theory predicts the states to lie twice as far apart from each other as observed. This is a rather serious error in the theory and one which, as we shall see in later chapters, seems to apply to molecules as well. Thus, as a *semiempirical* theory, the central-field approximation accurately rationalizes the ordering and separations among atomic terms, but it fails as an a priori theory to yield good term separation energies.

Since it is possible to separately minimize the energy of several different terms of a given configuration via the Hartree-Fock method, one can also calculate state energy differences by subtracting total Hartree-Fock atomic energies. For example, the total energies calculated for Cr^{3+} (d^3) by R. E. Watson (1959) are -1041.4718, -1041.3922, -1041.3910, and -1041.3644 au for the ^4F, ^4P, ^2G, and ^2P terms, respectively. Thus, with respect to the ^4F term, the energies are 0.0796, 0.0808, and 0.1074 au. Using the calculated $F_2 = 0.008113$, $F_4 = 0.0005649$ au ($B = 0.005288$, $C = 0.01977$ au), we predict (using Table 6.7) that the energies of the three excited states are 0.0793, 0.0805, and 0.1069 au, respectively. The two methods of obtaining excited state energies give much the same result in this case. Both methods give only qualitative agreement with the experimental values 0.0627, 0.0670, and 0.0623 au.

Assume for the moment that we can construct an approximate configurational wavefunction for a cation or anion by using the Roothaan SCF orbitals of the neutral atom, but including one less or one more electron in the configuration. This involves the rather drastic assumption that the radial wavefunctions will not be substantially changed by adding or subtracting an electron from the atom. As noted just above, detailed Hartree-Fock calculations on cations, atoms, and anions, or on any three states of ionization of an atom, show in fact that radial functions can change greatly with the state of ionization, but we ignore this. Doing so, one can prove that the highest occupied orbital energy of the neutral atom should be equal to the ionization potential of the atom, and the energy of the lowest empty (virtual) orbital should be equal to the electron affinity. Because of the drastic assumptions involved, we have no real reason to expect that either orbital energy should be close to the "experimental" value, and in fact experimental electron affinities are usually far from the corresponding Hartree-Fock orbital energies of the lowest virtual orbitals. On the other hand, in almost every case in which atomic ionization potentials have been compared to Hartree-Fock occupied orbital energies, one finds close agreement. The agreement seems to arise from a remarkable cancellation of errors, and not from any basic theoretical grounds, although the relationship between ionization potential and orbital energy is often referred to as Koopmans' theorem (Koopmans, 1923). The usual proof is developed in Sec. 10.3.

PROBLEMS

6.18 Verify Eq. (6.101), using Table 6.4. Note carefully the difference between spatial and spin orbitals, and the differences between integrals over such orbitals.

6.19 Verify the calculated energies of the excited states of Cr^{3+}, using $F_2 = 0.008113$, $F_4 = 0.0005649$ au, and Table 6.7.

6.20 Calculate the ratios B/C and F_2/F_4 for the $3d$ orbital of the hydrogen atom.

6.21 Obtain S in (6.93), and sketch the Slater $1s$ and $2s$ orbitals for helium. Graphically compare the orthogonalized $2s$ orbital with the ordinary nonorthogonal Slater $2s$ orbital.

6.22 Evaluate the integrals J_{11} and H_1 for helium using (6.111), and obtain E and ϵ_1.

6.9 CONFIGURATION INTERACTION

Even when taken to the Hartree-Fock limit, a configurational wavefunction is in no sense an eigenfunction of the Hamiltonian (6.1); as noted in Sec. 5.4, it ignores the instantaneous correlation of the electrons. Since this is true, it must be possible to set up *other* configurational wavefunctions, orthogonal to and linearly independent of the first, which have nonzero Hamiltonian matrix elements with the first wavefunction. Not just any such wavefunction will do, however; according to Eq. (6.12), it must belong to the same term or level as the first.

Consider the ^3P ground state of the carbon $(1s^22s^22p^2)$ atom. In the Hartree-Fock limit, we determine $1s$, $2s$, and $2p$ orbitals and a configurational function such that

$$E = \langle {}^3P(1s^22s^22p^2)|\mathfrak{IC}|{}^3P(1s^22s^22p^2)\rangle \tag{6.112}$$

is a minimum. Now consider an excited configuration such as $1s^22s^22p3p$ or $1s^22s2p^3$, both of which give rise to ^3P terms. In general, the matrix element

$$\langle {}^3P(1s^22s^22p^2)|\mathfrak{IC}|{}^3P(\text{other configuration})\rangle \tag{6.113}$$

is nonzero. In this way, by "mixing in" excited configurations belonging to the same term symbol, one can go beyond the Hartree-Fock limit and seek a more accurate solution to the Schrödinger equation. Using perturbation theory, or explicitly diagonalizing the entire matrix of excited configurations, one can construct a new wavefunction

$$|\Psi(^3P)\rangle = \sum_{i \text{ configurations}} c_i|\Psi_i(^3P) \text{ (various configurations)}\rangle \qquad \blacktriangleright(6.114)$$

If one mixes in a large number of such excited configurations, the function $|\Psi(^3P)\rangle$ will converge to the exact eigenfunction of $\mathcal{K}$. In practice, however, even an interaction with hundreds of excited configurations may produce an energy in noticeable disagreement with experiment. An order-of-magnitude estimate is almost impossible to give since the percent of energy improvement due to a given number of configurations depends on the number of electrons in the system. Because of the large number of configurations involved, the amount of work required is great, and the improvement is gained only at the expense of greatly increased complexity in the wavefunction. Furthermore, useful concepts such as orbitals and orbital energies are a casualty of the increased accuracy.

There are many ways to construct excited states suitable for configuration interaction (CI). One way which has great appeal is to use the *virtual* orbitals of the Roothaan treatment for the excited orbitals. Suppose we begin by considering a configurational wavefunction

$$|\Psi\rangle = |\varphi_1(1)\varphi_2(2) \cdots \varphi_n(n)\rangle \tag{6.115}$$

in which the φ_i are *spin orbitals* containing a total of n electrons. Let us suppose that the radial portions of the φ_i have been determined by the Roothaan method and that there is an additional set of virtual spin orbitals $\varphi_{n+1}, \ldots, \varphi_m$ which are also fixed by the Fock Hamiltonian. We now construct excited configurations by removing an electron from one or more of the occupied orbitals and placing it in one of the virtual orbitals. We then classify these excited configurations according to the number of such one-electron excitations. The first case occurs when *one* electron in φ_k ($k \leq n$) is transferred to φ_p ($p > n$); thus

$$|\Psi_{k \to p}\rangle = \mathcal{Q}|\varphi_1(1)\varphi_2(2) \cdots \varphi_p(k) \cdots \varphi_n(n)\rangle \qquad \blacktriangleright (6.116)$$

In the second case, we consider 2 one-electron excitations φ_k and φ_l ($k, l \leq n$) replaced by φ_p and φ_q ($p, q > n$). We then write

$$|\Psi_{k \to p, l \to p}\rangle = \mathcal{Q}|\varphi_1(1)\varphi_2(2) \cdots \varphi_p(k) \cdots \varphi_q(l) \cdots \varphi_n(n)\rangle$$
$$\blacktriangleright (6.117)$$

(In a third case, we could consider three such excitations, but this situation will not concern us.) Let us now consider the matrix element

$$\langle \Psi | \mathcal{K} | \Psi_{k \to p} \rangle = \langle \Psi | \mathbf{H} | \Psi_{k \to p} \rangle + \langle \Psi | \sum_{\substack{i,j \\ i<j}} \frac{e^2}{r_{ij}} | \Psi_{k \to p} \rangle \tag{6.118}$$

This is easily reduced to

$$\langle \Psi | \mathcal{K} | \Psi_{k \to p} \rangle = H_{kp} + \sum_{j=1}^{n} \left(2\langle \varphi_j \varphi_k \| \varphi_j \varphi_p \rangle - \langle \varphi_j \varphi_k \| \varphi_p \varphi_j \rangle \right) \tag{6.119}$$

We have assumed, for simplicity only, that the configuration described by Ψ is a closed shell. Using the results of the previous section, we see at once that

$$\langle \Psi | \mathfrak{IC} | \Psi_{k \to p} \rangle = F_{kp} = 0 \qquad \blacktriangleright (6.120)$$

where F_{kp} is the off-diagonal matrix element of the (diagonal) Fock matrix between orbitals.

We have shown that the SCF ground state has zero off-diagonal matrix elements with all configurations formed by one-electron excitations to virtual orbitals. This is known as Brillouin's theorem, and it is of paramount importance in discussing properties of Hartree-Fock wavefunctions. The theorem can be used to prove that Hartree-Fock predictions of the expectation value of all *one-electron* operators (such as the dipole moment operator) ought to be quite good despite the lack of correlation in the wavefunction.‡

We next consider the matrix element

$$\langle \Psi | \mathfrak{IC} | \Psi_{k \to p, l \to q} \rangle \qquad (6.121)$$

The one-electron term is zero, according to the discussion of Sec. 6.3. The two-electron portion is given by Eq. (6.33). Thus

$$\langle \Psi | \mathfrak{IC} | \Psi_{k \to p, l \to q} \rangle = \langle \varphi_k \varphi_p \| \varphi_l \varphi_q \rangle - \langle \varphi_k \varphi_p \| \varphi_q \varphi_l \rangle \qquad \blacktriangleright (6.122)$$

Furthermore, excitations involving more than two electrons have zero matrix elements with $|\Psi\rangle$, so the only excitations we need consider are the two-electron ones.

For the present, we will not consider any actual CI calculations. However, we will return to this method when we discuss molecules later. For specific examples of atomic calculations, see especially Watson's (1960*a*) calculation of Be and the calculation of Nesbet and Watson (1958) on helium.

6.10   THE VIRIAL THEOREM

In a classical planetary system, it is possible to show that the average value of the kinetic energy of the system, T, is equal to the negative of the total energy, $-E$. Hence, since $T + V = E$, we also have $T = -\frac{1}{2}V = -E$. Since the forces between charged particles obey an inverse square law, we suspect that the same relations may hold for atomic and molecular systems.

‡ M. Cohen and A. Delgarno (1961).

Consider the Schrödinger equation for an arbitrary system of n particles,

$$-\frac{1}{2} \sum_{j=1}^{3n} \frac{\hbar^2}{m_j} \frac{\partial^2 \Psi}{\partial x_j^2} + (V - E)\Psi = 0 \qquad (6.123)$$

where x_j stands for any of x, y, or z. It is possible to show‡ that

$$-2\langle T \rangle = -\left\langle \sum_{j=1}^{3n} x_j \frac{\partial V}{\partial x_j} \right\rangle \qquad (6.124)$$

where the bracket indicates the expectation value of the operator in the state Ψ. For an atom or a molecule, V is a homogeneous function of degree -1, so, according to Euler's theorem,

$$\sum_{j=1}^{3n} x_j \frac{\partial V}{\partial x_j} = -V \qquad (6.125)$$

The "planetary" form of the virial theorem is therefore obeyed, and

$$2\langle T \rangle = -\langle V \rangle = -2E \qquad \blacktriangleright (6.126)$$

Whether or not this theorem is obeyed for an arbitrary function Ψ which is an *approximate* solution to Schrödinger's equation depends on how Ψ is chosen. It is possible to show that the Hartree-Fock function does obey the virial theorem, but the less accurate Roothaan SCF function for a fixed basis need not. It is additionally possible to take any arbitrary function Ψ' which does not satisfy the virial theorem and construct from it a new function Ψ which does. The procedure by which this is done is called scaling. Suppose

$$\Psi' = \Psi'(r_1, r_2, \ldots, r_n) \qquad (6.127)$$

for an n-particle system. Let us *scale* this function by multiplying all distances r by a scaling factor η; thus

$$\Psi = \Psi(\eta r_1, \eta r_2, \ldots, \eta r_n) \qquad (6.128)$$

It can be shown§ that there exists an optimum η which will cause Ψ to satisfy the virial theorem. Furthermore, the energy of Ψ will be lower than the energy of Ψ', so the requirement that Ψ should obey the virial theorem is a necessary (but certainly not sufficient) condition for a minimization of the energy.

The virial theorem is of great importance in any discussion of chemical bonding, since all molecules, as well as the individual atoms of which they are composed, must obey the theorem. Since molecular forma-

‡ H. Eyring, J. Walter, and G. E. Kimball (1944, p. 355).
§ P. O. Löwdin (1959).

tion is accompanied by a lowering of the total energy, it is also accompanied by an increase in the kinetic energy of the system and a large compensating decrease in the potential energy.

PROBLEMS

6.23 Show that the wavefunction of Eq. (5.68) obeys the virial theorem.

6.24 Show that the correlation energy must obey the virial theorem and that improvement of the Hartree-Fock wavefunction through configuration interaction will *raise* the calculated kinetic energy of the system.

SUGGESTIONS FOR FURTHER READING

1. Condon, E. U., and G. H. Shortley: "The Theory of Atomic Spectra," Cambridge University Press, London, 1935.

   This text is still *the* standard reference on atomic electronic structure. The emphasis is on operator techniques, and the more modern subject of Hartree-Fock calculations is not included. There are many useful tables, and the discussions are invariably lucid.

2. Slater, J. C.: "Quantum Theory of Atomic Structure," vols. 1 and 2, McGraw-Hill Book Company, New York, 1966.

   These texts place more emphasis on modern developments in the theory. Especially useful are the many appendices and tables. For example, on p. 336 of Vol. II, a list of explicit wavefunctions for all terms of the d^n and p^n configurations is given. The second volume also contains a detailed explanation of the hole-particle relationships.

3. Griffith, J. S.: "The Theory of Transition Metal Ions," Cambridge University Press, London, 1961.

   Although primarily a text on ligand field theory, Chaps. 4 and 5 contain a lucid summary of the principal features of the theory of atomic structure.

The student who is interested in modern advances in the theory should consult the recent literature, beginning perhaps with the references cited in this chapter. At this writing, the problem of obtaining accurate atomic wavefunctions including electron correlation is just being solved, and papers in press by Frank Boys and R. K. Nesbet should prove of special interest.

7

THE ELECTRONIC STATES
OF DIATOMIC MOLECULES

I believe the chemical bond is
not so simple as some people
seem to think.
R. S. Mulliken

7.1 THE H_2^+ ION

We begin the discussion of molecular orbital theory with diatomic molecules, and in particular with homonuclear diatomic molecules. Historically, the MO treatment of the oxygen molecule was the first success of the method, and if it were not for the failure of valence-bond theory to predict a triplet ground state for O_2, molecular orbital theory might be far less important today. Indeed, the *conceptual* successes of molecular orbital theory are so firmly established that most first courses in chemistry at the college freshman level include a full discussion of molecular orbital binding in the first-row homonuclear diatomic. We will therefore pass over this more elementary material rather quickly; the usual treatment is developed at the beginning of Sec. 7.3. In keeping with the more advanced level of the present text, we begin our discussion with a relatively careful orbital treatment of the most simple conceivable molecule, H_2^+.

The hydrogen molecule ion actually exists. The equilibrium separation between the protons is 2.00 au (1.06 Å), and, relative to a separated hydrogen atom and proton, the dissociation energy is $D_0 = 2.64$ ev. Of course, the electronic dissociation energy D_e is larger, owing to the zero-point vibrational energy of the ion. Spectroscropic studies indicate $D_e = 2.78$ ev (0.103 au), corresponding to a total *electronic* energy of 1.103 au.

The Schrödinger equation for H_2^+ involves the relative coordinates of three particles. As in the classical three-body problem (or as in the case of helium), we do not anticipate an exact solution in closed form. However, in the present case, we find that the approach to a very good solution can be greatly simplified by assuming that the nuclear motion and the electronic motion are separable, such that

$$\Psi = \Psi_e \Psi_n = \Psi_{\text{electronic}} \Psi_{\text{nuclear}} \tag{7.1}$$

This is known as the Born-Oppenheimer approximation (1927). It can be physically rationalized on the grounds that the nuclei, being massive, move slowly compared to the electrons and may be considered at rest for a given electronic state. On purely mathematical grounds, it can be shown that (7.1) is a very good first-order approximation to the exact solution of Schrödinger's equation. The form of the higher-order corrections can be analyzed; an understanding of the "coupling" between nuclear and electronic motions is important in spectroscopy, where certain electronic selection rules are partially broken by the mixing of various electronic states through "vibronic" coupling. In the present text, however, we assume the Born-Oppenheimer approximation throughout and leave the vibronic problems to the spectroscopists. If we are interested in the motions of the *nuclei*, we will solve for Ψ_e (and E_e) at several values of R, the internuclear separation. We can then regard the nuclear motion as taking place in a potential well determined by the electronic energy and the nuclear repulsion energy. A plot of this energy as a function of R is simply the binding curve of the molecule (see Figs. 7.1 and 7.2 in which the binding curve is given for the neutral hydrogen molecule H_2).

With the nuclei regarded as at rest, the equation for H_2^+ reduces to a one-particle problem. The problem is not so easy as the hydrogen atom itself due to lack of the high symmetry which is present in all atoms. It can be simplified, however, by using such symmetry as is actually present. In the next chapter, we will show how molecular symmetry can be used to simplify the treatment of molecules in a systematic way, but for the present we restrict ourselves to a less general procedure in order to emphasize the physics of the situation.

The electronic Hamiltonian for H_2^+ in the Born-Oppenheimer

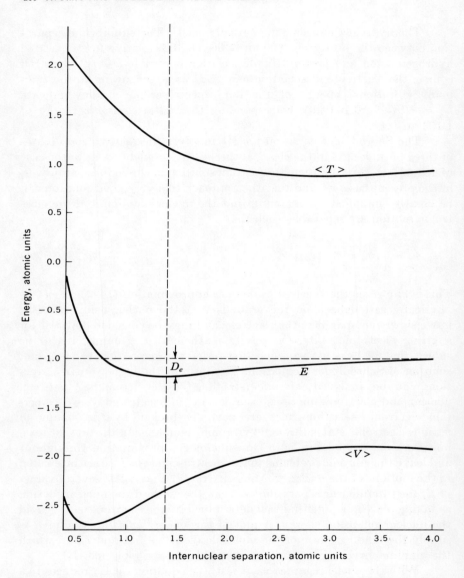

Fig. 7.1 Energy curves for the hydrogen molecule. The electronic energy E includes the nuclear-repulsion term. At $R = \infty$, $E = -1.0$, the energy of a pair of isolated hydrogen atoms. Note that the virial theorem is only satisfied at $R = \infty$ and at $R = R_e = 1.4$ au, the equilibrium internuclear separation. (*Data from Wolniewicz and Kolos*, 1964, 1965.)

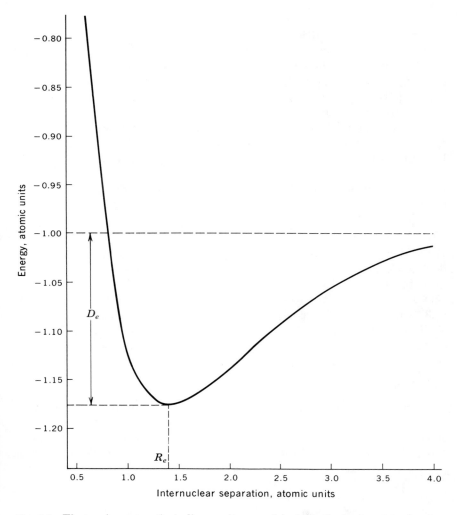

Fig. 7.2 Electronic energy (including nuclear repulsion) for the hydrogen molecule, enlarged.

approximation is

$$\mathfrak{IC} = -\frac{\hbar^2}{2m}\nabla^2 - \frac{e^2}{R_A} - \frac{e^2}{R_B} \tag{7.2}$$

where we have omitted the nuclear repulsion term e^2/R_{AB} since it is a constant. At the equilibrium internuclear separation, it contributes a repulsive energy of 0.500 au. The Schrödinger equation formed with this Hamiltonian is separable in ellipsoidal coordinates, and may be directly

solved by numerical methods. A study of this method is not very illuminating, however, since the techniques used in treating H_2^+ have not
proved useful in treating more complex molecules. However, we may
absorb a portion of the separability argument by considering the operator
$l_z = x\mathbf{p}_y - y\mathbf{p}_x = -i\hbar(\partial/\partial\varphi)$, where z is the internuclear axis and φ is
the azimuthal angle $(0 < \varphi < 2\pi)$ about the axis (Fig. 7.3). Just as is the
case in atoms, l_z commutes with ∇^2, but it is not so obvious that it commutes with the nuclear attraction terms. However, the proof is straightforward, and we leave it as an exercise (Prob. 7.1). Note that neither l_x nor
l_y commutes with $\mathcal{3C}$. The proof that l_z commutes with $\mathcal{3C}$ is essentially
equivalent to the statement that Ψ_e contains an angular part $\Phi(\varphi)$ as a
factor. As can be shown, the eigenvalues of l_z are integral multiples of $\hbar$,
so that

$$l_z\Psi_m = m\hbar\Psi_m \tag{7.3}$$

Eigenfunctions Ψ_m and Ψ_{-m} are degenerate since the energy does not
depend on the *sense* of the angular momentum.

It turns out that in the only bound state of H_2^+, $m = 0$. We will
therefore seek, as trial variational functions for Ψ, functions which satisfy
(7.3) with $m = 0$. A second symmetry operation which is useful is the
parity operator since H_2^+ (and any homonuclear diatomic molecule) has a
center of symmetry. Experimentally the ground state of H_2^+ has even
parity, so according to Eq. (4.72),

$$\mathbf{P}\Psi_0 = \Psi_0 \tag{7.4}$$

It is conventional to classify the states of diatomic molecules as Σ, Π, Δ, Γ,
. . . , according to whether M, the eigenvalue of $\mathbf{L}_z$, is 0, 1, 2, 3, . . . ;
thus the state we seek for H_2^+ may be given the "term symbol" $^2\Sigma_g$. The
superscript is equal to $2S + 1$, while subscripts g and u stand for even

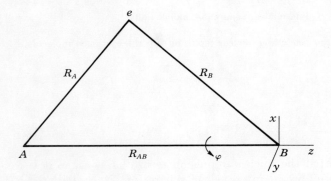

Fig. 7.3 Coordinate system for H_2^+. φ is the angle between
the projection of R_B (in the xy plane) and the x axis.

(German: *gerade*) and odd (*ungerade*) parity. We use capital Greek letters to denote the symmetries of *states*, and small Greek letters $(\sigma, \pi, \ldots)$ for the symmetries of *orbitals*. For the hydrogen molecule ion, either symbol can be used since there is just one electron. Our nomenclature thus parallels that for the hydrogen atom, which has a single electron in a $1s$ orbital and which therefore has a ^2S ground term.

It is possible to build up a very accurate variational function Ψ by using hydrogenic orbitals centered on the two nuclei. This technique is widely known as the LCAO (*linear combination of atomic orbitals*) method. Thus

$$\Psi = \sum_i [c_i \varphi^A_{n_i l_i m_i}(R_A) + c'_i \varphi^B_{n_i l_i m_i}(R_B)] \tag{7.5}$$

If Ψ is to be an eigenfunction of the parity operator, $c_i = \pm c'_i$. The reader should carefully verify this assertion. Secondly, with z the internuclear axis and with $1_{z_A} = 1_{z_B} = 1_z$, the eigenfunction of 1_z is m_i. Therefore

$$\Psi = \sum_i c_i [\varphi^A_{n_i l_i m_i}(R_A) \pm \varphi^B_{n_i l_i m_i}(R_B)] \tag{7.6}$$

The first few terms in the sum are of especial interest. For $m = 0$ and even parity, we have

$$\Psi(^2\Sigma_g) = c_{1s}(\varphi^A_{1s} + \varphi^B_{1s}) + c_{2s}(\varphi^A_{2s} + \varphi^B_{2s}) + c_{2p_0}(\varphi^A_{2p_0} - \varphi^B_{2p_0}) + \cdots \tag{7.7}$$

One can also define the excited (and unbound) state $^2\Pi_u$ as

$$\Psi(^2\Pi_u) = c_{2p}(\varphi^A_{2p_1} + \varphi^B_{2p_1}) + \cdots \tag{7.8}$$

This state has two-fold spatial degeneracy in addition to the two-fold spin degeneracy. One must be careful of the phases of the orbitals on the two centers. For example, if the $2p_0$ $(\equiv 2p_z)$ orbitals are directed along the z axis in the positive direction, the combination‡ $\varphi^A_{2p_0} - \varphi^B_{2p_0}$ has even parity, whereas $\varphi^A_{2p_0} + \varphi^B_{2p_0}$ has odd parity (note that the overlap integral $\langle \varphi^A_{2p_0} | \varphi^B_{2p_0} \rangle$ is negative).

Using Fig. 7.4, we can now see why the ground function has symmetry σ_g. In this orbital, the electron has the greatest probability of being between the two protons. The σ_u orbital has a nodal plane bisecting the line segment joining the two nuclei and therefore has a very low probability that the electron is between the two protons—low even compared to the idealized situation in which the electron has a probability distribution characteristic of an isolated hydrogen atom unaffected by the nearby

‡ One must beware of a variety of notations. Many authors write the even-parity sigma combination as $\varphi^A_{2p_\sigma} + \varphi^B_{2p_\sigma}$, where a $2p_\sigma$ orbital is defined with phase chosen such that the positive end points toward the center of the bond.

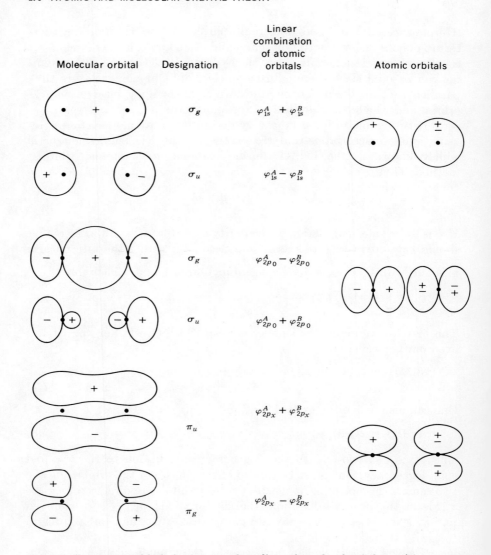

Fig. 7.4 Symmetry orbitals for homonuclear diatomic molecules (schematic).

proton. We call the σ_u orbital *antibonding*. Similarly, the π_u function is called *bonding* (although in the case of H_2^+ it is in fact not bonding) since the probability density between the protons is higher than in the non-bonding situation. If, corresponding to the real $2p_x$ and $2p_y$ functions formed by linear combinations of the complex $2p_1$ and $2p_{-1}$, we form functions π_x and π_y from linear combinations of $\varphi_{2p_x}^A$ and $\varphi_{2p_x}^B$ or $\varphi_{2p_y}^A$ and $\varphi_{2p_y}^B$, then we see at once that π_{ux} has a nodal plane *perpendicular* to the p_x

orbital and *including* the line joining the two nuclei (the yz plane). Despite this nodal plane, the π_u orbital still has a fair "amount of charge" in the region between the nuclei. On the other hand, the π_g function is antibonding since it has two nodal planes.

To actually solve for the ground-state function Ψ, we need to evaluate matrix elements of the form $\langle\varphi_i|\mathfrak{IC}|\varphi_j\rangle = \mathfrak{IC}_{ij}$ and $\langle\varphi_i|\varphi_j\rangle = S_{ij}$. Since Ψ is in linear parametric form, it is readily determined once the matrix elements are calculated. The symmetry we have imposed on the wavefunction will automatically appear when the secular equation is solved. This is because the symmetry is contained in the matrix elements themselves; for example, $\langle\varphi_{1s}^A|\varphi_{2p_z}^B\rangle = -\langle\varphi_{1s}^B|\varphi_{2p_z}^A\rangle$. However, it is possible to use the symmetry directly to aid in the solution of the secular equation. For example, suppose we use the limited basis set φ_{1s}^A, φ_{1s}^B, φ_{2s}^A, φ_{2s}^B, $\varphi_{2p_z}^A$, $\varphi_{2p_z}^B$. Instead of evaluating all matrix elements within this basis, we may instead use the basis $\chi_1 = \varphi_{1s}^A + \varphi_{1s}^B$, $\chi_2 = \varphi_{2s}^A + \varphi_{2s}^B$, $\chi_3 = \varphi_{2p_z}^A - \varphi_{2p_z}^B$, and evaluate only matrix elements between the new basis functions. This reduces the problem of a 6×6 secular determinant to a 3×3. The method is completely rigorous, since the function $\Psi(^2\Sigma_g)$ cannot contain σ_u combinations such as $\varphi_{1s}^A - \varphi_{1s}^B$. The χ are generally known as symmetry orbitals, and they are of even greater utility in large symmetric molecules. We will consider them from a more fundamental and rigorous point of view in the next chapter.

The reader may be bothered by the fact that the electron has equal probability of being near one nucleus or the other, regardless of inter-nuclear separation. This is a consequence of the stationary-state approximation. In practice, if H_2^+ is slowly pulled apart, the electron would end up on one proton or the other. Since we cannot say *which* proton it would be bound to, our equations must give it equal probability of being on either one. The energy of the function is, of course, not changed; the wavefunction $\Psi = (1/\sqrt{2})(\varphi_{1s}^A + \varphi_{1s}^B)$ is a perfectly valid quantum-mechanical description of a proton and a hydrogen atom separated from each other by a large distance. It is not necessarily a physically useful description, however.

PROBLEMS

7.1 Prove that l_z commutes with $\mathfrak{IC}$.

7.2 Verify the parity of the combination $\varphi_{2p_z}^A + \varphi_{2p_z}^B$. Is this function an eigenfunction of l_z?

7.3 As a trial function, let $\Psi = [1/\sqrt{2(1 + S)}] (\varphi_{1s}^A + \varphi_{1s}^B)$, where $\varphi_{1s}^A = Ne^{-\zeta r_A}$. Obtain‡ an expression for the energy, and optimize it with respect to ζ and R_{AB}. Optimally, $\zeta = 1.288$. What physical interpretation can be ascribed to this "negative shielding," if any?

‡ For a detailed discussion of approximate trial functions for the H_2^+ eigenfunctions, see Pauling and Wilson, p. 327.

7.2 THE HYDROGEN MOLECULE

With the H_2 molecule, the problem of interelectronic repulsion returns, and just as the helium atom is vastly more difficult to solve accurately than the hydrogen atom, so the hydrogen molecule presents difficulties far in excess of those presented by H_2^+. As in the case of helium, these difficulties can be avoided by explicitly including the interelectronic distance r_{12} in the wavefunction. This was first done by James and Coolidge (1933), whose treatment of H_2 is somewhat analogous to the Hylleraas treatment of He. As in the helium problem, the method of James and Coolidge‡ has not been successfully generalized to many-electron systems, and so we do not consider it further.

By analogy with the orbital treatment of helium, we may introduce a *molecular orbital* treatment of the hydrogen molecule. We may presume at first that we can place two electrons in a molecular orbital which is an eigenfunction of the H_2^+ Hamiltonian, just as we began our discussion of helium by placing two electrons in hydrogenic $1s$ orbitals. According to general orbital theory,

$$\Psi(^1\Sigma_g) = \mathcal{C}|\psi^+(1)\psi^-(2)\rangle = \psi(1)\psi(2)\{\alpha_1\beta_2 - \beta_1\alpha_2\}\frac{1}{\sqrt{2}} \tag{7.9}$$

where ψ is the molecular orbital. At the most elementary level,

$$\psi = N(\varphi_{1s}^A + \varphi_{1s}^B) \tag{7.10}$$

where

$$N = \sqrt{\frac{1}{2(1+S)}}$$
$$S = \langle \varphi_{1s}^A | \varphi_{1s}^B \rangle$$

In the most complex treatment, the molecular orbital ψ can be considered a linear combination of atomic functions (LCAO-MO) or, more specifically, of σ_g functions, such that

$$\psi = c_{1s}(\varphi_{1s}^A + \varphi_{1s}^B) + c_{2s}(\varphi_{2s}^A + \varphi_{2s}^B) + c_{2p}(\varphi_{2p_0}^A - \varphi_{2p_0}^B) + \cdots \tag{7.11}$$

as in (7.7). There is some choice in the orbital exponent for each atomic orbital, but within a given basis it is possible to determine the coefficients in the LCAO expansion by the Roothaan-Hartree-Fock method. Using a large basis (Kolos and Roothaan, 1960a, 1960b), one finds that the total electronic energy is -1.848 au at an internuclear separation of 1.4 au; this corresponds to a total energy of -1.134 au and a dissociation energy

---

‡ See also W. Kolos and C. C. J. Roothaan, *Rev. Mod. Phys.*, **32**:219 (1960). This issue of the Reviews of Modern Physics contains a large number of excellent articles which provide a useful summary of the state of molecular quantum mechanics prior to 1960.

$D_e = 0.134$ au. Experimentally, the dissociation energy is $D_e = 0.174$ au, so the total electronic energy in the Hartree-Fock limit is in error by only 0.04 au, or about 2 percent. Whether one considers this good or bad agreement depends on what one hopes to do with the results.

The accuracy of the wavefunction can be improved by configuration interaction. Rather than proceed in this direction, however, it is of interest to consider the physical interpretation of the molecular orbital picture a little more clearly. In order to do so, consider the MO function in its most elementary form, using (7.10). Expanding the total wavefunction, we obtain

$$\Psi_{MO} = \frac{1}{\sqrt{8(1 + S)}} [\varphi_{1s}^A \varphi_{1s}^A + \varphi_{1s}^B \varphi_{1s}^B + \varphi_{1s}^A \varphi_{1s}^B + \varphi_{1s}^B \varphi_{1s}^A](\alpha_1 \beta_2 - \beta_1 \alpha_2)$$

$$(7.12)$$

Examining the spatial part, we see that the first term represents an "ionic" function, with both electrons centered on φ_{1s}^A. The second term is also ionic, representing a bare proton at A adjacent to an H^- ion at B On the other hand, the third and fourth terms represent "covalent" structures, in which each center has a single electron. Clearly, in the hydrogen molecule, the molecular orbital method gives equal weight to covalent and ionic structures. This may seem unreasonable, and in fact a better function can be obtained by dropping the ionic terms. This gives rise to the following Heitler-London or valence-bond wavefunction for H_2:

$$\Psi_{VB} = \frac{1}{2\sqrt{1 + S^2}} [\varphi_{1s}^A \varphi_{1s}^B + \varphi_{1s}^B \varphi_{1s}^A](\alpha_1 \beta_2 - \beta_1 \alpha_2) \tag{7.13}$$

Evaluating the energy of Ψ_{MO}, we obtain

$$E_{MO} = 2\langle\psi|\mathbf{H}|\psi\rangle + \langle\psi(1)\psi(2)|\frac{1}{r_{12}}|\psi(1)\psi(2)\rangle + \frac{1}{R_{AB}} \tag{7.14}$$

where we have taken the Hamiltonian for H_2 as

$$\mathcal{H} = -\tfrac{1}{2}\nabla_1^2 - \tfrac{1}{2}\nabla_2^2 - \frac{1}{R_{A1}} - \frac{1}{R_{A2}} - \frac{1}{R_{B1}} - \frac{1}{R_{B2}} + \frac{1}{r_{12}}$$

$$+ \frac{1}{R_{AB}} = \mathbf{H}_1 + \mathbf{H}_2 + \frac{1}{r_{12}} + \frac{1}{R_{AB}} \tag{7.15}$$

H is a one-electron operator; its constituents include the kinetic energy and nuclear attractions of the electron. Note that we have omitted the fundamental constants $\hbar$, m, and e since they are all unity in the atomic system of units. Expanding the molecular orbital ψ, we have

$$
\begin{aligned}
E_{MO} &= \frac{2}{1+S}\left(\langle\varphi_{1s}^A|\mathbf{H}|\varphi_{1s}^A\rangle + \langle\varphi_{1s}^A|\mathbf{H}|\varphi_{1s}^B\rangle\right) \\
&+ \frac{1}{2(1+S)^2}\left[\langle\varphi_{1s}^A(1)\varphi_{1s}^A(2)|\,\frac{1}{r_{12}}\,|\varphi_{1s}^A(1)\varphi_{1s}^A(2)\rangle\right. \\
&+ \langle\varphi_{1s}^A(1)\varphi_{1s}^B(2)|\,\frac{1}{r_{12}}\,|\varphi_{1s}^A(1)\varphi_{1s}^B(2)\rangle + 4\langle\varphi_{1s}^A(1)\varphi_{1s}^A(2)|\,\frac{1}{r_{12}}\,|\varphi_{1s}^A(1)\varphi_{1s}^B(2)\rangle \\
&+ \left.2\langle\varphi_{1s}^A(1)\varphi_{1s}^B(2)|\,\frac{1}{r_{12}}\,|\varphi_{1s}^B(1)\varphi_{1s}^A(2)\rangle\right] + \frac{1}{R_{AB}} \\
&= \frac{2}{1+S}(H_{AA}+H_{AB}) + \frac{1}{2(1+S)^2}\left(\langle AA\|AA\rangle + \langle AB\|AB\rangle\right. \\
&\qquad\qquad\qquad\left. + 4\langle AA\|AB\rangle + 2\langle AB\|BA\rangle\right) + \frac{1}{R_{AB}} \qquad (7.16)
\end{aligned}
$$

Similarly, for Ψ_{VB},

$$
\begin{aligned}
E_{VB} &= \frac{2}{1+S^2}(H_{AA}+H_{AB}S) + \frac{1}{1+S^2}\left(\langle AB\|AB\rangle\right. \\
&\qquad\qquad\qquad\left. + \langle AB\|BA\rangle\right) + \frac{1}{R_{AB}} \qquad (7.17)
\end{aligned}
$$

When the orbitals φ_{1s} are hydrogen $1s$ eigenfunctions,

$$
H_{AA} = E_{1s} - \langle\varphi_{1s}^A|\,\frac{1}{R_B}\,|\varphi_{1s}^A\rangle = E_{1s} - \langle A|B|A\rangle
$$

Let us now investigate the behavior of E_{VB} and E_{MO} at large internuclear distances. As R_{AB} becomes large, $S \to 0$, $H_{AB} \to 0$, $\langle AB\|BA\rangle \to 0$, $\langle AA\|AB\rangle \to 0$, and $\langle AB\|AB\rangle$ and $\langle A|B|A\rangle \to 1/R_{AB}$. Thus,

$$
E_{VB} \to 2E_{1s} - \frac{2}{R_{AB}} + \frac{1}{R_{AB}} + \frac{1}{R_{AB}} = 2E_{1s} \qquad (7.18)
$$

On the other hand, for the molecular orbital function,

$$
\begin{aligned}
E_{MO} &\to 2E_{1s} - \frac{2}{R_{AB}} + \frac{1}{2R_{AB}} + \frac{1}{2}\langle AA\|AA\rangle + \frac{1}{R_{AB}} \\
&= 2E_{1s} - \frac{1}{2}\left(\frac{1}{R_{AB}} - \langle AA\|AA\rangle\right) \qquad (7.19)
\end{aligned}
$$

Thus the valence-bond function converges to the right limit as $R \to \infty$, whereas the molecular orbital function does not. This is essentially due to the symmetry requirement imposed on the molecular orbital—even at large internuclear separations, the ionic forms A^-B^+ and A^+B^- remain as probable as the covalent forms. Because the motion of the electrons is uncorrelated, pulling the protons apart in the MO method does not lead to two hydrogen atoms, but rather to a very extended hydrogen molecule. Therefore the valence-bond function has a great conceptual and calculational advantage over the molecular orbital function at large distances.

At smaller distances, the valence-bond function also gives lower energies. Using hydrogen atom eigenfunctions $\varphi_{1s} = Ne^{-r/a_0}$ ($\zeta = 1$), it is found that the valence-bond function gives a binding energy of 0.118 au, compared to the experimental 0.174 au and the Hartree-Fock limit 0.134 au. The MO function gives $D_e = 0.098$. If the basis functions are improved by letting ζ be a parameter, $D_e = 0.129$ au ($\zeta = 1.197$) for the molecular orbital function and $D_e = 0.139$ au ($\zeta = 1.166$) for the valence-bond function.

Considering that the VB method seems to give consistently better results than the MO method and provides a valid description even at large distances, the reader may wonder why we persist in using the less accurate method. The reasons are not obvious at this level, but they will become clearer as we proceed to larger and more complex molecules. The molecular orbital method, despite its inadequacies, retains considerable conceptual and practical mathematical simplicity in large molecules; the valence-bond method does not. Thus, although VB calculations are energetically more suitable for many small molecules, we will persist with the MO method as an introduction to more complex (and chemically more interesting) systems.

The single-configuration molecular orbital wavefunction can always be improved through configuration interaction. Since configurations produced by a one-electron excitation cannot mix with the ground configuration (see Sec. 6.9), the only configuration of importance in the present problem is that produced by two excitations to the antibonding orbital $\psi' = \sqrt{[1/2(1-S)]}(\varphi_{1s}^A - \varphi_{1s}^B)$. This configuration may be written

$$\Psi'_{\mathrm{MO}} = \alpha|\psi'^+(1)\psi'^-(2)\rangle \tag{7.20}$$

Expanding, we have

$$\Psi'_{\mathrm{MO}} = \frac{1}{\sqrt{8}\,(1-S)}\,(\varphi_{1s}^A\varphi_{1s}^A + \varphi_{1s}^B\varphi_{1s}^B - \varphi_{1s}^A\varphi_{1s}^B - \varphi_{1s}^B\varphi_{1s}^A)(\alpha_1\beta_2 - \beta_1\alpha_2)$$

$$\tag{7.21}$$

so the optimum function (within the φ_{1s} basis) is thus

$$\begin{aligned}\Psi &= a\Psi_{\mathrm{MO}} - b\Psi'_{\mathrm{MO}} \\ &= N[(\varphi_{1s}^A\varphi_{1s}^B + \varphi_{1s}^B\varphi_{1s}^A) + (C\varphi_{1s}^A\varphi_{1s}^A + \varphi_{1s}^B\varphi_{1s}^B)](\alpha_1\beta_2 - \beta_1\alpha_2)\end{aligned} \tag{7.22}$$

In this way we can clearly see that the molecular orbital function, including configuration interaction, is equivalent to the valence-bond function including the mixing of a certain amount of the ionic terms.

With a sufficiently large basis, both methods can be made to converge to the exact solution, although it would appear that the MO method of applying configuration interaction (CI) is somewhat easier to apply in a systematic and general way for large molecules.

Weinbaum (1933) was the first to apply Eq. (7.22). He found $\zeta = 1.193$ and $C = .256$ as the optimum parameters, yielding $D_e = 0.147$ au. Further improvement can only be obtained by expanding the basis set to include φ_{2s}, φ_{2p_0}, and so forth. It is customary to say that the hydrogen molecule has about 25 percent "ionic" character, but it must be realized that this is an approximate way of speaking and refers to a function having a very small and limited basis set.

PROBLEMS

7.4 Verify the normalizing factors given in Eqs. (7.10), (7.12), and (7.13).

7.5 Carefully verify Eqs. (7.14), (7.16), and (7.17). Equation (7.16) should be derived in two ways, starting from (7.14) and from (7.12).

7.6 Show that $H_{AA} = E_{1s} - \langle A|B|A \rangle$ when φ_{1s} is the hydrogen 1s eigenfunction. (Note that **H** is the one-electron operator $-\frac{1}{2}\nabla^2 - 1/R_A - 1/R_B$.)

7.7 Calculate the overlap integral $S = \langle \varphi_{1s}^A | \varphi_{1s}^B \rangle$ for arbitrary ζ. Show that $S = e^{-\zeta R}(1 + \zeta R + \zeta^2 R^2/3)$, where $R = R_{AB}$, the internuclear separation in atomic units. Find S for $\zeta = 1.2$, $R = 1.4$.

7.8 Calculate the overlap integral between the "ionic" and "covalent" functions given in the text. Are the two functions so very different? [Note: The covalent function is just Ψ_{VB}, whereas the ionic function is that multiplied by the constant C in Eq. (7.22)].

7.3 HOMONUCLEAR DIATOMIC MOLECULES

As noted in the first section, the lowest-energy molecular orbital has symmetry σ_g. Some authors write this as $\sigma_g 1s$ to indicate that in the simplest approximation it is a linear combination of 1s atomic orbitals. However, the notation $1\sigma_g$ is preferred, indicating only that this particular σ_g orbital is the lowest-energy (first) σ_g. In the Hartree-Fock limit, it can be formed from 1s, 2s, $2p_0$, . . . , orbitals. Another orbital which can be formed, in the simplest approximation, from 1s atomic orbitals, is the $1\sigma_u$. It is therefore the second-lowest orbital, since it is the only other molecular orbital which can be formed from the low-energy 1s atomic orbitals. According to this argument, the molecular orbital $2\sigma_g \sim (\varphi_{2s}^A + \varphi_{2s}^B)$ would be next in energy, followed by the $2\sigma_u \sim (\varphi_{2s}^A - \varphi_{2s}^B)$. The orbitals $3\sigma_g \sim (\varphi_{2p_0}^A - \varphi_{2p_0}^B)$ and $1\pi_u \sim (\varphi_{2p\pi}^A + \varphi_{2p\pi}^B)$ would appear to have similar energies. However, because of the interaction with the nearby $2\sigma_g$ (which will have an energy close to the $3\sigma_g$ because the 2s orbital of an atom has energy close to the 2p orbital), the $3\sigma_g$ orbital is ordinarily raised somewhat in energy. On the other hand, the $1\pi_u$ has a nodal surface along a plane containing both nuclei and so is somewhat less strongly binding than the $3\sigma_g$. These two factors compete so that the relative energy of the $1\pi_u$ and $3\sigma_g$ may vary from molecule to molecule; the two factors combine, on the other hand to ensure that the antibonding $1\pi_g$ always lies lower than the antibonding $3\sigma_u$. The most common ordering of molecular orbitals in homonuclear diatomic molecules is thus

$$1\sigma_g < 1\sigma_u < 2\sigma_g < 2\sigma_u < 1\pi_u \sim 3\sigma_g < 1\pi_g < 3\sigma_u \qquad (7.23)$$

We may now formulate a molecular aufbau principle entirely analogous to that for atoms, and we can write down configurations for all the first-row homonuclear diatomic molecules. For hydrogen, the configuration is simply $(1\sigma_g)^2$; for Li_2, $(1\sigma_g)^2(1\sigma_u)^2(2\sigma_g)^2$. Of course, since the σ_u orbital is antibonding (Table 7.1), the bonding effect of the $1\sigma_g$ orbital is canceled out, and the closed $1s$ shells contribute no net binding. Thus, for example, He_2 $[(1\sigma_g)^2(1\sigma_u)^2]$ is not stable. We may use the fact that occupied antibonding orbitals contribute a destabilization by defining the *bond order* as equal to half the *net* number of bonding electrons. Thus, the bond order for H_2 and Li_2 is unity, but for He_2 it is zero. Configurations and bond orders, together with experimental data, are given for the remainder of the first row homonuclear diatomic molecules in Table 7.2.

A number of predictions are immediate. According to the aufbau principle formulated above, the configuration of O_2 is $(1\sigma_g)^2(1\sigma_u)^2(2\sigma_g)^2$ $(2\sigma_u)^2(1\pi_u)^4(3\sigma_g)^2(1\pi_g)^2$. By Hund's rule, the ground state of O_2 must be a triplet, since the spatially degenerate π_g orbital can hold up to four electrons. This prediction was one of the earliest and most triumphant

Table 7.1 Characteristics of the symmetry orbitals for homonuclear diatomic molecules

Linear combination	Symbol	Bonding property
$\varphi_{1s}^A + \varphi_{1s}^B$	σ_g	Bonding
$\varphi_{1s}^A - \varphi_{1s}^B$	σ_u	Antibonding
$\varphi_{2s}^A + \varphi_{2s}^B$	σ_g	Bonding
$\varphi_{2s}^A - \varphi_{2s}^B$	σ_u	Antibonding
$\varphi_{2p_0}^A - \varphi_{2p_0}^B$	σ_g	Bonding
$\varphi_{2p_0}^A + \varphi_{2p_0}^B$	σ_u	Antibonding
$\varphi_{2p_1}^A + \varphi_{2p_1}^B$	π_u	Bonding
$\varphi_{2p_{-1}}^A + \varphi_{2p_{-1}}^B$	π_u	Bonding
$\varphi_{2p_1}^A - \varphi_{2p_1}^B$	π_g	Antibonding
$\varphi_{2p_{-1}}^A - \varphi_{2p_{-1}}^B$	π_g	Antibonding

Table 7.2   Molecular orbital configurations for the first-row homonuclear diatomic molecules (data from Herzberg, 1950)

	Configuration	Bond order	Ground state	D_0 ev	Bond length Å
H_2^+	$1\sigma_g$	$\frac{1}{2}$	$^2\Sigma_g^+$	2.65	1.06
H_2	$1\sigma_g{}^2$	1	$^1\Sigma_g^+$	4.48	0.74
He_2^+	$1\sigma_g{}^2 1\sigma_u{}^2$	$\frac{1}{2}$	$^2\Sigma_u^+$	3.1	1.08
He_2	$1\sigma_g{}^2 1\sigma_u{}^2$	0	Does not exist		
Li_2	$1\sigma_g{}^2 1\sigma_u{}^2 2\sigma_g{}^2$	1	$^1\Sigma_g^+$	1.1	2.67
Be_2	$1\sigma_g{}^2 1\sigma_u{}^2 2\sigma_g{}^2 2\sigma_u{}^2$	0	Does not exist		
B_2	$[Be_2]1\pi_u{}^2$	1	$^3\Sigma_g^-$	3.6	1.59
C_2	$[Be_2]1\pi_u{}^4$	2	$^1\Sigma_g^+$	5.0	1.24
N_2^+	$[Be_2]1\pi_u{}^4 3\sigma_g$	$2\frac{1}{2}$	$^2\Sigma_g^+$	8.72	1.12
N_2	$[Be_2]1\pi_u{}^4 3\sigma_g{}^2$	3	$^1\Sigma_g^+$	9.76	1.09
O_2^+	$[Be_2]1\pi_u{}^4 3\sigma_g{}^2 1\pi_g$	$2\frac{1}{2}$	$^2\Pi_g$	6.48	1.12
O_2	$[Be_2]1\pi_u{}^4 3\sigma_g{}^2 1\pi_g{}^2$	2	$^3\Sigma_g^-$	5.08	1.21
F_2	$[Be_2]1\pi_u{}^4 3\sigma_g{}^2 1\pi_g{}^4$	1	$^1\Sigma_g^+$	1.5	1.44

successes of molecular orbital theory. Note that O_2 has a bond order of 2, as expected from its Lewis structure. Similarly, N_2 has a triple bond, and hence a very short bond distance. The ion O_2^+ has a bond order $2\frac{1}{2}$, since an electron is removed from an antibonding orbital of O_2 to form the ion. We therefore expect that O_2^+ will have a shorter bond distance and larger dissociation energy than O_2.

Molecular orbital theory is capable of yielding a good deal of information on diatomic molecules for a relatively small investment of labor. This is essentially because of its heavy use of symmetry, rather than any deep fundamental significance in molecular orbitals. The orbitals arise, after all, from a one-electron treatment of a many-electron problem and have no real existence outside this approximation. It is perfectly possible that, as in the case of the $4s$ and $3d$ orbitals of the transition metal ions, the ordering of the molecular orbitals may depend on the state of ionization of the molecule. Nevertheless, because the molecular orbital method uses symmetry so directly, its predictions are clear and immediate, much in the way that certain predictions of atomic orbital theory are quite straightforward.

As in atomic theory, the ground state of each homonuclear diatomic molecule can be given a term symbol. For closed shells, the term symbol is‡ $^1\Sigma_g^+$, just as closed atomic shells have the term symbol ^1S. For a single electron outside a closed shell, the state has the symmetry of the odd electron, and so the ground state of O_2^+ is $^2\Pi_g$. The configuration $(\pi_g)^2$

‡ For the meaning of the plus sign in this symbol, see Sec. 8.7 and especially Table 8.1.

gives rise to three terms (just as does p^2 in the theory of atomic electronic structure), and these are labeled $^3\Sigma_g^-$, $^1\Sigma_g^+$, and $^1\Delta_g$. We can derive these terms as we did for atoms, using an implied terms table. Writing the possible Slater determinants in the $|m_1^{m_{s1}} m_2^{m_{s2}}\rangle$ notational scheme, we have six kets, $|1^+1^-\rangle$, $|1^+-1^+\rangle$, $|1^+-1^-\rangle$, $|1^--1^+\rangle$, $|1^--1^-\rangle$, and $|-1^+-1^-\rangle$. (Note that m_1 has only two values.) Each ket is an eigenfunction of $\mathbf{L}_z$ and $\mathbf{S}_z$. The ket with maximal M has $M = 2$, $M_S = 0$. There is just one such ket, so there must be a $^1\Delta$ term. Such a term implies the existence of another ket with $M = -2$, $M_S = 0$, and we may in fact write

$$|^1\Delta\ 0\ 2\rangle = |1^+1^-\rangle$$
$$|^1\Delta\ 0\ -2\rangle = |-1^+-1^-\rangle \tag{7.24}$$

This leaves four kets, all with $M = 0$, but one with maximal $M_S = 1$. Therefore, there exists a $^3\Sigma$ term, with three constituent states ($M_S = -1,0,1$). This finally leaves one ket, with $M = 0$, $M_S = 0$, hence a $^1\Sigma$ state. All functions have even parity since the constituent orbitals have even parity. The derivation is thus complete except for the labels $+$ and $-$ on the Σ terms. We ignore these for the present. One can find the allowed terms of other configurations in the same manner although, for complicated systems with several open shells or containing d electrons, the use of group theory is advisable.

PROBLEMS

7.9 Consider the molecule NO as if it were homonuclear. Find the configuration and term symbol for the ground state. Obtain the bond order. Compare NO and NO$^+$. Which has the stronger bond, and which the shorter bond distance?

7.10 Construct excited states for H_2 using the virtual orbitals determined by symmetry. What selection rules can you derive?

7.11 Find the linear combination of Slater determinants corresponding to the $|^3\Sigma\ 0\ 0\rangle$ state of the oxygen molecule.

7.12 Use the Hartree-Fock function (6.111) for helium and the formula developed in Prob. 7.7 to obtain a formula for the overlap integral between two helium atoms. Do the same with the single-ζ helium function. Calculate overlap integrals between a pair of helium atoms at internuclear separations of 3, 2, and 1 au.

7.4 HETERONUCLEAR DIATOMIC MOLECULES

In a few cases, it is possible to treat heteronuclear molecules as if they were homonuclear, as in Prob. 7.9. However, since heteronuclear diatomic molecules have no center of symmetry, the symbols g and u are inappropriate, and the coefficients of the primitive molecular orbitals cannot be determined by symmetry alone. It is possible to employ an aufbau principle of sorts, and, dropping the subscripts g and u from the corresponding homonuclear notational scheme and renumbering the orbitals, we have the order

$$1\sigma < 2\sigma < 3\sigma < 4\sigma < 1\pi < 5\sigma < \cdots \tag{7.25}$$

Thus, the configuration for LiH is $(1\sigma)^2(2\sigma)^2$. Unfortunately, we cannot definitely say, in general, which orbitals are bonding and which are antibonding, and so we cannot easily obtain bond orders.

One concept which is of some utility in discussing heteronuclear molecules is that of an isoelectronic series. The molecules CO, N_2, and NO^+ are isoelectronic; so are HF and Ne, and LiH and He_2. No sooner do we say this, however, than we recognize the limitations of this concept. LiH has an observed dissociation energy of about 2.5 ev, whereas He_2 does not form. Another possibility is to ignore the closed inner shells, which, although helpful, does not provide much more information about a molecule than the simple Lewis rules.

We thus see that the loss of symmetry in going from homonuclear to heteronuclear molecules makes it difficult to discuss the latter with the same qualitative accuracy with which we were able to treat the former. Only detailed calculations can provide the information we seek. For example, in LiH, we can form molecular orbitals by combining the $1s$ atomic orbitals of hydrogen with the $1s$ and $2s$ orbitals of lithium. We may suppose that the lowest molecular orbital is almost purely lithium $1s$ and is therefore *nonbonding;* thus

$$\psi_{1\sigma} \sim \varphi_{1s}^{Li} \tag{7.26}$$

The second orbital should be some linear combination of hydrogen $1s$ and lithium $2s$; thus

$$\psi_{2\sigma} \sim a\varphi_{1s}^{H} + b\varphi_{2s}^{Li} \tag{7.27}$$

It is found, however, that this function is grossly inadequate, and that the lithium $2p_0$ function must be added to the basis set—in this way we account for *polarization* of the lithium atom by the proton. With all orbital exponents optimized, Ransil (1960) has found

$$\begin{aligned}
\psi_{1\sigma} &= 0.996\varphi_{1s}^{Li} - 0.017\varphi_{2s}^{Li} + 0.007\varphi_{2p_0}^{Li} - 0.006\varphi_{1s}^{H} \\
\psi_{2\sigma} &= 0.141\varphi_{1s}^{Li} + 0.308\varphi_{2s}^{Li} + 0.211\varphi_{2p_0}^{Li} + 0.703\varphi_{1s}^{H}
\end{aligned} \tag{7.28}$$

We see that as expected, the $\psi_{1\sigma}$ orbital is nearly pure lithium $1s$, while in the second molecular orbital, the mixing of the $2p_0$ orbital is quite important. This mixing, usually referred to as *hybridization*, is quite important in molecules containing first-row atoms, which have $2s$ and $2p$ orbitals lying close together.

Ransil's calculation gives a total energy of -7.970 au for LiH. In the Hartree-Fock limit,‡ the total energy is -7.986 au, so our limited

‡ Kahalas and Nesbet (1963).

basis including the $2p_0$ would seem to do quite well. However, the total energy of lithium in the Hartree-Fock limit is -7.433 au, so the predicted electronic binding energy (D_e) of LiH is only 0.053 au, compared with the accurately calculated result (*vide infra*) 0.092 au. The exact total energy of LiH is -8.070; thus, although the Hartree-Fock method provides nearly 99 percent of the total energy, it does not accurately predict the binding energy. The situation is quite similar to that found in the hydrogen molecule. As in that case, we can improve matters through the use of configuration interaction. Browne and Matson (1964) have been able to determine the energy to within 0.2 percent, and this probably represents one of the most accurate calculations to date on a molecular system other than H_2. For practical purposes, the calculation of Browne and Matson has converged to the exact solution to Schrödinger's equation, and comparison of predicted and experimental properties using their wavefunction shows pleasing agreement. For example, the calculated dipole moment is only 1 percent greater than experimental, and the calculated vibrational frequency, $\omega_e = 1{,}438$ cm^{-1}, is in good agreement with the experimental 1,406 cm^{-1}. However, the calculation of Browne and Matson is in no sense based on "orbitals"; there is heavy configuration interaction, and the Hartree-Fock approach to excited configurations was *not* used. Instead, a more powerful method involving the interaction between *nonorthogonal* configurations was employed. Convergence starting with the Hartree-Fock orbitals and mixing excited configurations constructed from virtual orbitals is quite slow; such a calculation on LiH by Kahalas and Nesbet, involving about the same number (≈ 25) of configurations as that of Browne and Matson, gave only 99.34 percent of the experimental energy of LiH.

Lithium hydride is a four-electron system. To be able to treat larger systems than this to as high a degree of accuracy is clearly not feasible unless one is willing to expend a *great* deal of time and energy. Therefore, in the next section, we will consider the predictions of the Roothaan-Hartree-Fock method only, since this method *is* feasible for small- and intermediate-size systems. Later in the text, we will be using far *less* accurate methods, which often strive toward "Hartree-Fock accuracy." It is worthwhile seeing just how valid this goal is.

7.5 THE HARTREE–FOCK TREATMENT OF DIATOMIC MOLECULES

Very few SCF treatments which are of "Hartree-Fock accuracy" have been carried out for molecules. This may seem surprising since Hartree-Fock functions are available for most atoms. The reason that more molecular calculations have not been attempted lies in the integral evaluation problem. As should be clear from consideration of the atomic problem, a molecular calculation involves the one-electron integrals $\langle \varphi_1^A | 1/R_B | \varphi_2^C \rangle$, $\langle \varphi_1^A | \varphi_2^B \rangle$, and $\langle \varphi_1^A | -\frac{1}{2}\nabla^2 | \varphi_2^B \rangle$ and the two-electron integrals $\langle \varphi_1^A(1)\varphi_2^B(2) | 1/r_{12} | \varphi_3^C(1)\varphi_4^D(2) \rangle$. Once these integrals have been evaluated, the Roothaan procedure poses no difficulty and consumes little computation time. In fact, this part of the SCF problem is little different for atoms and molecules.

For the purpose of the discussion which follows, we shall assume that the basis functions are Slater-type orbitals. In point of fact, *gaussian-type* orbitals ($\varphi = Nr^n e^{-\zeta r^2} Y_{lm}$) are also widely used, and for these basis functions‡ the integral evaluation problem is quite easy (see Oohata et al., 1966). However, for a given degree of accuracy in the energy, it would appear that a factor of 5 more basis functions are required compared to a Slater basis. This would seem to negate any simplification, although the question is far from settled. Which basis is best will depend on the speed of integral evaluation and the storage problems associated with large numbers of integrals. The so-called *contracted* gaussian basis, consisting of a relatively small number of *fixed* linear combinations of basic gaussians, is sometimes used in place of the Slater basis; contracted gaussians can be chosen to mimic the radial behavior of a Slater-type orbital.

The integrals over basis functions may be classified into four types, according to the number of atoms involved in each integral. For example, for A, B, and C *separate* atomic centers, $\langle \varphi_1^A | 1/R_B | \varphi_2^C \rangle$ is referred to as a *three-center* nuclear attraction integral. The overlap and kinetic energy integrals can involve a maximum of two centers, whereas the most difficult to calculate is the four-center two-electron integral. (With a gaussian basis, four-center integrals are little more difficult than one-center integrals.) Actually, the two-center one-electron integrals pose no problem, and programs are available (Offenhartz, 1967) which will calculate several thousand of them per minute. Similarly, the one-center two-electron integrals are relatively simple, and for this reason the atomic Hartree-Fock problem is not excessively difficult. However, the three-center one-electron and certain of the two-center (and all of the three-center and four-center) two-electron integrals require considerably more work. Furthermore, there are an extremely large number of them; the reader may wish to verify that in a molecule with four atoms and 10 basis functions per atom, there are 820 overlap integrals, an equal number of kinetic energy integrals, 3,280 nuclear attraction integrals of which nearly half involve three centers, 30,000 four-center integrals, 186,000 three-center two-electron integrals, 114,450 two-center two-electron integrals, and 6,160 two-electron one-center integrals—341,630 integrals in all, most of which are difficult to calculate. Furthermore, 10 basis functions per

‡ Note that gaussian-type orbitals (GTOs) have r^2 rather than r in the exponential part. This makes a great difference in the shape of the orbital: GTOs die off rapidly at large distances and rise slowly and smoothly near the nucleus.

center is probably not sufficient for true Hartree-Fock accuracy even in molecules composed solely of first-row atoms.

Really accurate Hartree-Fock calculations are therefore out of the question for all but the simplest molecules.‡ At the present writing, only the first- and second-row hydrides have been systematically treated, together with a very few of the first-row diatomic molecules. A total of about 25 basis functions have been employed for the calculations. They have been carried out by Paul E. Cade, Winifred Huo, and A. C. Wahl at the University of Chicago and at Harvard, using programs§ developed by the "Chicago Group" under the direction of C. C. J. Roothaan. Although somewhat limited in scope (since we are interested in discovering just how good the predictions of Hartree-Fock theory are in general), they provide the best and most extensive series of accurate calculations to date.

We wish, therefore, to compare quantities calculated in the Hartree-Fock scheme with experiment. Since SCF calculations of Hartree-Fock accuracy are so very difficult to obtain, we wish also to compare the predictions of simpler, less accurate SCF functions, using limited basis sets.‖ First we must ask what experimental quantities are of interest, and of these, which are likely to be predicted accurately by an uncorrelated wavefunction.

We expect that the dipole moment will be accurately predicted since it is determined by a one-electron operator. (See the discussion and references given in Sec. 6.9.) There are many other one-electron operators whose expectation values have been calculated for the diatomic molecules, but they are somewhat esoteric, and we do not consider them further. They include the quadrapole moment, a number of field gradients, and the forces on the nuclei.

The quantities which are most naturally compared with experiment are those which can be inferred easily from spectroscopic measurements; such quantities are the internuclear distance R_e, the vibrational frequency ω_e, and the dissociation energy D_e. However, because we know that the Hartree-Fock equations *cannot* give the correct energy at infinite internuclear separation, we may expect some error in R_e and ω_e—just how

‡ There are no three or four center integrals present in a calculation on a diatomic molecule!

§ Wahl, Cade, and Roothaan, 1964.

‖ We will use the so-called minimum basis set, composed of one Slater-type orbital per occupied atomic orbital. A limited basis set of very nearly Hartree-Fock accuracy is the double-zeta basis, consisting of two STO's per atomic orbital. For F_2, there are 10 STO's in the minimal set and 20 in the double-zeta set. Note that the minimum-basis-set treatment of an intermediate-sized molecule (say seven first-row atoms and a few hydrogens) is easier than the four-atom double-ζ problem discussed above. It thus may be easier to do a calculation on benzene with a minimum basis than a calculation on propane with a basis allowing true Hartree-Fock accuracy.

much remains to be seen. Furthermore, because of correlation errors, we cannot expect a good prediction of D_e unless we compare the total Hartree-Fock energy of the molecule with the sum of the Hartree-Fock energies of the separated atoms. Another quantity which we may hope to predict is the ionization potential; we have a choice of calculating this either by subtracting the total energy of the ion from the total energy of the molecule, or by using Koopmans' theorem (see Secs. 6.8 and 10.3) and comparing the negative of the highest occupied orbital energy of the molecule with the ionization potential.

A detailed comparison of experimental data, minimum basis set SCF predictions, and accurate Hartree-Fock predictions for nine selected diatomic molecules is presented in Table 7.3. The agreement between predicted and experimental quantities is impressive in all cases except for the dissociation energy. The predicted internuclear distances are uniformly too small except for LiH, but the error is always less than 3 percent. Similarly, the agreement between predicted and experimental ω_e indicates that Hartree-Fock theory is capable of obtaining approximately the correct shape of the vibrational potential surface near R_e despite the fact that the prediction of the shape *must* be erroneous at large internuclear distances. Predictions of the dipole moment are in good agreement with experiment, although with the minimum basis set, there appears to be a random error of about 0.5 debye. This is serious only in carbon monoxide, where the Hartree-Fock prediction differs in sign from the prediction of the minimum basis set. Predictions of the ionization potentials, using Koopmans' theorem, are in remarkable agreement with experiment—remarkable considering the assumptions involved in the derivation. Numerically, the error is within 2 ev in all cases. Predictions using the total energies of molecules and ions (where Hartree-Fock energies are available for the ions) give about the same results. The minimum basis set does almost equally well, except for NH, which is in any case suspicious since at the experimental internuclear separation the molecule is predicted to have a negative dissociation energy.

One result not given in Table 7.3, but of great importance in spectroscopy and in the theory of molecular electronic structure, is the excitation energy from the ground state to the first excited state. This has been considered very carefully for carbon monoxide by Winifred Huo (1966), who obtained an excitation energy of 5.76 ev compared to the experimental 6.04 ev. The agreement here is quite good, in direct contrast with the results of Watson (1959) on transition metal ions. As discussed in Sec. 6.8, Watson found that predicted excitation energies were calculated to lie about twice as high as observed for the neutral atoms (the discrepancy is much less severe in the doubly and triply charged ions). Thus, the good agreement for CO is somewhat surprising and may be fortuitous. Clearly, a great deal more work needs to be done on the

Table 7.3 A comparison of calculated and experimental quantities for diatomic molecules

		ω_e, cm^{-1}	R_e, au	D_e, au*	μ, debyes	IP $E_{ion} - E_{mol}$	IP $-\epsilon_{mol}$	E_{mol}
CO	Minimum basis set†	—	—	0.228	-0.464	—	0.4842	-112.391
	Hartree-Fock†	2,431	2.081	0.290	+0.153(C$^+$O$^-$)	—	0.5505	-112.788
	Experimental†	2,170	2.132	0.413	-0.112	0.515	0.515	-113.377
BF	Minimum basis set†	—	—	0.206	1.40(B$^-$F$^+$)	—	0.3612	-123.646
	Hartree-Fock‡	1,496	2.354	0.227	1.04(B$^-$F$^+$)	—	0.4016	-124.166
	Experimental‡	1,402	2.391	0.315	—	0.4032	0.4032	-124.777
LiH	Minimum basis set§	—	—	0.052	5.92	—	0.2986	-7.970
	Hartree-Fock‡	1,433	3.034	0.055	6.002(Li$^+$H$^-$)	0.258	0.3017	-7.987
	Experimental‡	1,406	3.015	0.093	5.882	0.24 ± 0.02	0.24 ± 0.02	-8.070
BeH	Minimum basis set	—	—	0.080	0.282(Be$^+$H$^-$)	—	0.3128	-15.153
	Hartree-Fock‡	2,147	2.528	(0.096)	—	0.299	0.32 ± 0.01	-15.265
	Experimental‡	2,059	2.538	0.076	1.58	0.32 ± 0.01	0.3277	—
BH	Minimum basis set§	—	—	—	—	—	0.3484	-25.075
	Hartree-Fock‡	2,499	2.305	0.102	1.733(B$^-$H$^+$)	0.310	0.358	-25.131
	Experimental‡	2,368	2.336	0.132	—	0.358	—	-25.290
CH	Minimum basis set	—	—	—	—	—	0.4150	-38.279
	Hartree-Fock‡	3,053	2.086	0.091	1.570(C$^-$H$^+$)	0.370	0.391	-38.490
	Experimental‡	2,868	2.124	0.134	1.46	0.391	—	—
NH	Minimum basis set§	—	—	-0.44	2.01	—	0.3111	-54.325
	Hartree-Fock‡	3,556	1.923	0.077	1.627(N$^-$H$^+$)	0.471	0.5377	-54.978
	Experimental‡	(3,126)	1.961	0.14	—	0.481	0.481	-55.252
OH	Minimum basis set	—	—	0.111	—	—	0.5722	-75.421
	Hartree-Fock‡	4,062	1.795	0.170	1.780(O$^-$H$^+$)	0.420	0.492	-75.780
	Experimental‡	3,735	1.834	0.094	1.660	0.492	—	—
HF	Minimum basis set§	—	—	—	1.44	—	0.4686	-99.536
	Hartree-Fock‡	4,469	1.696	0.161	1.942(F$^-$H$^+$)	0.534	0.6501	-100.070
	Experimental‡	4,139	1.733	0.225	1.820	0.581	0.581	-100.530

* The dissociation energy for the minimum basis set is calculated by subtracting the SCF minimum-basis-set energy of the molecule from the corresponding energies of the separated atoms. Similarly, the Hartree-Fock dissociation energy uses the Hartree-Fock energy of the separated atoms as the zero of energy.

† W. Huo, *J. Chem. Phys.*, **43**:624 (1965).

‡ P. E. Cade and W. Huo, *J. Chem. Phys.*, **47**:614 (1967).

§ B. J. Ransil, *Rev. Mod. Phys.*, **32**:239 (1960).

predictions of the Hartree-Fock method for systems small enough for the method to be feasible.

It is difficult to draw pat conclusions from the comparisons of Table 7.3. We will later wish to know the accuracy of predictions made with approximate SCF functions calculated for large molecules. Clearly, we cannot hope to make good estimates of the dissociation energies because the correlation errors are too large. On the other hand, predictions of internuclear separations appear possible, at least if the basis set is adequate. Even with a small basis, *changes* and *trends* in internuclear separations may be calculable. Similarly, one may hope to obtain the dipole moment within 1 debye even with a small basis set and perhaps to predict trends. Similar hopes are in order for the ionization potential. In the end, however, one must bear in mind the inaccuracies inherent even in Hartree-Fock functions (due to the correlation error) and the difficulty of obtaining genuine Hartree-Fock functions by the Roothaan SCF procedure. Accurate Hartree-Fock functions for a molecule as large as benzene are still in the future, and correlated wavefunctions for even diatomic molecules are still to be produced. Therefore, for the foreseeable future, calculations on large molecules are going to be *very* approximate, and we should not expect too much of them. In particular, we should never expect a calculation on a large molecule to approach even the limited accuracy of the Hartree-Fock results of Table 7.3.

PROBLEMS

7.13 Write down the general formula for calculating the dipole moment of a diatomic molecule in terms of matrix elements over the basis set.

7.14 Derive a general formula for the total numbers of one- and two-electron integrals involved in a molecular calculation with a given basis size. How many integrals must be calculated if there are 100 basis functions?

SUGGESTIONS FOR FURTHER READING

1. Murrell, J. N., S. F. A. Kettle, and J. M. Tedder: "Valence Theory," John Wiley & Sons, Inc., New York, 1965.

   No student should learn about the molecular orbital theory of the electronic structure of diatomic molecules without learning something of the corresponding valence-bond treatment. An excellent discussion is given in Chaps. 11 and 12 of this book. Chapter 10 contains a first-rate discussion of molecular orbital theory for diatomic molecules, and the last part of Chap. 12 has a lucid description of Moffitt's *method of atoms in molecules*.

Two papers from the Chicago Group (Huo, 1965, and Bader, Henneker, and Cade, 1967) contain contour maps of the distribution of electronic charge in a number of diatomic molecules. The maps are quite revealing and lead to a novel and useful way of defining which orbitals are bonding and which are antibonding in a heteronuclear diatomic molecule.

A paper by Cade and Huo (1967) contains an extensive bibliography of previous calculations on diatomic molecules. The bibliography is *not* restricted to molecular orbital calculations.

8

GROUP THEORY AND
GROUP REPRESENTATIONS

Weyl protested that Dirac had said he would derive the results without the use of group theory, but, as Weyl said, all of Dirac's arguments were really applications of group theory. Dirac replied, "I said I would obtain the results without *previous knowledge* of group theory." Quoted by E. U. Condon and G. S. Shortley

8.1 APPLICATIONS AND DEFINITIONS

Of all the theorems developed in Chap. 2, the most useful have been the theorems on commuting operators. From (2.32), if two operators $\mathbf{O}_A$ and $\mathbf{O}_B$ commute, then there exists a set of simultaneous eigenstates $|O_A O_B\rangle$ such that

$$\begin{aligned}
\mathbf{O}_A|O_A O_B\rangle &= O_A|O_A O_B\rangle \\
\mathbf{O}_B|O_A O_B\rangle &= O_B|O_A O_B\rangle
\end{aligned} \tag{8.1}$$

Consequently, since the set $|O_A O_B\rangle$ is a complete orthonormal set, $\mathbf{O}_A$ is diagonal within the eigenstates of $\mathbf{O}_B$, and vice versa; thus

$$\begin{aligned}
\langle O_{A_1} O_{B_1}|\mathbf{O}_A|O_{A_1} O_{B_2}\rangle &= 0 \\
\langle O_{A_1} O_{B_1}|\mathbf{O}_B|O_{A_2} O_{B_1}\rangle &= 0
\end{aligned} \tag{8.2}$$

Furthermore, if the eigenstates of $\mathbf{O}_A$ are not known and those of $\mathbf{O}_B$ are, then $\mathbf{O}_A$ will be diagonal within the known eigenstates of $\mathbf{O}_B$; thus

$$\langle O_{B_1}|\mathbf{O}_A|O_{B_2}\rangle = 0 \qquad\qquad \blacktriangleright (8.3)$$

Owing to the power of the variation theorem, the most important operator in quantum mechanics is always the Hamiltonian. Thus, one of the most important problems in quantum theory is to find operators which commute with it. Having found such operators, we will wish to construct basis functions which are eigenfunctions of them and which can be used in a variational approach to the eigenfunctions of the Hamiltonian. Clearly, the larger the number of operators we can find which commute with $\mathfrak{IC}$, the simpler the variational problem becomes since fewer of the matrix elements of $\mathfrak{IC}$ will be nonzero.

Group theory provides us with a general way of finding operators which commute with $\mathfrak{IC}$. In fact, we could have derived the set of operators for atoms $(\mathbf{L}^2,\mathbf{S}^2,\mathbf{J}^2,\mathbf{J}_z)$ by using the full rotation group. This is an especially useful approach if one is interested in particle physics and in Lie groups.‡ However, much of the power of advanced group-theoretical methods is lost when they are applied to the simple groups common in molecular problems, and so we shall not develop these methods in the present text. Such methods may well be warranted in the case of octahedral§ symmetry or when special symmetry groups for "nonrigid molecules" are being considered,‖ but in any case they are outside the scope of treatment of a single chapter.

A group is a collection of elements—just what these elements are being of concern to the chemist and physicist, but not to the mathematician. An operation between any pair of these elements must be defined, and it is conventional to call this operation *group multiplication*, although it may be something unrelated to conventional multiplication and may in fact be addition. Let the elements be called A, B, C, ... (The number of elements in a group is called the *order* of the group. There exist groups of infinite order as well as finite groups.) For the set of these elements to form a group, the following requirements must be satisfied:

1. The set must be closed under the operation of multiplication, such that if X and Y are any two elements of the set, the element $Z = XY$ must also be a member of the set.

2. The operation of group multiplication must be associative, such that if X, Y, and Z are members of the set, $X(YZ) = (XY)Z = XYZ$.

‡ See Chap. 5 of M. Tinkham, 1964, for a discussion of the connection between the full rotation group and angular momentum. For an elementary introduction to Lie groups, using only the commutation rules discussed in Chap. 3, see the pleasantly lucid book by Harry J. Lipkin, 1966.
§ J. S. Griffith, 1962.
‖ H. C. Longuet-Higgins, 1963.

This is not as trivial a relation as it may seem; $1/(2/3) \neq (1/2)/3$, and hence $1/2/3$ is not defined in the ordinary division of numbers.

3. There exists a unit element in the set, usually designated E, such that $EX = XE = X$ for all elements X contained in the set. Note that if the operation of group multiplication were actually addition, the "unit element" would be, in effect, a zero.

4. For every element X in the set, there exists a inverse element X^{-1} in the set, such that $XX^{-1} = X^{-1}X = E$.

The reader should verify that the set of all integers under addition form an infinite group (the inverses are the negative integers) with 0 as the unit element. Similarly, the set of all rational numbers excluding zero form a group under ordinary multiplication, and the set of all possible 2×2 (or $n \times n$) matrices form an infinite group under group multiplication, provided that all matrices are nonsingular (have nonzero determinants) so their inverses are defined. The groups of most interest to us, however, are usually finite and in particular are connected with the symmetries of physical objects. These include rotations, reflections, and inversions; these are operations which take the object into a spatial position indistinguishable from the original position. For example, rotation of the benzene molecule about its center along an axis perpendicular to the plane of the molecule and by an angle $2\pi/6$ leaves the molecule in a position indistinguishable from the original one. This operation is usually written C_6, the letter C implying a rotation about some axis, and the subscript 6 implying a rotation by $\frac{1}{6}$ of 2π. In general, the symbol C_n^m implies a rotation about some axis by $2\pi m/n$, such that $C_6^2 = C_3$. The product of two operations is, physically, just another operation and corresponds to carrying out the two operations one at a time. As with operators in quantum mechanics, the operation is carried out from right to left, so that a sequence of operators XYZ implies that Z is performed first, followed by Y, followed by X. Thus, for benzene, C_6C_3 implies a rotation of $2\pi/3$ followed by a rotation of $2\pi/6$, and hence it is equivalent to C_2 about the same axis. It is clear that $C_2^2 = E$, and hence that $C_2^{-1} = C_2$.

Let us now consider a specific example of a finite group. We shall consider this group from three different points of view—as a purely mathematical abstract entity, as a series of geometrical manipulations, and as a set of matrices. We begin at the most abstract level. Consider the set of elements E, A, B, C, D, and F. In order to define the group, we need to list the inverse of each element and the product of every pair of elements. This is most directly done through the use of the *group multiplication table*. For example, in this particular group, we define $BA = F$, and in the multiplication table this is stated by having the symbol F in the row of B and the column of A. Similarly, we define $A^{-1} = A$, and in the table we place the symbol E in the row and column of A, since the statement

$A^{-1} = A$ is equivalent to the statement $A^2 = E$. Thus the group is entirely determined by its table, which we now define as

S_3	E	A	B	C	D	F
E	E	A	B	C	D	F
A	A	E	D	F	B	C
B	B	F	E	D	C	A
C	C	D	F	E	A	B
D	D	C	A	B	F	E
F	F	B	C	A	E	D

▶(8.4)

This group is conventionally given the name S_3, for reasons to be seen. We note from this table that the operation of multiplication is not always commutative, and, for example, $AB \neq BA$. A group for which group multiplication *is* always commutative is called *abelian*. The two groups involving the integers under addition and the rational numbers under multiplication are clearly abelian, but the group of all nonsingular 2×2 matrices is not. As is well known, matrix multiplication is, in general, noncommutative.

Our example group may be defined equally well as the set of symmetry operations of an equilateral triangle, as illustrated in Fig. 8.1. The elements A, B, and C are rotations by π (C_2 type) about the three different vertices and the center of the opposite edges. D and F are rotations of $2\pi/3$ in

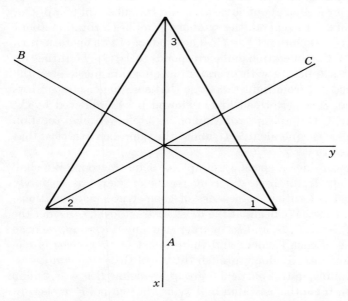

Fig. 8.1 Equilateral triangle and associated symmetry axes and coordinate system.

the plane of the triangle, clockwise and the counterclockwise, respectively. The reader should verify that a consideration of these symmetry elements leads to the same group multiplication table as given above. For example, *keeping the rotation axes fixed in space*, we see that operation A takes corner 3 into itself and corners 1 and 2 into each other. If we follow this operation by D, we find that DA takes corner 1 into 3, 3 into 1, and 2 into itself, which is equivalent to C; hence $C = DA$. In fact, because the set of symmetry operations of an equilateral triangle gives rise to all possible permutations of the corners, our example group can be viewed equally well as the *permutation group* of order 6. A permutation group of order $n!$ is the set of all possible permutations of n distinguishable things.‡ A convenient symbolic notation is

$$\begin{pmatrix} 1 & 2 & 3 & \cdots & n \\ a_1 & a_2 & a_3 & \cdots & a_n \end{pmatrix} \tag{8.5}$$

where $a_1, a_2, a_3, \ldots, a_n$ are the n different integers 1 through n except for the order. Thus, for example,

$$A = \begin{pmatrix} 1 & 2 & 3 \\ 2 & 1 & 3 \end{pmatrix} \qquad D = \begin{pmatrix} 1 & 2 & 3 \\ 2 & 3 & 1 \end{pmatrix} \tag{8.6}$$

or, for short,

$$A = (12)(3) \qquad D = (123) \qquad\qquad \blacktriangleright (8.7)$$

so that $DA = (123)(12)(3) = (13)(2) = C$. In the shortened notation, (13) indicates that 1 goes into 3 and 3 goes into 1. Similarly, (123) implies that 1 goes into 2, 2 into 3, and 3 into 1; (2) implies only that 2 goes into itself. Permutation groups are important in certain treatments of quantum theory, where the permutation symmetry of the identical particles is treated as fundamental. Such treatments have a way of running into enormous practical difficulties as the number of electrons becomes large, and so they will not be given here; we mention them only for the sake of completeness and refer the interested reader to the work of Goddard (1967). Permutation-group *nomenclature* is useful in itself, however, and we will often employ it when discussing the symmetries of physical objects as a means of bookkeeping. Thus, the group in our example is entirely specified either by Fig. 8.1 or by the permutation notations $E = (1)(2)(3)$, $A = (12)(3)$, $B = (1)(23)$, $C = (13)(2)$, $D = (123)$, and $F = (132)$.

A third, and still different, way of defining our group is by defining (or *representing*) each element in the group by a 2×2 matrix. Quite

‡ Permutation groups are given the general name S_n; hence the name S_3 in (8.4).

naturally, E is represented by the 2×2 unit matrix, and we have

$$E = \begin{pmatrix} 1 & 0 \\ 0 & 1 \end{pmatrix} \qquad\qquad A = \begin{pmatrix} 1 & 0 \\ 0 & -1 \end{pmatrix}$$

$$C = \begin{pmatrix} -\tfrac{1}{2} & -\tfrac{1}{2}\sqrt{3} \\ -\tfrac{1}{2}\sqrt{3} & \tfrac{1}{2} \end{pmatrix} \qquad B = \begin{pmatrix} -\tfrac{1}{2} & \tfrac{1}{2}\sqrt{3} \\ \tfrac{1}{2}\sqrt{3} & \tfrac{1}{2} \end{pmatrix}$$

$$F = \begin{pmatrix} -\tfrac{1}{2} & -\tfrac{1}{2}\sqrt{3} \\ \tfrac{1}{2}\sqrt{3} & -\tfrac{1}{2} \end{pmatrix} \qquad D = \begin{pmatrix} -\tfrac{1}{2} & \tfrac{1}{2}\sqrt{3} \\ -\tfrac{1}{2}\sqrt{3} & -\tfrac{1}{2} \end{pmatrix}$$

$$\blacktriangleright (8.8)$$

The reader may wish to verify (Prob. 8.1) that these matrices give the correct group multiplication table. However, the matrices in (8.8) are not the *only* possible matrix formulation of the group. Another different set of matrices representing our group can be written as

$$E = \begin{pmatrix} 1 & 0 & 0 \\ 0 & 1 & 0 \\ 0 & 0 & 1 \end{pmatrix} \qquad\qquad A = \begin{pmatrix} -1 & 0 & 0 \\ 0 & 1 & 0 \\ 0 & 0 & -1 \end{pmatrix}$$

$$B = \begin{pmatrix} -1 & 0 & 0 \\ 0 & -\tfrac{1}{2} & \tfrac{1}{2}\sqrt{3} \\ 0 & \tfrac{1}{2}\sqrt{3} & \tfrac{1}{2} \end{pmatrix} \quad C = \begin{pmatrix} -1 & 0 & 0 \\ 0 & -\tfrac{1}{2} & -\tfrac{1}{2}\sqrt{3} \\ 0 & -\tfrac{1}{2}\sqrt{3} & \tfrac{1}{2} \end{pmatrix}$$

$$D = \begin{pmatrix} 1 & 0 & 0 \\ 0 & -\tfrac{1}{2} & \tfrac{1}{2}\sqrt{3} \\ 0 & -\tfrac{1}{2}\sqrt{3} & -\tfrac{1}{2} \end{pmatrix} \quad F = \begin{pmatrix} 1 & 0 & 0 \\ 0 & -\tfrac{1}{2} & -\tfrac{1}{2}\sqrt{3} \\ 0 & \tfrac{1}{2}\sqrt{3} & -\tfrac{1}{2} \end{pmatrix}$$

$$(8.9)$$

The set (8.9) is a rather trivial extension of the set (8.8), formed by expanding the 2×2 matrices to 3×3 and by placing plus or minus unity in the first position of each new matrix and zeros in the $1 \times n$ and $n \times 1$ positions. We could equally have put only a plus unity in the first positions or expanded to 4×4 matrices, and so on; there are an infinite number of suitable sets of matrices isomorphic‡ to the group.

At this point the reader should have a fair intuitive conception of a group. In chemistry and physics we are interested in a particular kind of group, whose elements leave the Hamiltonian for a particular system unchanged. In mathematical language, we are interested in operators (or "elements") X such that $X\mathfrak{IC}$ and $\mathfrak{IC}$ are identical. For example, for X

---

‡ Isomorphic means in one-to-one correspondence, and therefore yielding the identical group multiplication table. See also Sec. 8.3.

(the permutation operator P_{12}) and $\mathcal{3C}$ (the Hamiltonian for helium), we have

$$\mathcal{3C} = -\tfrac{1}{2}\nabla_1{}^2 - \tfrac{1}{2}\nabla_2{}^2 - \frac{2}{R_1} - \frac{2}{R_2} + \frac{1}{r_{12}}$$

$$P_{12}\mathcal{3C} = -\tfrac{1}{2}\nabla_2{}^2 - \tfrac{1}{2}\nabla_1{}^2 - \frac{2}{R_2} - \frac{2}{R_1} + \frac{1}{r_{21}} = \mathcal{3C}$$

(8.10)

Thus, in this case, the Hamiltonian is invariant under the operation P_{12}. Since P_{12} leaves $\mathcal{3C}$ invariant, it commutes with $\mathcal{3C}$, and, in general, since any symmetry operation which leaves the Hamiltonian invariant commutes with it, the symmetry operations of a given Hamiltonian are just what we seek in the way of a complete commuting set. Of course, some of the symmetry operations may not commute with others. This turns out to be no obstacle, and it is, in fact, of great use in finding degenerate solutions to Schrödinger's equation for the system at hand; we shall shortly develop a theory for operators which do not mutually commute but which all commute with $\mathcal{3C}$.

The question then arises as to the maximum number of such operators for a given Hamiltonian. Surprisingly, there is no general way of knowing, and even today there is some debate as to whether the hydrogen atom has been classified under the largest possible set. It should be clear that any such set of operators will form a group, provided, of course, that the set is closed. The question is whether a *larger* group can be formed by adding elements to a smaller group, and this question cannot be answered systematically. Generally speaking, we ignore this difficulty and employ only rather simple groups having to do with rotational symmetry and inversion and mirror symmetry. These groups are small and relatively easy to handle, but they are in no sense the complete group of the molecule.

PROBLEMS

8.1 Verify that the matrices of (8.8) yield the group multiplication table of (8.4). Do the same for a few of the matrices of (8.9).

8.2 Write down the inverse of each element in the example group S_3. Check these inverses using the matrices of (8.8).

8.3 Prove that for any two elements X and Y of a group, $(XY)^{-1} = Y^{-1}X^{-1}$.

8.4 Consider the symmetries of a square. There are eight elements. By numbering the corners of the square, list the possible permutations of the corners, and write out the group multiplication table. It is conventional to let A and B represent C_2 rotations which bisect opposite sides of the square, C and D represent C_2 rotations through the corners, and F, G, and H represent clockwise rotations (C_4, C_4^2, and C_4^3, respectively) in the plane of the square.

8.2 SUBGROUPS AND CLASSES

A subgroup is a set of elements, all of which belong to some larger group, but which by themselves form a group. What is crucial here is that the subgroup must be closed. For example, E by itself always forms a subgroup. This is a rather trivial example since this is a subgroup of order 1; the elements E and A of our example group also form a subgroup. For that matter, E and B and E and C also form subgroups. E and D do not since $D^2 = F$, but E, D, and F form a subgroup. The groups with which we shall be dealing are, in general, subgroups of some larger "complete" group for the molecule. The major application of the study of subgroups, however, is in the treatment of systems which "nearly" have a certain symmetry, as in the case of *trans*-$Co(NH_3)_4Cl_2^+$, which nearly has octahedral symmetry and is more correctly classified under an octahedral subgroup which is also the group of the square.

The concept of a class is slightly more difficult. To understand it, we must first introduce the concept of a pair of conjugate group elements. Two elements A and B are conjugate if

$$A = X^{-1}BX \qquad\qquad \blacktriangleright(8.11)$$

for X *some* member of the group. Clearly, $B = XAX^{-1} = Y^{-1}AY$, so the relationship is reciprocal. Furthermore, if A and B are conjugate and A and C are conjugate, then B and C are conjugate (Prob. 8.5). Therefore, we can speak of several elements as being *mutually conjugate*. To discover whether two elements are conjugate, one must test the relation (8.11) with all members X of the group. When this has been done, one usually discovers that the elements break up into distinct *classes* of elements, the members of which are mutually conjugate. In our example group, A, B, and C belong to one class, and D and F to another. The element E always falls in a class by itself.

The concept of a class has an important physical significance—the members of a class are all the same *kind* of operation (C_2 rotation, C_3 rotation, mirror plane, etc.) and differ from each other only by choice of axes. In practice, this observation makes it easier to pick out the members of a class. For example, A, B, and C are all C_2 rotations about a vertex of the triangle, whereas D and F are C_3 rotations.

It is clear that a given element can belong to only one class and that the classes therefore exhaust the group. It is also possible to prove that the orders of the classes‡ must be integral divisors of the order of the group and that no two classes can have the same order. (Subgroups also have orders which are integral factors of the group order, but there can be more than one subgroup of a given order.) The most important theorem about classes, however, concerns the formulation of their elements as matrices.

‡ The order of a class is the number of elements in the class.

If we define the trace of a matrix as the sum of its diagonal elements, it is possible to prove that the traces of all elements of a given class must be equal. The proof is straightforward, and we give it primarily as an introduction to the nomenclature of group representations. Let $\Gamma(A)$ and $\Gamma(B)$ be two matrices chosen so that their multiplication is the same as the group multiplication $\Gamma(A)\Gamma(B) = \Gamma(AB)$. The trace of $\Gamma(A)$, written Tr $\Gamma(A)$, is then

$$\text{Tr } \Gamma(A) = \sum_i [\Gamma(A)]_{ii}$$

where $[\Gamma(A)]_{ij}$ are the elements of the matrix $\Gamma(A)$. Taking any element X, we find that the trace of a product of matrices is independent of the order of multiplication of the matrices, and

$$\begin{aligned}
\text{Tr } \Gamma(AX) &= \sum_i [\Gamma(AX)]_{ii} = \sum_i \sum_j [\Gamma(A)]_{ij}[\Gamma(X)]_{ji} \\
&= \sum_j \sum_i [\Gamma(X)]_{ji}[\Gamma(A)]_{ij} = \sum_j [\Gamma(XA)]_{jj} \\
&= \text{Tr } \Gamma(XA)
\end{aligned} \tag{8.12}$$

Therefore, Tr $(X^{-1}AX)$ = Tr $(XX^{-1}A)$ = Tr (A), and Tr (A) = Tr (B) if $B = X^{-1}AX$.

PROBLEMS

8.5 If two elements of a group A and B are conjugate and if A and C are conjugate, show that B and C are conjugate.

8.6 Show that in an abelian group each element is in a class by itself.

8.7 Verify that D and F are the only members of the class of order 2 in our example group S_3.

8.8 Find the classes and subgroups of the group of the square considered in Prob. 8.4.

8.3 MATRIX REPRESENTATIONS

In the first section we said that a group can be defined in terms of matrices instead of a group multiplication table, provided that the multiplication of the matrices properly gives the group multiplication table. This leads to the more general concept of an *isomorphism* between two groups. Two groups are isomorphic if they have essentially the same multiplication table; more precisely, groups G and G' are isomorphic if there is a one-to-one correspondence between their elements, such that if A and A' and B and B' correspond, then AB and $A'B'$ correspond for all A, B and A', B'

in the two groups. To give a trivial example, the group defined by the multiplication table

	E'	A'	B'	C'	D'	F'
E'	E'	A'	B'	C'	D'	F'
A'	A'	E'	D'	F'	B'	C'
B'	B'	F'	C'	E'	A'	D'
C'	C'	D'	E'	B'	F'	A'
D'	D'	C'	F'	A'	E'	B'
F'	F'	B'	A'	D'	C'	E'

$$(8.13)$$

is isomorphic to the example group S_3 defined by (8.4) under the correspondence $E \rightleftarrows E'$, $A \rightleftarrows D'$, $B \rightleftarrows A'$, $C \rightleftarrows F'$, $D \rightleftarrows C'$, and $F \rightleftarrows B'$. Clearly, in an isomorphic relation, the groups are identical except for a possible change of nomenclature.

In a second kind of relation, called a homomorphism, *many* elements of one group may correspond to a single element of another group. This is a *many-to-one* correspondence. Mathematically, if A corresponds to A' and B to B', AB must still correspond to $A'B'$. But in a homomorphism, A and C may both correspond to A', in which case both A^2 and AC must correspond to A'^2. For example, consider the trivial group with just one element E'. *All* groups are homomorphic to the trivial group. Similarly, consider the group with just two elements E' and I', with $I'^2 = E$. Our example group is homomorphic to this group, with $(E,D,F) \rightarrow E'$ and $(A,B,C) \rightarrow I'$. The reader may wish to verify this assertion (Prob. 8.9).

A *matrix representation* of a group is defined to be a *homomorphic* relation between group elements and matrices. Since an isomorphism is but a special case of a homomorphism, an isomorphic representation is possible, and such a representation is called *true*, or *faithful*. The matrix representation given in (8.8) is clearly a faithful representation of our example group. Another representation, arising out of the homomorphism of our example group with the trivial group, is that in which all elements correspond to $n \times n$ unit matrices. It is conventional to choose the 1×1 matrices, so every element is represented by unity. In general, this is called the *identical representation*. By analogy with the homomorphism of S_3 with the group of order 2, we also have a one-dimensional representation in which E, D, and F are represented by $+1$ and A, B, and C are represented by -1. Note that in this example, members of the same class have the same representative, a relationship which is always true of one-dimensional representations since the traces of the matrices representing the elements of a class must be the same.

As noted earlier, there are an infinite number of different matrix representations for a given group. We therefore require some systematic method of limiting and classifying them. Clearly, there are only a limited number of different one-dimensional representations, so these pose little difficulty. On the other hand, one can construct an infinite number of two-dimensional representations. Not all of these are *essentially* different, however, in the sense that many different two-dimensional representations may give the same *type* of multiplication scheme, and indeed some of them may give the same *type* of multiplication scheme as a one-dimensional representation. For example, the two-dimensional representation in which all elements are represented by the 2×2 unity matrix is essentially the same as the identical representation.

We clarify these somewhat vague ideas as follows. Suppose we have a 2×2 (or $n \times n$) representation in which the group elements A, B, C, . . . are represented by the matrices $\mathbf{\Gamma}(A)$, $\mathbf{\Gamma}(B)$, $\mathbf{\Gamma}(C)$, ... Suppose we are given a matrix $\mathbf{U}$ and form a new set of matrices

$$\mathbf{\Gamma}'(A) = \mathbf{U}^{-1}\mathbf{\Gamma}(A)\mathbf{U}, \; \mathbf{\Gamma}'(B) = \mathbf{U}^{-1}\mathbf{\Gamma}(B)\mathbf{U}, \; \ldots$$

Then we can very easily prove (Prob. 8.10) that

$$\mathbf{\Gamma}'(A)\mathbf{\Gamma}'(B) = \mathbf{\Gamma}'(AB) \tag{8.14}$$

so the transformed matrices $\mathbf{\Gamma}'(X)$ *have the same multiplication properties* as $\mathbf{\Gamma}(X)$ and also form a representation of the group. The two representations, however, are said to be *equivalent* via a *similarity transformation*,‡ and it is obvious that since they have the same multiplication table, they are essentially the same. We will therefore seek to find the *inequivalent* representations of a given group.

Only representations of the same dimensionality (degree) can be equivalent. However, as noted above, all $n \times n$ representations in which all group elements are represented by the unity matrix are essentially the same. Such representations are considered *reducible*. In general, a representation is defined to be reducible if *all* the matrices of the representation are in the *same* block form

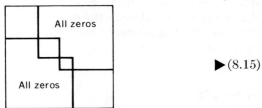

$$\blacktriangleright(8.15)$$

The matrices representing A and E in (8.8) are not reducible since not all of the matrices representing the elements of our example group are in diagonal form. Clearly, all one-dimensional representations are *irreducible*. A representation of degree n is irreducible if there does not exist *any*

‡ If $\mathbf{U}$ is a unitary matrix, this is called a *unitary transformation*.

unitary matrix which brings *all* the matrices of the representation into some simpler block form. Thus, a two-dimensional representation is reducible if and only if there exists a similarity transformation which simultaneously diagonalizes all of the matrices of the representation. The identical representation of degree 2 is clearly reducible, and writing $\Gamma^{(2)}$ for this representation and $\Gamma^{(1)}$ for the identical representation of degree 1, we obtain $\Gamma^{(2)} = 2\Gamma^{(1)}$. This is a somewhat peculiar notation; it does *not* refer to matrix addition but to a kind of matrix decomposition. In general, we write a reducible representation as a sum of irreducible ones; thus

$$\Gamma = \sum_i c_i \Gamma^{(i)} \qquad\qquad \blacktriangleright (8.16)$$

where the $\Gamma^{(i)}$ are the irreducible representations and the c_i are integers. The superscripts number the irreducible representations and do not necessarily indicate their degree. The decomposition (8.16) involves, of course, similarity transformations, but we need not indicate this in the equation.

Because the block form may easily be hidden through a unitary transformation, it will rarely be obvious whether a given representation is reducible. Furthermore, we need some way of finding out how many irreducible representations are possible for a given group, and we also need a standard way of forming them since, for a given irreducible representation of degree 2 or greater, there will be an infinite number of *equivalent* representations, which will have a different appearance. All will have the same trace, however (Prob. 8.11).

Before proceeding further with the theory, it is advisable to have a look at some of the physical implications of an irreducible representation. It turns out that there is a very close relationship between the degeneracy of an energy level and the degree of an irreducible representation. Furthermore, it turns out that an abelian group has only one-dimensional irreducible representations, so there is an intimate connection between spatial degeneracies and whether the symmetry operations which commute with the Hamiltonian mutually commute. We develop this connection in the next section.

PROBLEMS

8.9 Show that there is a homomorphism between S_3 and the group of two elements, E' and I', $I'^2 = E'$, such that $(E, D, F) \to E'$ and $(A, B, C) \to I'$. Show that this leads to a one-dimensional irreducible representation for S_3 with E, D, and F represented by $+1$ and A, B, and C represented by -1.

8.10 Verify Eq. (8.14).

8.11 Prove that all equivalent representations of a given symmetry operation have the same trace.

8.4   EIGENFUNCTIONS, BASIS SETS, AND IRREDUCIBLE REPRESENTATIONS

In the eigenvalue equation

$$\mathcal{3C}\Psi_i = E_i\Psi_i \tag{8.17}$$

there are an infinite number of linearly independent solutions Ψ_i. For a given energy level E_i there may be a number of degenerate (but independent) solutions, and we indicate this by explicitly writing

$$\mathcal{3C}\Psi_{ik} = E_i\Psi_{ik} \qquad k = 1, 2, \ldots, m \tag{8.18}$$

where there are m degenerate functions Ψ_{ik}. Without loss of generality we may assume that these functions are orthonormal; thus

$$\langle \Psi_{ik} | \Psi_{jl} \rangle = \delta_{ij}\delta_{kl} \tag{8.19}$$

We shall now show that there is a close connection between the number of degenerate functions belonging to a given energy level and the degree of a certain irreducible representation.

Consider a group of operators R which commute with the Hamiltonian. Then

$$R\mathcal{3C}\Psi_{ik} = \mathcal{3C}R\Psi_{ik} = E_iR\Psi_{ik} \qquad \blacktriangleright (8.20)$$

so $R\Psi_{ik}$ is an eigenfunction of $\mathcal{3C}$ with eigenvalue E_i. We may therefore conclude that any wavefunction $R\Psi_{ik}$ obtained by applying any symmetry operation R which leaves the Hamiltonian invariant to any eigenfunction Ψ_{ik} is also an eigenfunction of $\mathcal{3C}$. This function will always have the same energy as Ψ_{ik}. Therefore, given any eigenfunction, we ought to be able to *generate* all others degenerate with it by application of the symmetry operators which leave the Hamiltonian invariant. Whether we can in practice do this depends on the operators R—we may have chosen a group too small to generate all functions. If we choose the full group of the Hamiltonian, which by definition includes all possible operations which leave it invariant, then it is an article of faith with physicists that this group will generate all degenerate functions. We exclude "accidental" degeneracies which may occur because of a peculiarity in the experimental conditions; for example, two atomic terms might become degenerate in an electric field just exactly of the right size. This field need not have any symmetry, but only a very special magnitude. Clearly such degeneracies fall outside the domain of group theory, and we exclude them. By the same token, the degeneracy between the $2s$ and $2p$ levels in the hydrogen atom is *not* accidental, and it can be explained by considering a suitably large group.

If the function Ψ_{ik} is nondegenerate ($m = 1$), then, dropping the subscript k, we have

$$R\Psi_i = e^{ir}\Psi_i \tag{8.21}$$

where e^{ir} is a phase factor. If we apply each operation $(A, B, \ldots)$ of the group to this function, we generate a set of phase factors $e^{ia}, e^{ib}, \ldots$. If we apply two operations R and S successively, then

$$SR\Psi_i = e^{ir}e^{is}\Psi_i \qquad (8.22)$$

and thus the phase factors form a representation of the group. The representation is clearly irreducible since it is one-dimensional.

Let us now consider the complex conjugate of Eq. (8.17). Since both $\mathfrak{IC}$ and E_i are real,

$$\mathfrak{IC}\Psi_i^* = E_i\Psi_i^* \qquad (8.23)$$

Applying any symmetry operator R,

$$R\mathfrak{IC}\Psi_i^* = \mathfrak{IC}R\Psi_i^* = E_i R\Psi_i^* \qquad (8.24)$$

This equation, however, is just the complex conjugate of Eq. (8.20), since R is real. Therefore

$$R\Psi_i^* = e^{-ir}\Psi_i^* \qquad (8.25)$$

We have assumed that E_i is nondegenerate, which implies that Ψ_i^* and Ψ_i are *not* linearly independent. They therefore differ at most by a phase factor, and we thus obtain

$$R\Psi_i = e^{-ir}\Psi_i \qquad (8.26)$$

This equation is in conflict with (8.21) unless all the phase factors are real, so that $e^{ir} = \pm 1$ for all R. We have thus proved that a nondegenerate eigenfunction of $\mathfrak{IC}$ *provides a basis* for a *real* irreducible representation of the group of operators which leave $\mathfrak{IC}$ invariant. We commonly say that Ψ_i *transforms as* the irreducible representation, and if, for example, we give the name $\mathbf{\Gamma}^{(1)}$ (or A_1 or A_2 or B_1, etc.) to the irreducible representation, we say that Ψ_i *transforms as* $\mathbf{\Gamma}^{(1)}$ or *belongs to* $\mathbf{\Gamma}^{(1)}$.

Suppose now that Ψ_i and Ψ_i^* are linearly independent functions, so that E_i is doubly degenerate. In this case Ψ_i and Ψ_i^* can still belong to one-dimensional irreducible representations in certain circumstances. According to (8.21), the phase factors e^{ir} form a one-dimensional representation. Equally, the phase factors e^{-ir} form a one-dimensional representation which is the complex conjugate of the first. Thus, in general, if a complex irreducible representation exists, the conjugate representations always exist in pairs, and in the character tables (*vide infra*) and other group tabulations they are bracketed together and given a nomenclature appropriate to an irreducible representation of degree 2. Conversely, if a given symmetry group has such complex one-dimensional representations, we anticipate that a Hamiltonian commuting with all elements of the group may have some doubly degenerate eigenfunctions which are complex conjugates of each other and which transform as the two irreducible representations.

Complex representations are somewhat uncommon. Usually a doubly degenerate energy level will belong to an irreducible representation of degree 2. Consider the general case of degeneracy in which m functions Ψ_{ik} ($k = 1, 2, \ldots , m$) belong to the same energy level E_i. A symmetry operator R will then transform a given Ψ_{ik} into a linear combination‡ of the degenerate set; thus

$$R\Psi_{ik} = \sum_j [\boldsymbol{\Gamma}(R)]_{jk} \Psi_{ij} \tag{8.27}$$

where the coefficients $[\boldsymbol{\Gamma}(R)]_{jk}$ form a matrix when the effect of R on all Ψ_{ik} ($k = 1, 2, \ldots , m$) is considered. For other operations $S, \ldots$, we have

$$S\Psi_{ik} = \sum_j [\boldsymbol{\Gamma}(S)]_{jk} \Psi_{ij} \tag{8.28}$$

and so we associate a matrix with each operator. The matrices rather obviously form a representation of the group since

$$\begin{aligned}
SR\Psi_{ik} &= S \sum_j [\boldsymbol{\Gamma}(R)]_{jk} \Psi_{ij} \\
&= \sum_j \sum_l [\boldsymbol{\Gamma}(R)]_{jk} [\boldsymbol{\Gamma}(S)]_{lj} \Psi_{il} \\
&= \sum_l [\boldsymbol{\Gamma}(S) \boldsymbol{\Gamma}(R)]_{lk} \Psi_{il} \\
&= \sum_l [\boldsymbol{\Gamma}(SR)]_{lk} \Psi_{il} \tag{8.29}
\end{aligned}$$

The crucial question is whether or not the representation $\boldsymbol{\Gamma}$ is irreducible. Suppose it is reducible. Then we can divide the functions Ψ_{ik} into two sets $\Psi_{i1}, \Psi_{i2}, \ldots , \Psi_{ip}$ and $\Psi_{ip+1}, \Psi_{ip+2}, \ldots , \Psi_{im}$, such that the symmetry operators carry each set into itself; for example,

$$R\Psi_{il} = \sum_{j=1}^{p} [\boldsymbol{\Gamma}'(R)]_{jl} \Psi_{ij} \tag{8.30}$$

for all R in the group. If this division into two sets is possible, it is hard to see any reason why the two sets should have the same energy. It is thus a fundamental postulate of physics that the representation formed from the basis functions Ψ_{ik} is in fact irreducible (except for the possibility of complex one-dimensional representations, as mentioned above). Equivalently, it is assumed there is no degeneracy, other than accidental, which is not fundamentally due to symmetry.

‡ In the groups with complex representations, none of the symmetry operations transforms Ψ_i into Ψ_i^*, hence the one-dimensional representations.

As noted, the functions Ψ_{ik} for all i,k form bases for the irreducible representations of symmetry groups whose elements commute with $\mathfrak{IC}$. Since, in general, the Ψ_{ik} are unknown, it is desirable to have other functions which similarly provide a basis for the irreducible representations. Such functions are usually suggested by the physical nature of the group. For example, consider our triangle in Fig. 8.1. Let the coordinate z be the high-symmetry axis. Then

$$
\begin{aligned}
Ez &= z \\
Az &= -z \\
Bz &= -z \\
Cz &= -z \\
Dz &= z \\
Fz &= z
\end{aligned}
\tag{8.31}
$$

so z provides a basis for the one-dimensional irreducible representation mentioned in Prob. 8.9. Furthermore, z^2 provides a basis for the one-dimensional identity representation. Finally, the axes x and y in the plane of a triangle form a basis for the irreducible representation given in (8.8). With the x and y axes placed as noted on the figure, we can easily derive $Ax = x$, $Ay = -y$, $Bx = -\frac{1}{2}x + (\frac{1}{2}\sqrt{3})y$, $By = (\frac{1}{2}\sqrt{3})x + \frac{1}{2}y$, $Cx = -\frac{1}{2}x - (\frac{1}{2}\sqrt{3})y$, $Cy = -(\frac{1}{2}\sqrt{3})x + \frac{1}{2}y$. We can view these transformations as rotations of the triangle or rotations of the coordinate system. When we come to the rotations D and F, it is important which direction we choose as standard. Conventionally, we rotate the coordinates counterclockwise, which is equivalent to a *clockwise* rotation of the object. The reader should verify (Prob. 8.12) that the matrices of (8.8) are obtained with this convention. For example, note that $Dx = -\frac{1}{2}x + \frac{1}{2}\sqrt{3}\,y$. More generally we can write $\mathbf{r'} = \mathbf{\Gamma}(R)\mathbf{r}$, where $\mathbf{r'}$ and $\mathbf{r}$ are *column vectors*. The vector $\mathbf{r'}(x',y')$ specifies the coordinates x',y' of the new vector in the fixed, unrotated coordinate system.

The definitions above completely define the matrices $\mathbf{\Gamma}(D)$ and $\mathbf{\Gamma}(F)$. However, in the fundamental equation (8.27), the functions Ψ_{ik} are best written as *row vectors* $\mathbf{\Psi}_i = (\Psi_{i1}\Psi_{i2}\cdots\Psi_{im})$. Then $R\mathbf{\Psi}_i = \mathbf{\Psi}_i\mathbf{\Gamma}(R)$, according to (8.27). This definition *conflicts* with what we have just written above, as may be seen by substituting $R = D$, $\mathbf{\Psi}_i = (x,y)$. The conflict may be removed by assuming that we are dealing with two isomorphic groups, one of which (the one we have been discussing) rotates *coordinates* and the other of which rotates *functions* in such a way as to reverse the effect of the coordinate rotation. The matrices of one group are the inverses of the matrices of the other. Rather than introduce a second set of nomenclature for this second group, we will merely write, for example, $Dx = -\frac{1}{2}x + \frac{1}{2}\sqrt{3}\,y$, always implying coordinate rotation. Should we actually need the matrices $\mathbf{\Gamma}(R)$ in the sense of (8.27), we can use the inverses of the coordinate-rotation matrices. However, the reader should be aware that this simple approach contradicts the usual nomenclature in the literature, wherein it is customary to specifically define an operator P_R, such that $P_R\mathbf{\Psi}_i = \mathbf{\Psi}_i\mathbf{\Gamma}(R)$ is satisfied simultaneously with $\mathbf{r}' = \mathbf{\Gamma}(R)\mathbf{r}$. P_R is then a function rotating operator.‡

In Sec. 8.6 we will be dealing with products of basis functions. We can introduce the subject at this point by considering the transformation properties of $xy - yx$. For example, since $Ax = x$, etc., we have $A(xy - yx) = -xy + yx = -(xy - yx)$. Similarly, since $Dy = -\frac{1}{2}\sqrt{3}\, x - \frac{1}{2}y$, $Dxy = \frac{1}{4}\sqrt{3}\, (x^2 - y^2) + \frac{1}{4}xy - \frac{3}{4}yx$, and $Dyx = \frac{1}{4}\sqrt{3}\, (x^2 - y^2) - \frac{3}{4}xy + \frac{1}{4}yx$, so $D(xy - yx) = xy - yx$. Thus D and A (and E) have the same effect on $xy - yx$ as they have on z. Similarly, $xy - yx$ behaves like z under the operators B, C, and F. (Only enough operators sufficient to generate the entire group via consideration of all possible operator products need to be considered.) Therefore we say that $xy - yx$ *transforms* as z in the group S_3; $xy - yx$ is essentially the same basis as z.

PROBLEMS

8.12 Consider the triangle and associated coordinate system of Fig. 8.1. The operation D rotates the triangle clockwise by $2\pi/3$ and rotates the coordinates counterclockwise. Show that this transformation gives rise to the matrix $\mathbf{D}$ of (8.8) so that $\mathbf{r}' = \mathbf{D}\mathbf{r}$ with $\mathbf{r}'$ and $\mathbf{r}$ column vectors. Do the same for the other matrices.

8.13 Show that the matrix representations formed by using degenerate eigenfunctions as a basis are unitary. Show in addition that any group representation is equivalent (through a similarity transformation) to a representation by unitary matrices.

8.5 CHARACTER TABLES

The character of a matrix representation of a given symmetry operation is defined as the trace of the matrix; that is,

$$\chi^{(i)}(R) = \mathrm{Tr}\mathbf{\Gamma}^{(i)}(R) = \sum_{j}^{l_i} [\mathbf{\Gamma}^{(i)}(R)]_{jj} \qquad \blacktriangleright (8.32)$$

The integer l_i is the dimension of the representation, which is labeled by the superscript i. Since a similarity transformation leaves the trace of a matrix unchanged, all equivalent representations have the same character, and in this sense the trace *characterizes* an irreducible representation.

‡ See M. Tinkham, 1964, p. 32.

Furthermore, all members of the same class have the same character in a given representation. Therefore the complete information about the characters of the irreducible representations of a given group may be specified by listing the character for each class and each irreducible representation. Such a listing is called a character table; for S_3, the character table is

S_3	E	$3C_2$	$2C_3$	
$\Gamma^{(1)}$	1	1	1	(8.33)
$\Gamma^{(2)}$	1	-1	1	
$\Gamma^{(3)}$	2	0	-1	

assuming that the three irreducible representations we have found so far are the only ones which exist. The notation $3C_2$ implies a class (A,B,C) of three C_2 operations, and similarly the notation $2C_3$ stands for the operations D and F.

For a *reducible* representation, it is clear that the similarity transformation which brings the representation into block form does not change the character of the representation. Therefore $\chi(R)$ for a reducible representation is given by

$$\chi(R) = \sum_i c_i \chi^{(i)}(R) \qquad \blacktriangleright (8.34)$$

where $\chi^{(i)}(R)$ is the character of R in the ith irreducible representation and c_i is the number of times the irreducible representation $\Gamma^{(i)}$ occurs in the reducible representation Γ. It is in this sense that (8.16) is true.

We still have no way of knowing how many irreducible representations exist for a given group. The answer is that *the number of irreducible representations is equal to the number of classes in the group*. The proof is rather involved, and so we omit it and refer the interested reader to the standard works on group theory. However, if the result is assumed and if we assume the truth of the following proposition, known as the *great orthogonality theorem*, we can derive a number of useful results. We first state without proof (but see Prob. 8.13) that any irreducible representation is equivalent through a similarity transformation to a representation by unitary matrices. Then, considering all the different inequivalent, irreducible unitary representations of a group, the great orthogonality theorem states that

$$\sum_R [\mathbf{\Gamma}^{(i)}(R)]^*_{pq}[\mathbf{\Gamma}^{(j)}(R)]_{ab} = \frac{h}{l_i}\delta_{ij}\delta_{pa}\delta_{qb} \qquad \blacktriangleright (8.35)$$

where R runs over all elements of the group, l_i is the degree of $\Gamma^{(i)}$, and h is the order of the group.

From (8.35) we can immediately derive a number of important rules and relationships. For example, letting $p = q$ and $a = b$, we obtain

$$\sum_R [\mathbf{\Gamma}^{(i)}(R)]^*_{pp}[\mathbf{\Gamma}^{(j)}(R)]_{aa} = \frac{h}{l_i} \delta_{ij}\delta_{pa} \tag{8.36}$$

If we now sum over p, we obtain at once

$$\sum_R \chi^{(i)}(R)^*\chi^{(j)}(R) = h\delta_{ij} \tag{8.37}$$

If h_k is the number of elements in class k and there are n classes (and n irreducible representations), then

$$\sum_{k=1}^{n} h_k\chi^{(i)}(R_k)^*\chi^{(j)}(R_k) = h\delta_{ij} \tag{8.38a}$$

or

$$\sum_{k=1}^{n} \left(\frac{h_k}{h}\right)^{1/2} \chi^{(i)}(R_k)^* \left(\frac{h_k}{h}\right)^{1/2} \chi^{(j)}(R_k) = \delta_{ij} \tag{8.38b}$$

where the sum extends only over one element R_k in each class. The numbers $(h_k/h)^{1/2}\chi^{(i)}(R_k)$ therefore form a unitary matrix, which implies that the relationship

$$\sum_{k=1}^{n} \left(\frac{h_i}{h}\right)^{1/2} \chi^{(k)}(R_i)^* \left(\frac{h_j}{h}\right)^{1/2} \chi^{(k)}(R_j) = \delta_{ij} \tag{8.39}$$

is also true. Therefore, setting $i = j$, we find

$$\sum_{k=1}^{n} [\chi^{(k)}(R_i)]^2 = \frac{h}{h_i} \tag{8.40}$$

or, since $\chi^{(k)}(E) = l_k$,

$$\sum_{k=1}^{n} l_k^2 = h \qquad \blacktriangleright(8.41)$$

Starting with the great orthogonality theorem [Eq. (8.35)], and the assumption that the number of irreducible representations is equal to the number of classes, we have thus obtained a number of useful rules which we list here in order of importance. The first and most important rule is one of our two assumptions; the rest are obtained from Eqs. (8.38) to (8.41):

1. The number of irreducible representations of a group is equal to the number of classes.

2. The sum of the squares of the dimensions of the irreducible representations is equal to the order of the group [Eq. (8.41)].

3. The columns of the character table form orthogonal vectors normalized to h/h_i [Eqs. (8.39) and (8.40)].

4. The rows of the table, when weighted by $h_k^{1/2}$, are orthogonal and normalized to h [Eq. (8.38)].

From these rules we can obtain the character tables of most groups without reference to any specific matrix representations. For example, for S_3, with three classes and six elements, we know at once from rules 1 and 2 that the representations have degrees 1, 1, and 2. Thus the "first draft" of the character table for S_3 is

S_3	E	$3C_2$	$2C_3$
$\Gamma^{(1)}$	1	1	1
$\Gamma^{(2)}$	1	a	b
$\Gamma^{(3)}$	2	c	d

where the first row represents the identical representation and is the same for all groups. Using rule 3, we have the two relations $1 + a + 2c = 0$ and $1 + a^2 + c^2 = 2$. These have the roots $c = 0$, $a = -1$, and $c = -\frac{4}{5}$, $a = \frac{3}{5}$. Using rule 4, we have two additional relations $1 + 3a + 2b = 0$ and $1 + 3a^2 + 2b^2 = 6$, which have the roots $a = -1$, $b = 1$ and $a = \frac{3}{5}$, $b = -\frac{7}{5}$. Similarly, one may show that $d = -1$ or $d = \frac{1}{5}$. Additional rules are necessary to prove that only the integral solutions are satisfactory, but apparently it is usually sufficient to simply discard the nonintegral solutions.

In this way we may ordinarily derive the character table for a given group without explicit knowledge of the irreducible representations. In practice, one looks up the character tables in standard texts on group theory. A full list of such tables is not included in the present text, in the belief that any student going further in the study of orbital theory ought to own at least one book on group theory and its applications to quantum mechanics. Character tables for the most common groups in chemistry are given in Table 8.1.

It is customary in tabulating a character table to list suitable basis functions for the irreducible representations. Most authors list simple functions (x,y,z) in one column and products of simple functions $(x^2, xy,$ etc.) in a separate column. These are obviously not the only possible kinds of basis functions, however, and other authors use spherical harmonics, the special vectors S_x, S_y, and S_z which behave like x, y, and z but which are symmetric with respect to inversion through the origin, and the special vectors R_x, R_y, and R_z which represent rotation axes. Another notation important in character tables concerns the labeling of the irreducible representations. In the system of nomenclature devised by Mulliken, one-dimensional representations are labeled A or B, two-dimensional representations E, and three-dimensional representations T. Generally speaking, A is symmetric with respect to the high-symmetry axis, whereas B is antisymmetric. Subscripts 1 and 2 are added to A and B to distinguish representations which are symmetric or antisymmetric with respect to a C_2 operation perpendicular to the high-symmetry axis or, if none exists, with respect to a plane of symmetry containing the high-symmetry axis. Subscripts 1 and 2 can also be assigned to E and T, but here the rules are more complex. When the group contains a center of symmetry, subscripts g and u are added as well; hence the notation, in the octahedral group, of T_{2g} for one of the three-dimensional irreducible representations.

Applying these rules to our example group S_3, we obtain the final character table

S_3	E	$3C_2$	$2C_3$		
A_1	1	1	1		$z^2,\ x^2 + y^2,\ x^2 + y^2 + z^2$
A_2	1	-1	1	z	
E	2	0	-1	(x,y)	$(yz,\ xz),\ (x^2 - y^2,\ xy)$

$\blacktriangleright$(8.42)

The information given in the table is thus sufficient to completely determine the irreducible representations in their entirety, from consideration of how each basis function behaves under the group operations.

Let us now consider an important yet elementary application of the character table of our example group. Let us suppose there exists a molecule, which we designate A_3, having three identical nuclei located at the corners of an equilateral triangle. We shall try to construct molecular orbitals of $2p$ atomic orbitals on each atom; for the moment we neglect the $1s$ and $2s$ orbitals. We label the atomic orbitals by their center and by their direction, and write, for example, φ_x^1, φ_y^1, and φ_z^1 for the three $2p$ orbitals centered on atom 1. These nine orbitals can then be considered to form a nine-element vector, which we write as

$$\Phi = (\varphi_x^1, \varphi_y^1, \varphi_z^1, \varphi_x^2, \varphi_y^2, \varphi_z^2, \varphi_x^3, \varphi_y^3, \varphi_z^3) \tag{8.43}$$

We use this vector as a basis for a reducible representation. For example, it is clear that $E\Phi = \Phi$, and so the character of E in the basis Φ is 9. Similarly, the C_2 operation through center 1 takes φ_x^1 into itself, φ_z^1 into $-\varphi_z^1$, φ_y^1 into $-\varphi_y^1$, and the orbitals of center 2 into those of center 3, and vice versa. Therefore, the character of a C_2 operation in this basis is -1 since the matrix representing C_2 has all zeros along the diagonal except for one $+1$ and two -1 elements. Finally, any C_3 operation takes all orbitals of a given center into the orbitals of a different center, so the character of a C_3 operation in the basis Φ is zero. Explicitly,

$$\chi_\Phi = (E) = 9 \qquad \chi_\Phi(C_2) = -1 \qquad \chi_\Phi(C_3) = 0 \qquad (8.44)$$

However, from Eqs. (8.34) and (8.42) we know that $\chi_\Phi(E) = c_1 + c_2 + 2c_3$, $\chi_\Phi(C_2) = c_1 - c_2$, and $\chi_\Phi(C_3) = c_1 + c_2 - c_3$, where c_1, c_2, and c_3 are integers. Thus, Φ transforms as $A_1 + 2A_2 + 3E$. Furthermore, it is easily shown that the $2p_z$ orbitals themselves transform as $A_2 + E$, as can be seen by considering the three-element vector composed only of these orbitals. We may best interpret this result by considering an SCF-Roothaan Hamiltonian in which the basis set is restricted to the $2p$ orbitals of the three atoms. All symmetry operations leave this Hamiltonian invariant, and so the molecular orbitals, which are eigenfunctions of this Hamiltonian, form bases for various irreducible representations of S_3. For example, there is but one A_2 molecular orbital which can be formed from the $2p_z$ basis. Therefore, within this basis the normalized linear combination $\varphi_z^1 + \varphi_z^2 + \varphi_z^3$ must be an eigenfunction of the SCF Hamiltonian. (The reader should verify that this combination transforms as A_2.) Similarly, the combinations $\varphi_z^2 - \varphi_z^1$ and $\varphi_z^1 + \varphi_z^2 - 2\varphi_z^3$ transform as x and y, that is, as two basis functions for the irreducible representation E. In this way the molecular orbitals composed of $2p_z$ basis functions are completely determined by symmetry. In Chap. 11 we shall see an even more dramatic case, in which the $\pi(2p_z)$ molecular orbitals of benzene are completely determined by symmetry. Of course, the basis functions must be restricted to the $2p_z$ orbitals of the individual atoms for this to be true.

The p_x and p_y orbitals of the atoms also form bases for irreducible representations of S_3, in this case $A_1 + A_2 + 2E$. There is more than one molecular orbital belonging to each E irreducible representation, so the form of the orbital would appear to be no longer completely determined by symmetry.

However, let us consider a set of orbitals defined by the relations

$$\sigma_1 = -\tfrac{1}{2}\varphi_x^1 - \tfrac{1}{2}\sqrt{3}\,\varphi_y^1 \qquad \sigma_2 = -\tfrac{1}{2}\varphi_x^2 + \tfrac{1}{2}\sqrt{3}\,\varphi_y^2 \qquad \sigma_3 = \varphi_x^3$$

$$(8.45)$$

$$\pi_1 = -\tfrac{1}{2}\sqrt{3}\,\varphi_x^1 + \tfrac{1}{2}\varphi_y^1 \qquad \pi_2 = \tfrac{1}{2}\sqrt{3}\,\varphi_x^2 + \tfrac{1}{2}\varphi_y^2 \qquad \pi_3 = -\varphi_y^3$$

These orbitals are sigma and pi with respect to a point in the center of the triangle. It is clear that all symmetry operations carry the sigma orbitals into themselves, and all pi orbitals into themselves, just as all operations carried the $2p_z$ orbitals into one another. Using Fig. 8.2, it is easily shown that the sigma orbitals transform as $A_1 + E$, whereas the pi orbitals transform as $A_2 + E$. Thus the three different E representations which occur in the reduction of Φ are completely independent, and all molecular orbitals within the $2p$ basis are determined entirely by sym-

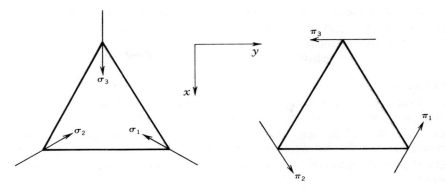

Fig. 8.2 Sigma and pi orbitals in the equilateral triangle.

metry. As is easily shown, the appropriate A_1 orbital is proportional to $\sigma_1 + \sigma_2 + \sigma_3$, whereas the A_2 orbital is proportional to $\pi_1 + \pi_2 + \pi_3$. The two E orbitals (sigma and pi) are only slightly more difficult to find, and we leave this as an exercise (Prob. 8.14).

It may strike the reader as somewhat curious that three different kinds of orbitals, sigma, pi, and $2p_z$, should all transform according to the E representation. One might expect that such different combinations would have rather different symmetry behavior. Actually, we have classified our A_3 molecule under a subgroup (that is, S_3) of a larger and more useful group known as D_{3h}. The latter group includes reflection symmetries, and we shall consider it later. For the present, the classification under S_3 is sufficient because it does allow us to completely determine by symmetry alone all molecular orbitals within the $2p$ basis.

Just as we have used capital letters to represent states (S,P,D,Ψ) and small letters to represent orbitals (s,p,d,ψ), it is customary to use small letters to specify the irreducible representations to which the molecular orbitals belong. Thus, for example, we write ψ_{a_1} for a molecular orbital belonging to the identical representation. If there is more than one such orbital, we may write ψ_{1a_1}, ψ_{2a_1}, and so on, exactly as we did when we classified the molecular orbitals of the homonuclear diatomic molecules.

Since orbitals and states can both belong to irreducible representations, it is clear that we must formulate a means of combining orbitals together to form states. For example, if, in a three-electron system, two electrons occupy the a_1 orbital and one occupies the a_2 orbital, we can determine the symmetry of the state of the system. The way in which we do this is similar to the way in which we determined the possible states of the p^3 configuration (the configuration here being $a_1^2 a_2$). In the course of considering the transformation properties of such products of orbitals, we shall also discuss how state and orbital selection rules can be derived from a knowledge of state and orbital symmetry. Problem 8.15 hints at one of the applications.

PROBLEMS

8.14 Find linear combinations of sigma and pi orbitals [i.e., those of Eq. (8.45)] which transform as x and y components of the E representation of S_3.

8.15 Show that all the symmetry orbitals determined for the A_3 molecule are orthogonal to each other. Can you suggest a reason why?

8.16 Show that the $2s$ orbitals of the A_3 molecule transform as $A_1 + E$, as do the sigma orbitals.

8.6 DIRECT PRODUCTS AND SELECTION RULES

In this section we consider the following closely related problems: building up state functions (and determining the irreducible representations to which they belong) by considering the symmetries of the molecular orbitals, and finding selection rules for matrix elements of the general form $\langle \Psi | O | \Psi' \rangle$ or $\langle \Psi | \Psi' \rangle$. In both problems we consider products of basis function for irreducible representations and ask the following general question: To what representation does the product belong? For example, in the character table (8.42) for S_3, z forms a basis for A_2 and (x,y) form a basis for E. The products z^2, x^2, y^2, xy, xz, and yz belong to a variety of different representations as well, and our general problem is to find how such products transform and to what extent products will form bases for irreducible representations.

Consider two irreducible representations $\Gamma^{(p)}$ and $\Gamma^{(q)}$ (not necessarily different) with bases

$$\alpha_1, \alpha_2, \alpha_3, \ldots, \alpha_n$$
$$\beta_1, \beta_2, \beta_3, \ldots, \beta_m$$

where n and m are the degrees of the two representations. If we form all possible product functions $\alpha_i \beta_j$, these functions form a representation (usually reducible) for our group. The proof follows by acting on the product functions $\alpha_i \beta_j$ by an arbitrary operator R contained in the group. Then

$$R\alpha_i\beta_j = (R\alpha_i)(R\beta_j) = \sum_{k=1}^{n} [\mathbf{\Gamma}^{(p)}(R)]_{ki}\alpha_k \sum_{l=1}^{m} [\mathbf{\Gamma}^{(q)}(R)]_{lj}\beta_l$$

$$= \sum_{k=1}^{m} \sum_{l=1}^{n} \alpha_k\beta_l[\mathbf{\Gamma}(R)]_{k \cdot l, i \cdot j} \tag{8.46}$$

where‡ $\Gamma(R) = \Gamma^{(p)}(R) \times \Gamma^{(q)}(R)$. It is easily shown that

$$\Gamma(R)\Gamma(S) = \Gamma(RS)$$

so the product functions $\alpha_i\beta_j$ indeed form a basis for a representation. The character of this representation is

$$\chi(R) = \sum_{k=1}^{n} \sum_{l=1}^{m} [\Gamma(R)]_{k \cdot l, k \cdot l} = \sum_{k=1}^{n} [\Gamma^{(p)}(R)]_{kk} \sum_{l=1}^{m} [\Gamma^{(q)}(R)]_{ll}$$

$$= \chi^{(p)}(R)\chi^{(q)}(R) \qquad (8.47)$$

and so the representation is readily reduced as follows: $\Gamma = \sum_i c_i \Gamma^{(i)}$.

If we take S_3 once again as our example, it is easily shown (Prob. 8.17) that

$$
\begin{aligned}
A_1 \times A_1 &= A_1 \\
A_1 \times A_2 &= A_2 \\
A_1 \times E &= E \\
A_2 \times A_2 &= A_1 \\
A_2 \times E &= E \\
E \times E &= A_1 + A_2 + E
\end{aligned}
\qquad (8.48)
$$

From this information we can construct the following table of direct product reductions for S_3:

S_3	A_1	A_2	E	
A_1	A_1	A_2	E	(8.49)
A_2	A_2	A_1	E	
E	E	E	$A_1 + A_2 + E$	

Furthermore, we can show that the product xz transforms as the y part of the E representation, whereas yz transforms as $-x$ (Prob. 8.18). In fact, it is possible to define coupling coefficients just as we did for the addition of angular momentum, and we may write

$$|\Gamma^{(p)}\Gamma^{(q)}\Gamma\gamma_k\rangle = \sum_{i,j} |\Gamma^{(p)}\Gamma^{(q)}\alpha_i\beta_j\rangle\langle\Gamma^{(p)}\Gamma^{(q)}\alpha_i\beta_j|\Gamma^{(p)}\Gamma^{(q)}\Gamma\gamma_k\rangle \qquad \blacktriangleright(8.50)$$

which is totally analogous to the formula

$$|l_1 l_2 L M\rangle = \sum_{m_{l_1} + m_{l_2} = M} |l_1 l_2 m_{l_1} m_{l_2}\rangle\langle l_1 l_2 m_{l_1} m_{l_2}|l_1 l_2 L M\rangle \qquad (8.51)$$

given in Chap. 3. The coupling coefficients $\langle\Gamma^{(p)}\Gamma^{(q)}\alpha_i\beta_j|\Gamma^{(p)}\Gamma^{(q)}\Gamma\gamma_k\rangle$ are just numbers, with α_i, β_j, and γ_k defining certain basis functions for the

‡ $\Gamma(R)$ is known as the direct product matrix of $\Gamma^{(p)}(R)$ and $\Gamma^{(q)}(R)$. It has dimension $m \cdot n$ with one element for each possible product of elements of the two component matrices.

three irreducible representations. For S_3, we have the rather simple set of coefficients

$$\langle A_1 A_1 | A_1 A_1 A_1 \rangle = 1$$
$$\langle A_2 A_2 | A_2 A_2 A_1 \rangle = 1$$
$$\langle A_1 A_2 | A_1 A_2 A_2 \rangle = 1$$
$$\langle A_1 E x | A_1 E E x \rangle = \langle A_1 E y | A_1 E E y \rangle = 1$$
$$\langle A_2 E x | A_2 E E y \rangle = -\langle A_2 E y | A_2 E E x \rangle = 1 \qquad (8.52)$$
$$\langle E E x x | E E A_1 \rangle = \langle E E y y | E E A_1 \rangle = \sqrt{\tfrac{1}{2}}$$
$$\langle E E x x | E E E x \rangle = -\langle E E y y | E E E x \rangle = \sqrt{\tfrac{1}{2}}$$
$$\langle E E x y | E E E y \rangle = \langle E E y x | E E E y \rangle = \sqrt{\tfrac{1}{2}}$$
$$\langle E E x y | E E A_2 \rangle = -\langle E E y x | E E A_2 \rangle = \sqrt{\tfrac{1}{2}}$$

All other coefficients are zero. Note that the last two coefficients imply that the combination $\sqrt{\tfrac{1}{2}}\,(xy - yx)$ forms a basis for A_2. [The proper combination transforming as y is not actually xy, but $\sqrt{\tfrac{1}{2}}\,(xy - yx)$]. The reader should carefully verify all coefficients, in order to make sure he understands the principle involved. There are, of course, arbitrary phase factors as part of the definitions, but unfortunately there are no general rules for choosing them.

The analogy between a function "forming a basis for an irreducible representation" and "belonging to a given atomic term" is thus very close. This is because the functions belonging to a given term in fact form a basis for an irreducible representation of the full rotation group, and it is possible to develop the theory of atomic electronic structure from this point of view. The S, P, and D terms thus correspond to irreducible representations of degree 1, 3, and 5.

Using the vector coupling coefficients and the direct product table for the group, we can determine the symmetry of a configurational state function composed of molecular orbitals. We begin by considering the coupling of 2 one-electron functions. If the first function ψ_1 belongs to $\Gamma^{(p)}\alpha_i$ and the second to $\Gamma^{(q)}\beta_j$, then the antisymmetrized product

$$\Psi = \frac{1}{\sqrt{2}} [\psi_1(1)\psi_2(2) - \psi_2(1)\psi_1(2)] \qquad (8.53)$$

does *not* necessarily form a basis for a direct product representation. However, by using the coupling coefficients, it is possible to determine the symmetries of all possible two-electron states. For example, in S_3, the configurations a_1^2 and a_2^2 (both of which are closed shells) give rise to functions of the form

$$\Psi = \frac{1}{\sqrt{2}} [\psi_{a_1}(1)\psi_{a_1}(2)](\alpha_1\beta_2 - \beta_1\alpha_2) \qquad (8.54)$$

so the resultant state is a singlet A_1, which we write as $^1A_1$. The configurations a_1e and a_2e give rise to both singlet and triplet E states, and we have, for example,

$$\Psi(^3E, y, M_S = 1) = \frac{1}{\sqrt{2}}[\psi_{a_2}(1)\psi_{e_x}(2) - \psi_{e_x}(1)\psi_{a_2}(2)]\alpha_1\alpha_2$$

$$\Psi(^1E, y) = \frac{1}{2}[\psi_{a_2}(1)\psi_{e_x}(2) + \psi_{e_x}(1)\psi_{a_2}(2)](\alpha_1\beta_2 - \beta_1\alpha_2)$$

(8.55)

Finally, the e^2 configuration is somewhat more difficult, since the Pauli principle must be carefully applied. There is but one triplet state, which is clearly

$$\Psi(^3A_2, M_S = 1) = \frac{1}{\sqrt{2}}[\psi_{e_x}(1)\psi_{e_y}(2) - \psi_{e_y}(1)\psi_{e_x}(2)]\alpha_1\alpha_2$$

(8.56)

The other two possibilities are a $^1A_1$ and a ^1E as follows:

$$\begin{aligned}
\Psi(^1A_1) &= \frac{1}{2}[\psi_{e_x}(1)\psi_{e_x}(2) + \psi_{e_y}(1)\psi_{e_y}(2)](\alpha_1\beta_2 - \beta_1\alpha_2) \\
\Psi(^1E, x) &= \frac{1}{2}[\psi_{e_x}(1)\psi_{e_x}(2) - \psi_{e_y}(1)\psi_{e_y}(2)](\alpha_1\beta_2 - \beta_1\alpha_2) \\
\Psi(^1E, y) &= \frac{1}{2}[\psi_{e_x}(1)\psi_{e_y}(2) + \psi_{e_y}(1)\psi_{e_x}(2)](\alpha_1\beta_2 - \beta_1\alpha_2)
\end{aligned}$$

(8.57)

In this way the Pauli principle limits the possible coupling of orbitals, and we must use implied terms tables as in Chap. 6 to determine which states are possible.‡

Functions for n-electron states may be built up by adding one electron at a time to the two-electron functions. There are certain obvious simplifications for a filled shell; only one Slater determinant is possible, in which all basis functions belonging to a given irreducible representation occur twice. We may write§

$$\Psi(\text{closed shell}) = (\alpha_1^2\alpha_2^2 \cdots \alpha_m^2)$$

(8.58)

and it is not difficult to show (Prob. 8.20) that the spatial part $\alpha_1^2\alpha_2^2 \cdots \alpha_m^2$ belongs to the identical representation. Thus the only state for the configuration e^4 is $^1A_1$. For a single electron outside a closed shell, the state has the symmetry of the orbital in which the odd electron is placed.

The techniques for coupling one-electron functions together in this way are really only interesting and useful when the group contains one or more two- (or three-) dimensional irreducible representations. Actually, very little work has been done along these lines except in the area of ligand field theory, where the group of interest is usually octahedral. Most of the ordinary applications of group theory to quantum chemistry concern the prediction of selection rules, and we now turn to this second use.

‡ We assume that ψ_{e_x} and ψ_{e_y} are two *degenerate* molecular orbitals. If they belong to two different energy levels, we can form the $^1A_2$, $^3A_1$, and ^3E states as well.
§ The reader should be careful to distinguish the two following (conventional) uses of the symbol α: as a basis label [Eq. (8.58)] and as a spin function [(8.54) to (8.57)].

Consider the matrix element $\langle\psi_1|\psi_2\rangle = \langle\psi_1\Gamma^{(p)}\alpha_i|\psi_2\Gamma^{(q)}\beta_j\rangle$ where ψ_1 forms a basis α_i for $\Gamma^{(p)}$ and ψ_2 forms a basis β_j for $\Gamma^{(q)}$. The matrix element is just a number and so is unchanged by any group operation applied to it; we have

$$R\langle\psi_1\Gamma^{(p)}\alpha_i|\psi_2\Gamma^{(q)}\beta_j\rangle = \langle\psi_1\Gamma^{(p)}\alpha_i|\psi_2\Gamma^{(q)}\beta_j\rangle \tag{8.59}$$

However, the operator R applied to ψ_1 transforms it into a linear combination of other functions belonging to $\Gamma^{(p)}$ and forming bases $\alpha_1, \alpha_2, \ldots, \alpha_m$. A similar statement is true for the effect of R on ψ_2. Let us abbreviate these transformed functions simply as $\alpha_1, \alpha_2, \ldots, \alpha_m$, $\beta_1, \beta_2, \ldots, \beta_n$. Then in this notation

$$R\langle\Gamma^{(p)}\alpha_i|\Gamma^{(q)}\beta_j\rangle = \langle\Gamma^{(p)}\alpha_i|\Gamma^{(q)}\beta_j\rangle \qquad \blacktriangleright (8.60)$$

We now expand the left-hand side of (8.60), and sum over R so as to be able to use (8.35). The result is

$$h\langle\Gamma^{(p)}\alpha_i|\Gamma^{(q)}\beta_j\rangle = \sum_{k=1}^{m}\sum_{l=1}^{n}\sum_{R}[\mathbf{\Gamma}^{(p)}(R)]_{ki}^{*}[\Gamma^{(q)}(R)]_{lj}\langle\Gamma^{(p)}\alpha_k|\Gamma^{(q)}\beta_l\rangle$$

$$= \frac{h}{m}\,\delta_{ij}\delta_{pq}\sum_{k=1}^{m}\sum_{l=1}^{n}\delta_{kl}\langle\Gamma^{(p)}\alpha_k|\Gamma^{(q)}\beta_l\rangle \tag{8.61}$$

We therefore see at once that the integral $\langle\psi_1\Gamma^{(p)}\alpha_i|\psi_2\Gamma^{(q)}\beta_j\rangle$ is zero unless $\Gamma^{(p)} = \Gamma^{(q)}$ *and* $\alpha_i = \beta_j$. This equation is of the highest importance and we state it separately as the following fundamental rule:

$$\langle\psi_1\Gamma^{(p)}\alpha_i|\psi_2\Gamma^{(q)}\beta_j\rangle = \delta_{pq}\delta_{ij} \qquad \blacktriangleright (8.62)$$

The reader should compare this result with Prob. 8.15. We can simply state the rule by saying that all functions belonging to different irreducible representations or different components of the same irreducible representation are orthogonal.

A corollary to this rule which is of equal practical importance concerns the matrix element $\langle\psi_1\Gamma^{(p)}\alpha_i|\mathcal{3C}|\psi_2\Gamma^{(q)}\beta_j\rangle$. Since $R\mathcal{3C} = \mathcal{3C}$, it is clear that a similar derivation will result in the rule

$$\langle\psi_1\Gamma^{(p)}\alpha_i|\mathcal{3C}|\psi_2\Gamma^{(q)}\beta_j\rangle = 0 \quad \text{unless } \Gamma^{(p)} = \Gamma^{(q)} \text{ and } \alpha_i = \beta_j \quad \blacktriangleright (8.63)$$

Furthermore, when $\Gamma^{(p)} = \Gamma^{(q)}$ and $\alpha_i = \beta_j$, we find that

$$\langle\Gamma^{(p)}\alpha_i|\mathcal{3C}|\Gamma^{(p)}\alpha_i\rangle = H \tag{8.64}$$

independent of the value of α_i. [The derivation follows by using an equation corresponding to (8.61).] This is the exact analog of the fact that the matrix element $\langle^{2S+1}LM|\mathcal{3C}|^{2S+1}LM\rangle$ is independent of the value of M. It shows that a collection of functions belonging to a given irreducible representation is indeed degenerate.

A third selection rule applies to operators other than $\mathcal{K}$. Suppose a given operator O is not left invariant by all operations of the group, but transforms as a given irreducible representation Γ. (The Hamiltonian $\mathcal{K}$ transforms according to the identical representation. If we write this representation as A_1, we always have $A_1 \times \Gamma^{(i)} = \Gamma^{(i)}$.) Then $O|\psi_2\Gamma^{(q)}\beta_j\rangle$ transforms as $\Gamma \times \Gamma^{(q)}$ which may be reducible. In this way we deduce the rule

$$\langle\psi_1\Gamma^{(p)}\alpha_i|O(\Gamma)|\psi_2\Gamma^{(q)}\beta_j\rangle = 0 \qquad \text{unless } \Gamma^{(q)} \text{ is contained in}$$
$$\text{the product } \Gamma \times \Gamma^{(p)} \quad \blacktriangleright (8.65)$$

We can illustrate this last rule with reference to S_3. Consider the dipole moment operator $\mathbf{\mu} = \mathbf{r}e = e(x \; y \; z)$. We can deduce selection rules for transitions between states by considering matrix elements of the form $\langle\psi_1|\mathbf{r}|\psi_2\rangle$. When $\mathbf{r} = z$, the dipole operator transforms as A_2, and we have the rule that transitions $A_1 \to A_2$, $A_2 \to A_1$, and $E \to E$ are allowed, but transitions $A_1 \to A_1$, $A_2 \to A_2$, $E \to A_1$, $E \to A_2$, $A_1 \to E$, and $A_2 \to E$ are forbidden. When $\mathbf{\mu} \propto (x \; y)$, the dipole operator transforms as E, and only transitions $A_1 \to E$, $E \to A_1$, $A_2 \to E$, $E \to A_2$, and $E \to E$ are allowed. We therefore say that the transitions $A_1 \to A_2$, and $A_2 \to A_1$ are z-polarized, whereas the transitions $A_1 \to E$, $E \to A_1$, $A_2 \to E$, and $E \to A_2$ are (x,y)-polarized. The transition $E \to E$ has no fixed sense of polarization, and all other transitions are forbidden. The usefulness of these selection rules depends on whether we can find a molecular system belonging to S_3; as we shall show in the next section, a molecule of the type A_3 previously mentioned belongs to a larger group containing reflections.

All three selection rules may be summarized by noting that if

$$\Gamma^{(p)} \times \Gamma^{(q)} = \Gamma = \sum_i c_i\Gamma^{(i)} \qquad \blacktriangleright (8.66)$$

then the reducible Γ contains the identical representation (say, A_1) if and only if $\Gamma^{(p)} = \Gamma^{(q)}$. Furthermore, $c_{A_1} = 1$. The proof follows from (8.37), the orthogonality theorem for characters. We have first

$$\chi(R) = \chi^{(p)}(R)\chi^{(q)}(R) = \sum_i c_i\chi^{(i)}(R) \qquad (8.67)$$

so

$$\sum_R \chi(R)\chi^{(j)}(R) = \sum_i \sum_R c_i\chi^{(i)}(R)\chi^{(j)}(R)$$
$$= \sum_R c_j[\chi^{(j)}(R)]^2$$
$$= hc_j \qquad (8.68)$$

Thus

$$c_j = \frac{1}{h}\sum_R \chi(R)\chi^{(j)}(R) \qquad (8.69)$$

From this equation we obtain

$$c_{A_1} = \frac{1}{h} \sum_R \chi(R) \tag{8.70}$$

since $\chi^{A_1} = 1$ for all R for the identical representation. Therefore

$$c_{A_1} = \frac{1}{h} \sum_R \chi^{(p)}(R)\chi^{(q)}(R) = 1 \tag{8.71}$$

which completes the proof. Thus we have formulated the general selection rule

$$\langle \psi_1 \Gamma^{(p)}\alpha_i | \mathbf{O}(\Gamma) | \psi_2 \Gamma^{(q)}\beta_j \rangle = 0 \qquad \text{unless } \Gamma^{(p)} \times \Gamma^{(q)} \times \Gamma \text{ contains the identical representation } \blacktriangleright (8.72)$$

When $\mathbf{O} = 1$ or $\mathcal{3C}$, $\Gamma =$ the identical representation, and $\Gamma^{(p)} \times \Gamma^{(q)}$ must contain the identical representation for the matrix element to be nonzero.

PROBLEMS

8.17 Verify the direct product reduction table for S_3, Eq. (8.48).

8.18 By explicitly considering the effect of several group operations on x, y, and z, show that xz transforms as y while yz transforms as $-x$.

8.19 Verify the coupling coefficients of (8.52).

8.20 Consider the m basis functions $\alpha_1, \alpha_2, \ldots, \alpha_m$ for an irreducible representation. Show that the product function $\alpha_1^2 \alpha_2^2 \cdots \alpha_m^2$ forms a basis for the total symmetric representation.

8.7 COMMON SPATIAL SYMMETRY GROUPS

There are an infinite number of point groups, that is, groups which leave a point in space unmoved. In solid-state physics, one is interested in those point groups which are consistent with the translational symmetry of the crystal (there are only 32 of these), but for the study of single molecules, any point group may be applicable.

We first consider the symmetries of regular polygons with two different sides (so that C_2 rotations perpendicular to the high-symmetry axis are *not* included in the group). There is an infinite sequence of such groups, and we label them C_n. For example, C_3 is the rotation group of an equilateral triangle having inequivalent sides. (The reader should not confuse the *group* C_3 with the *operator* C_3.) The C_n groups for $n > 2$ have complex characters. A second infinite sequence of groups is the series of rotations which leave a regular polygon with *equivalent* sides invariant. These groups are labeled D_3. Each group includes n different C_2 rotations perpendicular to the high-symmetry axis. The group D_3 is isomorphic with S_3, and it is conventional to use the label D_3 instead of S_3 for the triangle.

The groups C_n and D_n are called rotation groups since these are the only kinds of operations included in the set. It turns out that there are only three other distinct rotation groups, labeled T, O, and K (or I). These are the rotation groups of the tetrahedron, octahedron (or cube) and icosahedron (or dodecahedron). The icosahedron is the most complicated, and we omit it; it is a regular polygon whose 20 faces are triangles. The icosahedral group is essentially identical with the group of the dodecahedron, whose 12 faces are pentagons. The octahedral group is essentially identical with the group of a cube, and an octahedron is conveniently drawn (Fig. 8.3) by placing each vertex in the center of a face of a cube. Character tables for T and O are given in Table 8.1. The C_3 operations of O are rotations about the center of a triangular face. The six C_4 operations are rotations about the vertices, as are the three C_2 operations. The remaining class of six C_2 operations are rotations about the midpoint of a line joining two adjacent vertices.

The tetrahedron is most easily drawn by reference to the cube (Fig. 8.4). The eight C_3 rotations about the centers of faces break up into two distinct classes (hence the group has complex characters). There are three C_2 axes, R_x, R_y, and R_z. The group is simple enough that the reader will find it a useful exercise (Prob. 8.21) to write down all the operations (best done by considering how they permute the tetrahedral vertices) and work out the group multiplication table. The same should be done for the octahedron if the reader plans to study ligand field theory in Chap. 9.

Having exhausted the rotation groups, we next consider other symmetry operations. These include reflections in planes (which contain the origin) and the inversion, which takes $\mathbf{r}$ into $-\mathbf{r}$. A typical reflection is the xy mirror plane which takes x into x, y into y, and z into $-z$. The

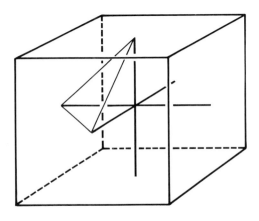

Fig. 8.3 The six corners of the octahedron as inscribed in a cube. One of the eight triangular faces is shown.

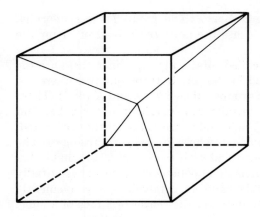

Fig. 8.4 The four corners of the tetrahedron as inscribed in a cube.

reflections are classified by the kinds of planes in which they lie. The notation σ_h is used for a reflection perpendicular to the *h*igh-symmetry axis. Similarly, σ_v represents a *v*ertical reflection in which the mirror plane *contains* the high-symmetry axis. Finally, σ_d is a special kind of σ_v plane which bisects the angle between two C_2 axes perpendicular to the high-symmetry axis. Another kind of operation, closely related to the reflection planes, is the improper rotation S_n. This may be thought of as an ordinary rotation C_n followed by a reflection σ_h in the plane perpendicular to the C_n axis. The reader should verify that the operation S_2 is equivalent to the inversion operator i.

We can now form a new set of groups from C_n and D_n by adding mirror planes. For *even* n, the D_n groups contain a center of symmetry. We may therefore create a new group by considering all elements R contained in D_n together with all elements $i \times R$ formed by combining the rotations with the inversion. The new group, written D_{nh}, is said to be a *direct product* of the groups S_2 (elements E and $i = S_2$) and D_n. (Again, do not confuse the operation S_2 with the group S_2.) We write this

$$D_{nh} = D_n \times S_2$$

The relationship is closely analogous to the relationship for the direct product of irreducible representations, and D_{nh} (for *even* n) contains twice as many elements as D_n. The notation D_{nh} arises from the fact that this group contains σ_h. We also write D_{nh} for odd n, but here there is no center of symmetry. However, for odd n, D_{nh} is the direct product of D_n and C_{1h}, where C_{1h} is the two-element group consisting of E and σ_h. It also has twice as many elements as D_n. Note that D_{nh} also always contains incidental σ_v reflections.

Another group which can be formed from D_n is D_{nd} (sometimes written D_{nv}) which contains σ_d. D_{nd} is the group of the regular n antiprism,

Table 8.1 Character tables for the most common groups in chemistry

C_{2v}	E	C_2	$\sigma_v(xz)$	$\sigma_v'(yz)$		
A_1	1	1	1	1	z	x^2, y^2, z^2
A_2	1	1	-1	-1	R_z	xy
B_1	1	-1	1	-1	x, R_y	xz
B_2	1	-1	-1	1	y, R_x	yz

C_{2h}	E	C_2	σ_h	i		
A_g	1	1	1	1	R_z	x^2, y^2, z^2, xy
A_u	1	1	-1	-1	z	
B_g	1	-1	-1	1	R_x, R_y	xz, yz
B_u	1	-1	1	-1	x, y	

D_{3h}	E	$2C_3$	$3C_2$	σ_h	$2S_3$	$3\sigma_v$		
A_1'	1	1	1	1	1	1		x^2+y^2, z^2
A_2'	1	1	-1	1	1	-1	R_z	
E'	2	-1	0	2	-1	0	(x,y)	(x^2-y^2, xy)
A_1''	1	1	1	-1	-1	-1		
A_2''	1	1	-1	-1	-1	1	z	
E''	2	-1	0	-2	1	0	(R_x, R_y)	(xz, yz)

D_{4h}	E	$2C_4$	C_2	$2C_2'$	$2C_2''$	i	$2S_4$	σ_h	$2\sigma_v$	$2\sigma_d$		
A_{1g}	1	1	1	1	1	1	1	1	1	1		x^2+y^2, z^2
A_{2g}	1	1	1	-1	-1	1	1	1	-1	-1	R_z	
B_{1g}	1	-1	1	1	-1	1	-1	1	1	-1		x^2-y^2
B_{2g}	1	-1	1	-1	1	1	-1	1	-1	1		xy
E_g	2	0	-2	0	0	2	0	-2	0	0	(R_x, R_y)	(xz, yz)
A_{1u}	1	1	1	1	1	-1	-1	-1	-1	-1		
A_{2u}	1	1	1	-1	-1	-1	-1	-1	1	1	z	
B_{1u}	1	-1	1	1	-1	-1	1	-1	-1	1		
B_{2u}	1	-1	1	-1	1	-1	1	-1	1	-1		
E_u	2	0	-2	0	0	-2	0	2	0	0	(x,y)	

D_{6h}	E	$2C_6$	$2C_3$	C_2	$3C_2'$	$3C_2''$	i	$2S_3$	$2S_6$	σ_h	$3\sigma_d$	$3\sigma_d$		
A_{1g}	1	1	1	1	1	1	1	1	1	1	1	1		x^2+y^2, z^2
A_{2g}	1	1	1	1	-1	-1	1	1	1	1	-1	-1	R_z	
B_{1g}	1	-1	1	-1	1	-1	1	-1	1	-1	1	-1		
B_{2g}	1	-1	1	-1	-1	1	1	-1	1	-1	-1	1		
E_{1g}	2	1	-1	-2	0	0	2	1	-1	-2	0	0	(R_x, R_y)	(xz, yz)
E_{2g}	2	-1	-1	2	0	0	2	-1	-1	2	0	0		(x^2-y^2, xy)
A_{1u}	1	1	1	1	1	1	-1	-1	-1	-1	-1	-1		
A_{2u}	1	1	1	1	-1	-1	-1	-1	-1	-1	1	1	z	
B_{1u}	1	-1	1	-1	1	-1	-1	1	-1	1	-1	1		
B_{2u}	1	-1	1	-1	-1	1	-1	1	-1	1	1	-1		
E_{1u}	2	1	-1	-2	0	0	-2	-1	1	2	0	0	(x,y)	
E_{2u}	2	-1	-1	2	0	0	-2	1	1	-2	0	0		

O	E	$8C_3$	$3C_2$	$6C_2'$	$6C_4$		
A_1	1	1	1	1	1		
A_2	1	1	1	-1	-1		
E	2	-1	2	0	0		$(x^2-y^2, 3z^2-r^2)$
T_1	3	0	-1	-1	1	(x,y,z) (R_x, R_y, R_z)	
T_2	3	0	-1	1	-1		(xy, yz, zx)

Table 8.1 Character tables for the most common groups in chemistry (*Continued*)

O_h	E	$8C_3$	$6C_2$	$6C_4$	$3C_2$	i	$6S_4$	$8S_6$	$3\sigma_h$	$6\sigma_d$		
A_{1g}	1	1	1	1	1	1	1	1	1	1		$x^2 + y^2 + z^2$
A_{2g}	1	1	-1	-1	1	1	-1	1	1	-1		
E_g	2	-1	0	0	2	2	0	-1	2	0		$(x^2 - y^2, 3z^2 - r^2)$
T_{1g}	3	0	-1	1	-1	3	1	0	-1	-1	(R_x, R_y, R_z)	
T_{2g}	3	0	1	-1	-1	3	-1	0	-1	1		(xy, xz, yz)
A_{1u}	1	1	1	1	1	-1	-1	-1	-1	-1		
A_{2u}	1	1	-1	-1	1	-1	1	-1	-1	1		
E_u	2	-1	0	0	2	-2	0	1	-2	0		
T_{1u}	3	0	-1	1	-1	-3	-1	0	1	1	(x, y, z)	
T_{2u}	3	0	1	-1	-1	-3	1	0	1	-1		

T	E	$3C_2$	$4C_3$	$4C_3'$		
A	1	1	1	1		
E	1	1	ϵ	ϵ^2		$\epsilon = e^{2\pi i/3}$
	1	1	ϵ^2	ϵ		
T	3	-1	0	0	(x, y, z)	
					(R_x, R_y, R_z)	

T_d	E	$8C_3$	$3C_2$	$6S_4$	$6\sigma_d$		
A_1	1	1	1	1	1		$x^2 + y^2 + z^2$
A_2	1	1	1	-1	-1		
E	2	-1	2	0	0		$(x^2 - y^2, 3z^2 - r^2)$
T_1	3	0	-1	1	-1	(R_x, R_y, R_z)	
T_2	3	0	-1	-1	1	(x, y, z)	(xy, xz, yz)

$C_{\infty v}$	E	$2C_\infty^\varphi$	$\cdots$	$\infty \sigma_v$		
Σ^+	1	1	$\cdots$	1	z	$x^2 + y^2, z^2$
Σ^-	1	1	$\cdots$	-1	R_z	
Π	2	$2\cos\varphi$	$\cdots$	0	(x, y)	(xz, yz)
					(R_x, R_y)	
Δ	2	$2\cos 2\varphi$	$\cdots$	0		$(x^2 - y^2, xy)$
Φ	2	$2\cos 3\varphi$	$\cdots$	0		
$\cdots$						

$D_{\infty h}$	E	$2C_\infty^\varphi$	$\cdots$	$\infty \sigma_v$	i	$2S_\infty^\varphi$	$\cdots$	∞C_2		
Σ_g^+	1	1	$\cdots$	1	1	1	$\cdots$	1		$x^2 + y^2, z^2$
Σ_g^-	1	1	$\cdots$	-1	1	1	$\cdots$	-1	R_z	
Π_g	2	$2\cos\varphi$	$\cdots$	0	2	$-2\cos\varphi$	$\cdots$	0	(R_x, R_y)	(xz, yz)
Δ_g	2	$2\cos 2\varphi$	$\cdots$	0	2	$2\cos 2\varphi$	$\cdots$	0		$(x^2 - y^2, xy)$
$\cdots$										
Σ_u^+	1	1	$\cdots$	1	-1	-1	$\cdots$	1	z	
Σ_u^-	1	1	$\cdots$	-1	-1	-1	$\cdots$	1		
Π_u	2	$2\cos\varphi$	$\cdots$	0	-2	$2\cos\varphi$	$\cdots$	0	(x, y)	
Δ_u	2	$2\cos 2\varphi$	$\cdots$	0	-2	$-2\cos 2\varphi$	$\cdots$	0		
$\cdots$										

just as D_{nh} is the group of the n prism. Thus D_{nd} contains the inversion for *odd* n, and in fact $D_{nd} = D_n \times S_2$ in this case.

Two sets of groups can be formed from C_n. The group C_{nv} contains n σ_v planes and is isomorphic to D_n for all n. The group C_{nh} contains σ_h, and it is clear that for even n, $C_{nh} = C_n \times S_2$. In fact, for odd n, $C_{nh} = C_n \times C_{1h}$. Another set of groups are the improper rotation groups written S_n or S_{2n}, consisting of the improper rotation *operator* S_n or S_{2n} and its multiples S_n^2, S_n^3, etc. For n *odd* these groups are identical with C_{nh} and so are not usually considered separately. Hence the notation S_{2n}. One should not confuse the improper rotation groups S_{2n} with the permutation group S_n, although they share the same symbolic nomenclature.

Four new groups can be formed by combining T, O, and K with reflection and inversion symmetry. The icosahedron has a center of symmetry, and we can form the direct product group K_i (or I_h) $= K \times S_2$. Similarly, the octahedron has a center of symmetry, and we have the full octahedral group $O_h = O \times S_2$. It has 48 elements. From the rotation group of the tetrahedron we can form both T_d and T_h. T_d consists of all possible rotations and reflections of the tetrahedron. It has 24 elements and is isomorphic to O. In addition, T_d is isomorphic to the permutation group S_4, as can be seen from the fact that a combination of rotations and reflections will make possible all permutations of the four tetrahedral vertices. T_h is a rather curious group which is the direct product of T and S_2. Griffith (1961, p. 144) has given this group an interesting discussion.

The groups C_n, D_n, C_{nh}, C_{nv}, D_{nh}, D_{nd}, S_{2n}, T, O, K, T_d, T_h, O_h, and K_i exhaust all the possible *finite* point symmetry groups. In addition to the full rotation group, there are two *infinite* groups which contain a C_∞ axis. $C_{\infty v}$, the group of the general linear molecule, has axial symmetry plus reflection symmetry in *any* vertical plane containing the high-symmetry axis. $D_{\infty h}$ also has a reflection plane perpendicular to the high-symmetry axis. Although the groups are infinite, one can work out character tables for them with little difficulty. These are given in Table 8.1. The meaning of the superscript plus and minus symbols on the Σ states introduced in the previous chapter should now be clear.

Most texts on group theory contain character tables for many of the groups mentioned above. Rather than duplicate this work, we give tables only for C_{2v}, C_{2h}, D_{3h}, D_{4h}, D_{6h}, O, O_h, T, T_d, $C_{\infty v}$, and $D_{\infty h}$. These are by far the most common groups in chemistry and will suffice for the purposes of this text.

Using the character table for $D_{\infty h}$, we can repeat in a somewhat more elegant fashion the discussion in Sec. 7.3 on the states for the configuration π_g^2 (or π_u^2) of the homonuclear diatomic molecules. From Table 8.1 we see that $\pi \times \pi = \Delta_g + \Sigma_g^+ + \Sigma_g^-$, since $2 \cos^2 \varphi = \cos 2\varphi + 1$. The basis functions for Δ_g are spatially symmetric ($x^2 - y^2$ and $xy + yx$) so the spin part must be the antisymmetric singlet function. The state is therefore $^1\Delta_g$. Similarly, the basis for Σ_g^+ is $x^2 + y^2$, and we have

$$\Psi(^1\Sigma_g^+) = \tfrac{1}{2}[\psi_{\pi x}(1)\psi_{\pi x}(2) + \psi_{\pi y}(1)\psi_{\pi y}(2)](\alpha_1\beta_2 - \beta_1\alpha_2) \qquad (8.73)$$

Finally, $xy - yx$ forms an antisymmetric basis for the Σ_g^- representation and we obtain

$$\Psi(^3\Sigma_g^-) = \tfrac{1}{2}[\psi_{\pi x}(1)\psi_{\pi y}(2) - \psi_{\pi y}(1)\psi_{\pi x}(2)]\alpha_1\alpha_2 \qquad (8.74)$$

for the $M_S = 1$ component. The reader should compare this result with the discussion of Sec. 7.3.

PROBLEMS

8.21 Work out the group multiplication table for the tetrahedron. Determine the classes and the character table.

8.22 Verify that CH_4 belongs to T_h, FeF_6^{3-} to O_h, CO_3^{2-} (planar) to D_{3h}, NH_3 to C_{3v}, H_2O to C_{2v}, benzene to D_{6h}, LiH to $C_{\infty v}$, and H_2 to $D_{\infty h}$. The largest group to which the A_3 molecule of our example can belong is D_{3h}.

8.23 Obtain the correct selection rules for the A_3 molecule by constructing the direct product reduction table for D_{3h}.

SUGGESTIONS FOR FURTHER READING

Presently there are a rather large number of books on group theory and its applications to quantum mechanics and quantum chemistry. I mention only two favorites, which I like for essentially opposite reasons.

1. Tinkham, Michael: "Group Theory and Quantum Mechanics," McGraw-Hill Book Company, New York, 1964.

   This book is precise, terse, and extraordinarily lucid. Chapters 1 to 4 plus Chap. 7 contain about all a chemist need know, and the first three chapters provide an excellent introduction. For a more leisurely and less mathematical development of some of the same material, see

2. Cotton, F. A.: "Chemical Applications of Group Theory," Interscience Publishers, New York, 1963.

   Cotton is often deliberately verbose, which may be what the beginning student needs.

3. Griffith, J. S.: "The Irreducible Tensor Method For Molecular Symmetry Groups," Prentice-Hall, Inc., Englewood Cliffs, N.J., 1962.

   This book contains the most advanced applications of the theory of *finite* groups to problems in quantum chemistry. The reader should not attempt this book until he fully understands the use of Wigner and Clebsch-Gordon coefficients and tensor operators in the full rotation group. Griffith's text on ligand field theory (1961) also contains a lucid and *relatively* elementary introduction to group theory, as well as many useful appendixes for the octahedral group.

   Most chemists are primarily interested in applications of group theory, rather than in group theory per se. In orbital theory, there are two major kinds of uses, both involving group tables. In the first instance, character tables are employed to classify wavefunctions and operators according to different irreducible representations of a molecular symmetry group. In the second instance, direct product tables are used to determine selection rules, and coupling coefficients are used to build up many-electron wavefunctions from orbitals. The reader who understands these applications probably does not need further training in group theory at this level. In order to achieve this level of understanding, a careful reading of Secs. 8.1 to 8.3, together with the discussion following Eq. (8.42) (to the end of Sec. 8.5) and the discussion following Eq. (8.47) (to the end of Sec. 8.6) may be sufficient. Furthermore, a number of extended examples are worked out in Sec. 11.1. The student who can follow the details of these examples should have no trouble following the usual applications of group theory to be found in the literature.

9

LIGAND FIELD THEORY

The peculiar interest of the
transition metals is generally
agreed to be connected with
their ability to form compounds
in which the outermost set of
stable *d* electron orbitals is only
partially filled.
L. E. Orgel

9.1 COORDINATION COMPLEXES

Transition metal ions react in aqueous solution with small molecules or
ions, called ligands, to produce coordination complexes. Commonly, six
such ligands react with a single metal ion to produce a complex ion with
at least approximately octahedral symmetry. Complexes with four ligands,
either tetrahedrally bound or in a square planar array, are also found. The
chemistry of these complexes has been studied in enormous detail, starting
with the work of S. M. Jorgensen and Alfred Werner around the turn of
the century and continuing through the present day. In solution, the complexes are "weak," in the sense that one ligand can easily be removed and
replaced by another; on the other hand, the replacement reactions often
take place relatively slowly, so the kinetics of replacement are easily
observed. The complexes may be crystallized, together with suitable
counterions, and the detailed structures of many of these crystals are
known from x-ray analysis. As in organic chemistry, the compounds can

show optical activity and have various kinds of isomers, and hence they provide the kind of broadly intriguing behavior likely to attract a chemist.

Almost all of the transition metal complexes are highly colored. This is, of course, the major reason they received early attention; the striking blue color produced when copper chloride is added to an ammonia solution is clear proof that some sort of chemical reaction has occurred. From a theoretical point of view these colors are interesting because they suggest that complex formation in some way produces a set of low-lying excited states to which transitions can occur. One can show that the transitions to these states are just the forbidden d-d transitions of the open d shell of the transition metal ion, perturbed somewhat in energy by the fact of complex formation and enhanced in intensity through vibronic coupling with the ligands. When this theoretical interpretation was first suggested, in the 1930s, by Bethe and Van Vleck, the ligands were regarded as producing a purely electrostatic perturbation on the d-shell levels of the metal. The theory was therefore at first called "crystal field theory." It was soon recognized, however, that the same *kinds* of perturbations could be produced by any sort of interaction, whether electrostatic or not, provided it had the proper symmetry behavior. It is therefore customary to distinguish a modern approach, based on molecular orbitals, as "ligand field theory," and reserve the original name for the electrostatic theory. Nevertheless, it is the *symmetry* of the field which is basic to ligand field theory, whereas the *nature* of field hardly enters at all into consideration.‡ Furthermore, the application of symmetry concepts is not limited to a discussion of spectra but is readily extended to magnetic properties, including spin resonance, and to certain thermodynamic properties.

Ligand field theory, then, arises from a marriage of group theory with the theory of atomic electronic structure developed in Chap. 6. Consider, for example, the complex $[Ti(H_2O)_6]^{3+}$, produced by dissolving a titanium $(+3)$ salt in water. To at least a rough approximation, the ion has octahedral symmetry, with the oxygen atoms located at the six corners. (Strict octahedral symmetry is lacking because of the presumably random locations of the protons and the vibrations of the metal-water bonds.) Assuming that the nature of the d orbitals is not greatly changed by the fact of complex formation, the titanium ion retains its uncomplexed electronic configuration $1s^2 2s^2 2p^6 3s^2 3p^6 3d^1$. The salient feature here is the single d electron; all other shells are closed, including the shells of the water molecules. The free ion has ground state ^2D, and within the d^n configurational framework this is the only possible state. In the complex, the ^2D state splits into two states, classified ^{2}§T_{2g} and ^2E_g. The $^2T_{2g}$ state is

‡ It is assumed that the ligand field directly affects only the spatial degeneracy, and not the spin degeneracy, of a given atomic term. Nevertheless, the ligand field can have a very important indirect effect on the spin of a system, that is, *vide infra*.

§ Since all d^n states have even parity, the subscript g is often dropped.

the ground state; a transition from this state to the ^2E_g can be observed in $[Ti(H_2O)_6]^{3+}$ at about 20,000 cm^{-1}. The splitting itself (but not its magnitude) can be predicted from a knowledge of how basis functions of the full rotation group transform when the symmetry is reduced to octahedral; a list of such reductions is given in Table 9.1. Since we have not studied the full rotation group, this approach is somewhat awkward. Alternatively, we can consider how the d orbitals themselves transform when the symmetry is reduced. Consulting the character table for O_h given in the previous chapter, we see that the functions xz, yz, and xy form a basis for the representation T_{2g}. Therefore, the orbitals d_{xz}, d_{yz}, and d_{xy} transform‡ as t_{2g}. Similarly, $d_{x^2-y^2}$ and d_{z^2} ($\equiv d_{3z^2-r^2}$) transform as e_g. Thus, from an orbital point of view, the ligand field splits the five-fold degenerate d orbitals into a three-fold degenerate t_{2g} and a doubly degenerate e_g. The t_{2g} orbital lies lower, and so the ground state of the titanium complex is $^2T_{2g}$, and the only excited state is the ^2E_g produced when the electron is excited into the e_g orbital. *In all octahedral complexes observed, the t_{2g} orbital lies lower.* The distance between the t_{2g} and e_g orbitals is conventionally called $10Dq$ or Δ (Fig. 9.1). The name $10Dq$ arises from attempted calculations of this quantity using crystalline field theory; if one assumes that the splitting of the d orbitals is caused by the electrostatic field of the (negative) ligands, one can calculate that the e_g orbital lies higher than the t_{2g} by an amount that was called $10Dq$. Originally, then, $10Dq$ was a purely theoretical quantity. Many authors therefore refer to the *experimental* separation as Δ, which is thus a purely empirical quantity. However, Δ is a rather general sort of symbol, and in this text we will reserve the name $10Dq$ for the empirical quantity, the original meaning having by now no importance.

In the d^1 case, it does not matter whether we say that the ^2D state is split into two states, according to Table 9.1, or that the five d-orbitals are split into two sets. Since there is but one d electron, the state and

‡ As is customary, we use small letters for orbitals and capital letters for states.

Table 9.1 Transformation properties of atomic terms in an octahedral field

L	Term	Irreducible representations of O
0	S	A_1
1	P	T_1
2	D	$E + T_2$
3	F	$A_2 + T_1 + T_2$
4	G	$A_1 + E + T_1 + T_2$
5	H	$E + 2T_1 + T_2$
6	I	$A_1 + A_2 + E + T_1 + 2T_2$

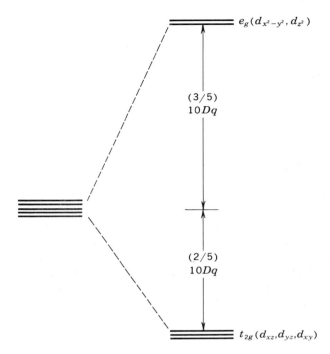

Fig. 9.1 Splitting of the d orbitals in an octahedral field. It is conventional to keep the center of gravity of the orbitals unchanged, so the t_{2g} orbitals are lowered by $(\frac{2}{5})10Dq$ and the e_g orbitals are raised by $(\frac{3}{5})10Dq$.

orbital pictures are identical. When we consider the d^2 case, however, it becomes clear that we must think more carefully about the exact nature of the ligand field splitting. For example, the ground state of the d^2 configuration is a ^3F. According to Table 9.1, this will be split into the states $^3A_2$, $^3T_1$, and $^3T_2$. We wish to know the relative energies of these three states, and in particular we need to know which is the ground state of the complex. This question can only be answered by calculating matrix elements of the form

$$\langle {}^3FM_SM_L|\mathbf{V}|{}^3FM'_SM'_L\rangle \tag{9.1}$$

To do so, we must consider the behavior of the ligand field operator $\mathbf{V}$ in greater detail.

Let us first write

$$\mathbf{V} = \mathbf{V}_0 + \mathbf{V}' \qquad\qquad \blacktriangleright (9.2)$$

where $\mathbf{V}'$ is a *splitting* operator which does not change the center of gravity of a set of orbitals or states. The latter sort of change is lumped in $\mathbf{V}_0$. Thus, for example, the overall change in d-orbital energy in the formation of a complex from Cr^{3+} and six F^- anions is enormous, whereas the *splitting* of the d orbitals is quite small. Then, by the previous discussion,

$$
\begin{aligned}
\langle d(e_g\gamma)|\mathbf{V}'|d(e_g\gamma')\rangle &= \delta_{\gamma\gamma'}\tfrac{3}{5}(10Dq) \\
\langle d(t_{2g}\gamma)|\mathbf{V}'|d(t_{2g}\gamma')\rangle &= -\,\delta_{\gamma\gamma'}\tfrac{2}{5}(10Dq) \\
\langle d(t_{2g}\gamma)|\mathbf{V}'|d(e_g\gamma')\rangle &= 0 \\
\langle d(\Gamma\gamma)|\mathbf{V}_0|d(\Gamma'\gamma')\rangle &= \delta_{\Gamma\Gamma'}\delta_{\gamma\gamma'}C
\end{aligned}
$$

▶(9.3)

Here $d(\Gamma\gamma)$ stands for any one of the five d orbitals labeled by the irreducible representation and the appropriate component to which it belongs. The delta functions arise from the symmetry selection rules (8.63). However, this form for the matrix elements of $\mathbf{V}$ is not entirely convenient, since they are expressed over the *real* orbitals d_{xy}, d_{xz}, d_{yx}, $d_{x^2-y^2}$, and d_{z^2}. On the other hand, matrix elements such as those of (9.1) are inherently in terms of the complex d orbitals involving the Y_{lm}. For example, take the matrix element

$$\langle {}^3F\,1\,3|\mathbf{V}'|{}^3\mathbf{F}\,1\,3\rangle \tag{9.4}$$

From Chap. 6, we know that $|{}^3F\,1\,3\rangle = |d_2^+\,d_1^+\rangle = |2^+\,1^+\rangle$. In order to evaluate Eq. (9.4), we thus need to know matrix elements of the form (9.3) over complex orbitals.

From Table 3.1 we may write

$$
\begin{aligned}
d_{z^2} &= d_0 = |0\rangle \\[4pt]
d_{x^2-y^2} &= -\frac{1}{\sqrt{2}}(d_2 + d_{-2}) = -\frac{1}{\sqrt{2}}(|2\rangle + |-2\rangle) \\[4pt]
d_{xz} &= -\frac{1}{\sqrt{2}}(d_1 - d_{-1}) = -\frac{1}{\sqrt{2}}(|1\rangle - |-1\rangle) \\[4pt]
d_{yz} &= \frac{i}{\sqrt{2}}(d_1 + d_{-1}) = \frac{i}{\sqrt{2}}(|1\rangle + |-1\rangle) \\[4pt]
d_{xy} &= -\frac{i}{\sqrt{2}}(d_2 - d_{-2}) = -\frac{i}{\sqrt{2}}(|2\rangle - |-2\rangle)
\end{aligned}
$$

▶(9.5)

Together with the results of (9.3), we can derive (Prob. 9.1) the matrix elements of Table 9.2. If we then assume that $\mathbf{V}$ is a one-electron operator,‡

‡ Throughout ligand field theory, one assumes that $\mathbf{V}$ is a one-electron operator. This assumption cannot be strictly justified theoretically, although it is supported by a great deal of empirical evidence.

Table 9.2 Matrix elements of the ligand field splitting V' for a single d electron in units of $10Dq$

m_1	2	1	0	-1	-2
2	$\frac{1}{10}$	0	0	0	$\frac{1}{2}$
1	0	$-\frac{2}{5}$	0	0	0
0	0	0	$\frac{3}{5}$	0	0
-1	0	0	0	$-\frac{2}{5}$	0
-2	$\frac{1}{2}$	0	0	0	$\frac{1}{10}$

we can use Eqs. (6.20) and (6.21) to derive the matrix elements (9.4). For example,

$$\langle {}^3F\ 1\ 3|V'|{}^3F\ 1\ 3\rangle = \langle 2^+\ 1^+|V'|2^+\ 1^+\rangle = \langle 2|V'|2\rangle + \langle 1|V'|1\rangle$$
$$= -\tfrac{3}{10}(10Dq) \quad (9.6)$$

In order to find the energies of the three states ($^3A_2$, $^3T_1$, and $^3T_2$), we must evaluate all such elements and diagonalize the resulting matrix. The process is simplified somewhat by the assumption that V' is independent of the spin, so in the present case we need only evaluate matrix elements with $M_S = 1$. However, even if we followed such a procedure [which is at best tedious since we have $(2L + 1)^2 = 49$ matrix elements to evaluate], it might not be at all clear *which* energy level belonged to a given symmetry state. (The $^3A_2$ state could be distinguished easily in this case since it is spatially nondegenerate. But the two T states could be confused with each other.) What is required is a method of determining which combinations of components in the $|{}^{2S+1}LM_SM_L\rangle$ scheme belong to components in the $|{}^{2S+1}\Gamma M_S\gamma\rangle$ scheme. Thus, for the $^3T_2$ state, we require coefficients in the expansion

$$|{}^3T_2\ 1\ \gamma\rangle = \sum_{M_L} c_{M_L}|{}^3F\ 1\ M_L\rangle \quad (9.7)$$

This is again a group-theoretical problem, although not requiring much knowledge of the full rotation group. The coefficients c_{M_L} are, of course, independent of the spin, and we may equally write

$$|T_2\ \gamma\rangle = \sum_{M_L} c_{M_L}|3\ M_L\rangle \quad (9.8)$$

In order to derive these coefficients, we first need to develop a number of group-theoretical relations for the octahedral group. This we do in the next section.

PROBLEM

9.1 Derive the matrix elements of Table 9.2.

9.2 THEORETICAL RELATIONS FOR THE OCTAHEDRAL GROUP‡

We have already met the character tables for the groups O and O_h in the previous chapter. It remains to determine the direct product reduction tables and the tables of coupling coefficients, both of which are necessary for a discussion of how the states of the free ion are affected by the ligand field.

It is easily verified that $T_1 \times T_1 = T_2 \times T_2 = A_1 + E + T_1 + T_2$, and that $T_1 \times T_2 = A_2 + E + T_1 + T_2$. The remaining relations are even easier to confirm and are given in Table 9.3.

Since $g \times g = u \times u = g$ and $g \times u = u$, it is entirely unnecessary to give the corresponding table for O_h.

In order to determine how products of individual components transform, it is necessary to obtain coupling coefficients in the form of Eq. (8.50). In order to do so it is, of course, first necessary to adopt a definite choice of basis functions for the irreducible representations. We take them to be the conventional real orbital forms. Thus, since p orbitals transform as T_1 in octahedral symmetry (Table 9.1), we choose bases x, y, and z (analogous to p_x, p_y, and p_z) for the T_1 representation. Similarly, we choose bases yz, xz, and xy for the T_2 representation and take $x^2 - y^2(d_{x^2-y^2})$ and $3z^2 - r^2(d_{z^2})$ for the bases of the E representation. In the A_1 and A_2 representations there is no real choice involved since each representation has but one component. In order to actually work

‡ The reader who is not interested in the theoretical details can skip this section, although he should have a quick look at Eqs. (9.9) and (9.10), Table 9.4, and Table 9.5. It is also possible to skim over Secs. 9.3 and 9.4 without seriously affecting the continuity of the text.

Table 9.3   Direct product reduction table for the octahedral group

O	A_1	A_2	E	T_1	T_2
A_1	A_1	A_2	E	T_1	T_2
A_2	A_2	A_1	E	T_2	T_1
E	E	E	$A_1 + A_2 + E$	$T_1 + T_2$	$T_1 + T_2$
T_1	T_1	T_2	$T_1 + T_2$	$A_1 + E + T_1 + T_2$	$A_2 + E + T_1 + T_2$
T_2	T_2	T_1	$T_1 + T_2$	$A_2 + E + T_1 + T_2$	$A_1 + E + T_1 + T_2$

out the coefficients, however, it is useful to note that xyz transforms as A_2 and that $(xyz)^2$ and $x^2 + y^2 + z^2$ both transform as A_1. Two small points of difficulty arise from these choices, however. Basis functions such as xy and xyz are not entirely suitable since in matrix multiplication we distinguish between the products xy and yx and between xyz, yxz, etc. We therefore actually take the bases for T_2 as $(yz + zy)$, $(xz + zx)$, and $(xy + yx)$, and similarly choose the basis for A_2 as the symmetrized product of x, y, and z. A second difficulty arises in the fact that our bases must all be normalized to the same value. If we define the bases x, y, and z to be normalized, we must then take the bases of T_2 as $(1/\sqrt{2})(yz + zy)$, etc., the basis of A_2 as $(1/\sqrt{6})(xyz + yxz + zxy + xzy + zyx + yzx)$, and the two bases of E as $(1/\sqrt{2})(x^2 - y^2)$ and $(1/\sqrt{6})(3z^2 - r^2)$. It is inconvenient to keep repeating the normalization factors, and so we adopt the *specific definitions*

$$A_1 a_1 = \frac{1}{\sqrt{3}} (x^2 + y^2 + z^2)$$

$$A_2 a_2 = \frac{1}{\sqrt{6}} (xyz + yxz + xzy + zxy + zyx + yzx)$$

$$T_2 x = \frac{1}{\sqrt{2}} (yz + zy)$$

$$T_2 y = \frac{1}{\sqrt{2}} (xz + zx) \qquad\qquad\blacktriangleright (9.9)$$

$$T_2 z = \frac{1}{\sqrt{2}} (xy + yx)$$

$$E\theta = \frac{1}{\sqrt{6}} (3z^2 - r^2)$$

$$E\epsilon = \frac{1}{\sqrt{2}} (x^2 - y^2)$$

The Japanese school of ligand field theorists uses u and v for θ and ϵ, but Griffith's notation‡ is to be preferred since it is easy to remember that θ is the polar function. The symbols ξ, η, and ζ are often used in place of $T_2 x$, $T_2 y$, and $T_2 z$.

To consider how *products* of basis functions transform, we must first determine how the individual basis functions themselves behave under all operations of the group. The task is greatly simplified by the fact that all operations of the octahedral group can be generated as the product of just three elements,§ C_4^z, C_4^x, and C_3^{111}. The latter operator takes $(x\ y\ z)$ into

‡ J. S. Griffith, 1961, pp. 164 and 226.
§ J. S. Griffith, 1961, pp. 164 and 390.

$(y\ z\ x)$. Thus the following generating table completely determines the behavior of all basis functions in O:

	C_4^z	C_4^x	C_3^{111}
A_1a_1	a_1	a_1	a_1
A_2a_2	$-a_2$	$-a_2$	a_2
T_1x	y	x	y
y	$-x$	z	z
z	z	$-y$	x
T_2x	$-y$	$-x$	y
y	x	$-z$	z
z	$-z$	y	x
$E\theta$	θ	$-\tfrac{1}{2}\theta - \dfrac{\sqrt{3}}{2}\epsilon$	$-\tfrac{1}{2}\theta + \dfrac{\sqrt{3}}{2}\epsilon$
ϵ	$-\epsilon$	$-\dfrac{\sqrt{3}}{2}\theta + \tfrac{1}{2}\epsilon$	$-\dfrac{\sqrt{3}}{2}\theta - \tfrac{1}{2}\epsilon$

$\blacktriangleright$ (9.10)

From the above table it is clear, for example, that $\pm(T_1x \cdot T_1y + T_1y \cdot T_1x)$ transforms as T_2z, and it is easy to verify that $\pm(T_1x \cdot T_1y - T_1y \cdot T_1x)$ transforms as T_1z. Similarly, $\pm(T_1x \cdot T_1x - T_1y \cdot T_1y)$ transforms as $E\epsilon$, whereas $\pm(T_1x \cdot T_1x + T_1y \cdot T_1y + T_1z \cdot T_1z)$ transforms as A_1 and $\pm(2T_1z \cdot T_1z - T_1x \cdot T_1x - T_1y \cdot T_1y)$ transforms as $E\theta$. Of course, the functions must be normalized. We always have an arbitrary phase choice to contend with, and no simple prescriptions are available for this. We follow Griffith's *original* choice of phase,‡ which differs both from that of Tanabe and Sugano and from his own choice in his later book. The reason we do this is that Griffith's comprehensive tables in his first book are somewhat difficult to use unless one sticks to the original phase choice, and these tables are the best source of detailed information in ligand field theory. A table of relative phases is given in his second book (1964).

As a further example of the way in which coupling coefficients are determined, we leave it to the reader to show that

$$|EEA_1a_1\rangle = \pm(1/\sqrt{2})(\theta\theta + \epsilon\epsilon)$$
$$|EEA_2a_2\rangle = \pm(1/\sqrt{2})(\theta\epsilon - \epsilon\theta)$$
$$|EEE\epsilon\rangle = \pm(1/\sqrt{2})(\theta\epsilon + \epsilon\theta)$$
$$|EEE\theta\rangle = \mp(1/\sqrt{2})(\theta\theta - \epsilon\epsilon)$$

The complete set of coupling coefficients is given in Table 9.4.

Next we must find the coefficients which determine the linear combinations of atomic states which transform as irreducible representations of the octahedral group. For $L = 0$, 1 and 2, the problem is trivial, since

‡ J. S. Griffith, 1961, p. 396.

Table 9.4 Coupling coefficients for the octahedral group. We write

$$|\Gamma^{(p)}\Gamma^{(q)}\Gamma\gamma_k\rangle = \sum_{\alpha_i,\beta_j} \langle\Gamma^{(p)}\Gamma^{(q)}\alpha_i\beta_j|\Gamma^{(p)}\Gamma^{(q)}\Gamma\gamma_k\rangle|\Gamma^{(p)}\Gamma^{(q)}\alpha_i\beta_j\rangle$$

and define

$$\langle\Gamma^{(q)}\Gamma^{(p)}\beta_j\alpha_i|\Gamma^{(q)}\Gamma^{(p)}\Gamma\gamma_k\rangle = \langle\Gamma^{(p)}\Gamma^{(q)}\alpha_i\beta_j|\Gamma^{(p)}\Gamma^{(q)}\Gamma\gamma_k\rangle$$

for $\Gamma^{(p)} \neq \Gamma^{(q)}$. T_i stands for either T_1 or T_2

Γ	γ_k	$\Gamma^{(p)}$	$\Gamma^{(q)}$						
A_1	a_1	A_2	A_2	a_2a_2					
		E	E	$\frac{1}{\sqrt{2}}(\theta\theta + \epsilon\epsilon)$					
		T_i	T_i	$\frac{1}{\sqrt{3}}(xx + yy + zz)$					
A_2	a_2	E	E	$\frac{1}{\sqrt{2}}(\theta\epsilon - \epsilon\theta)$					
		T_1	T_2	$\frac{1}{\sqrt{3}}(xx + yy + zz)$					
E	θ	A_2	E	$a_2\epsilon$					
		E	E	$\frac{1}{\sqrt{2}}(\epsilon\epsilon - \theta\theta)$					
		T_i	T_i	$-\frac{1}{\sqrt{6}}(2zz - xx - yy)$					
		T_1	T_2	$\frac{1}{\sqrt{2}}(yy - xx)$					
E	ϵ	A_2	E	$-a_2\theta$					
		E	E	$\frac{1}{\sqrt{2}}(\theta\epsilon + \epsilon\theta)$					
		T_i	T_i	$\frac{1}{\sqrt{2}}(yy - xx)$					
		T_1	T_2	$\frac{1}{\sqrt{6}}(2zz - yy - xx)$					
T_1	x	A_2	T_2	a_2x	T_2	x	A_2	T_1	a_2x
		E	T_1	$\frac{1}{2}\sqrt{3}\,\epsilon x - \frac{1}{2}\theta x$			E	T_2	$\frac{1}{2}\sqrt{3}\,\epsilon x - \frac{1}{2}\theta x$
		E	T_2	$-\frac{1}{2}\sqrt{3}\,\theta x - \frac{1}{2}\epsilon x$			E	T_1	$-\frac{1}{2}\sqrt{3}\,\theta x - \frac{1}{2}\epsilon x$
		T_i	T_i	$\frac{1}{\sqrt{2}}(zy - yz)$			T_i	T_i	$-\frac{1}{\sqrt{2}}(zy + yz)$
		T_1	T_2	$-\frac{1}{\sqrt{2}}(yz + zy)$			T_1	T_2	$\frac{1}{\sqrt{2}}(zy - yz)$
T_1	y	A_2	T_2	a_2y	T_2	y	A_2	T_1	a_2y
		E	T_1	$-\frac{1}{2}\theta y - \frac{1}{2}\sqrt{3}\,\epsilon y$			E	T_2	$-\frac{1}{2}\theta y - \frac{1}{2}\sqrt{3}\,\epsilon y$
		E	T_2	$\frac{1}{2}\sqrt{3}\,\theta y - \frac{1}{2}\epsilon y$			E	T_1	$\frac{1}{2}\sqrt{3}\,\theta y - \frac{1}{2}\epsilon y$
		T_i	T_i	$\frac{1}{\sqrt{2}}(xz - zx)$			T_i	T_i	$-\frac{1}{\sqrt{2}}(xz + zx)$
		T_1	T_2	$-\frac{1}{\sqrt{2}}(xz + zx)$			T_1	T_2	$\frac{1}{\sqrt{2}}(xz - zx)$
T_1	z	A_2	T_2	a_2z	T_2	z	A_2	T_1	a_2z
		E	T_2	ϵz			E	T_1	ϵz
		E	T_1	θz			E	T_2	θz
		T_i	T_i	$-\frac{1}{\sqrt{2}}(xy - yx)$			T_i	T_i	$-\frac{1}{\sqrt{2}}(xy + yx)$
		T_1	T_2	$-\frac{1}{\sqrt{2}}(xy + yx)$			T_1	T_2	$-\frac{1}{\sqrt{2}}(xy - yx)$

we have chosen the basis functions for A_1, T_1, T_2, and E to be the normalized combinations $|LM\rangle \pm |L-M\rangle$. For $L = 3$ (and higher L) the problem is more difficult. The combinations $|LM\rangle \pm |L-M\rangle$ do not, in general, form a basis for a representation. However, by considering the transformation properties of such real combinations we can find the desired coefficients. For example, $|3\ 3\rangle - |3\ -3\rangle \sim \sqrt{5}\, x(3y^2 - x^2)$, $|3\ 3\rangle + |3\ -3\rangle \sim i\sqrt{5}y\,(y^2 - 3x^2)$, $|3\ 2\rangle - |3\ -2\rangle \sim i\sqrt{120}\,xyz$, $|3\ 2\rangle + |3\ -2\rangle \sim \sqrt{30}\,z(x^2 - y^2)$, $|3\ -1\rangle - |3\ -1\rangle \sim -\sqrt{3}\,x(5z^2 - r^2)$, $|3\ 1\rangle + |3\ -1\rangle \sim -i\sqrt{3}\,y(5z^2 - r^2)$, $|3\ 0\rangle \sim z(5z^2 - 3r^2)$. Therefore, the combination $(1/\sqrt{2})(|3\ 2\rangle - |3\ -2\rangle)$ forms a basis for A_2, and the combina-

Table 9.5 Linear combinations of atomic states $|LM\rangle$ (with $L \leq 3$) transforming as irreducible representations of the octahedral group

$L = 0$	$\|A_1 a_1\rangle = \|0\ 0\rangle$
$L = 1$	$\|T_1 x\rangle = -\dfrac{1}{\sqrt{2}}\,(\|1\ 1\rangle - \|1\ -1\rangle)$
	$\|T_1 y\rangle = \dfrac{i}{\sqrt{2}}\,(\|1\ 1\rangle + \|1\ -1\rangle)$
	$\|T_1 z\rangle = \|1\ 0\rangle$
$L = 2$	$\|E\theta\rangle = \|2\ 0\rangle$
	$\|E\epsilon\rangle = \dfrac{1}{\sqrt{2}}\,(\|2\ 2\rangle + \|2\ -2\rangle)$
	$\|T_2 x\rangle = \dfrac{i}{\sqrt{2}}\,(\|2\ 1\rangle + \|2\ -1\rangle)$
	$\|T_2 y\rangle = -\dfrac{1}{\sqrt{2}}\,(\|2\ 1\rangle - \|2\ -1\rangle)$
	$\|T_2 z\rangle = -\dfrac{i}{\sqrt{2}}\,(\|2\ 2\rangle - \|2\ -2\rangle)$
$L = 3$	$\|A_2 a_2\rangle = -\dfrac{i}{\sqrt{2}}\,(\|3\ 2\rangle - \|3\ -2\rangle)$
	$\|T_1 x\rangle = \dfrac{\sqrt{3}}{4}\,(\|3\ 1\rangle - \|3\ -1\rangle) - \dfrac{\sqrt{5}}{4}\,(\|3\ 3\rangle - \|3\ -3\rangle)$
	$\|T_1 y\rangle = -i\dfrac{\sqrt{3}}{4}\,(\|3\ 1\rangle + \|3\ -1\rangle) - i\dfrac{\sqrt{5}}{4}\,(\|3\ 3\rangle + \|3\ -3\rangle)$
	$\|T_1 z\rangle = \|3\ 0\rangle$
	$\|T_2 x\rangle = \dfrac{\sqrt{5}}{4}\,(\|3\ 1\rangle - \|3\ -1\rangle) + \dfrac{\sqrt{3}}{4}\,(\|3\ 3\rangle - \|3\ -3\rangle)$
	$\|T_2 y\rangle = i\dfrac{\sqrt{5}}{4}\,(\|3\ 1\rangle + \|3\ -1\rangle) - i\dfrac{\sqrt{3}}{4}\,(\|3\ 3\rangle + \|3\ -3\rangle)$
	$\|T_2 z\rangle = \dfrac{1}{\sqrt{2}}\,(\|3\ 2\rangle + \|3\ -2\rangle)$

tion $(1/\sqrt{2})(|3\ 2\rangle + |3\ -2\rangle)$ forms a basis for $T_2 z$ since, from Table 9.4, ϵz transforms as $T_2 z$. Similarly, one can show that the combinations $\frac{1}{2}\sqrt{3}(|3\ 1\rangle - |3\ -1\rangle) - \frac{1}{2}\sqrt{5}(|3\ 3\rangle - |3\ -3\rangle)$, $-i\ \frac{1}{2}\sqrt{3}(|3\ 1\rangle + |3\ -1\rangle) - \frac{1}{2}\sqrt{5}(|3\ 3\rangle + |3\ -3\rangle)$, and $|3\ 0\rangle$ transform as $T_1 x$, $T_1 y$, and $T_1 z$, respectively. Coefficients for $L \leq 3$ are given in Table 9.5; Griffith‡ has derived the corresponding coefficients for $L = 4$, 5, and 6.

9.3 CALCULATIONS IN THE WEAK–FIELD LIMIT

We are now in a position to evaluate the matrix elements of the ligand field-splitting operator within any atomic term and determine the energy of any given state. For example, consider the ^3F term of the d^2 configuration. According to Table 9.1, we must evaluate the matrix elements $\langle ^3F\ A_2\ a_2|\mathbf{V}'|^3F\ A_2\ a_2\rangle$, $\langle ^3F\ T_1\ z|\mathbf{V}'|^3F\ T_1\ z\rangle$, and $\langle ^3F\ T_2\ z|\mathbf{V}'|^3F\ T_2\ z\rangle$, where we have chosen the z components of T_1 and T_2 to facilitate the calculation. Substituting from Table 9.5, we obtain at once§

$$
\begin{aligned}
E(^3A_2) &= E(^3F) + \tfrac{1}{2}\langle ^3F\ 1\ 2|\mathbf{V}'|^3F\ 1\ 2\rangle \\
&\quad + \tfrac{1}{2}\langle ^3F\ 1\ -2|\mathbf{V}'|^3F\ 1\ -2\rangle - \langle ^3F\ 1\ 2|\mathbf{V}'|^3F\ 1\ -2\rangle \\
E(^3T_2) &= E(^3F) + \tfrac{1}{2}\langle ^3F\ 1\ 2|\mathbf{V}'|^3F\ 1\ 2\rangle \\
&\quad + \tfrac{1}{2}\langle ^3F\ 1\ -2|\mathbf{V}'|^3F\ 1\ -2\rangle + \langle ^3F\ 1\ 2|\mathbf{V}'|^3F\ 1\ -2\rangle \\
E(^3T_1) &= E(^3F) + \langle ^3F\ 1\ 0|\mathbf{V}'|^3F\ 1\ 0\rangle
\end{aligned}
\tag{9.11}
$$

where we have chosen the $M_S = 1$ component for simplicity. Using the shift operator $\mathbf{L}^-$ beginning with the ket $|^3F\ 1\ 3\rangle = |2^+\ 1^+\rangle$, we derive the following expansions of each ket in terms of the constituent Slater determinants:

$$
\begin{aligned}
|^3F\ 1\ 2\rangle &= |2^+\ 0^+\rangle \\
|^3F\ 1\ 1\rangle &= \sqrt{\tfrac{2}{5}}\,|1^+\ 0^+\rangle + \sqrt{\tfrac{3}{5}}\,|2^+\ -1^+\rangle \\
|^3F\ 1\ 0\rangle &= \sqrt{\tfrac{4}{5}}\,|1^+\ -1^+\rangle + \sqrt{\tfrac{1}{5}}\,|2^+\ -2^+\rangle \\
|^3F\ 1\ -1\rangle &= \sqrt{\tfrac{2}{5}}\,|0^+\ -1^+\rangle + \sqrt{\tfrac{3}{5}}\,|1^+\ -2^+\rangle \\
|^3F\ 1\ -2\rangle &= |0^+\ -2^+\rangle
\end{aligned}
\tag{9.12}
$$

---

‡ J. S. Griffith, 1961, p. 393.
§ There should be no problem carrying out this process even if the reader has omitted the previous section.

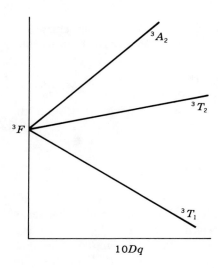

Fig. 9.2 Energy levels arising from the ^3F term of d^2 in the weak-field limit.

Thus, using Table 9.2, we find

$$\langle {}^3F\ 1\ 2|\mathbf{V}'|{}^3F\ 1\ 2\rangle = \tfrac{7}{10}(10Dq) = \langle {}^3F\ 1\ -2|\mathbf{V}'|{}^3F\ 1\ -2\rangle$$
$$\langle {}^3F\ 1\ 2|\mathbf{V}'|{}^3F\ 1\ -2\rangle = -\tfrac{1}{2}(10Dq) \qquad (9.13)$$
$$\langle {}^3F\ 1\ 0|\mathbf{V}'|{}^3F\ 1\ 0\rangle = -\tfrac{3}{5}(10Dq)$$

Combining results, we finally obtain

$$E({}^3A_2) = E({}^3F) + \tfrac{6}{5}(10Dq)$$
$$E({}^3T_2) = E({}^3F) + \tfrac{1}{5}(10Dq) \qquad (9.14)$$
$$E({}^3T_1) = E({}^3F) - \tfrac{3}{5}(10Dq)$$

The energy levels are plotted as functions of $10Dq$ in Fig. 9.2. The $^3T_1$ state is the ground state for all values of $10Dq$. Relative to zero ligand field strength, the ground state is stabilized by $\tfrac{3}{5}(10Dq)$, and it would appear that this gives rise to observable thermodynamic effects.‡

The $^3T_1$ ground state arising from the ^3F atomic term is not the only $^3T_1$ state in the d^2 configuration. In addition, the ^3P term also becomes $^3T_1$ in octahedral systems. These two states can then interact, and we must solve a 2×2 secular equation to obtain the correct energies. However, when $10Dq$ is small compared to the separation between the two atomic terms ($15B$ from Table 6.7), the interaction can be neglected. This is the so-called *weak-field limit*. Here we treat the ligand field splitting as a small perturbation on the atomic energy levels. However, for larger $10Dq$ it may be necessary to consider explicitly the interaction between the two states, and we illustrate how this can be done. By orthogonality with the $|{}^3F\ 1\ 0\rangle$ state, we may write $|{}^3P\ 1\ 0\rangle = \sqrt{\tfrac{4}{5}}\ |2+$

‡ P. George and D. S. McClure, 1959.

$-2^+\rangle - \sqrt{\frac{1}{5}} |1^+ -1^+\rangle$. Since, from Table 9.5, the $|{}^3T_1 \; 1 \; z\rangle$ component transforms as $|{}^3P \; 1 \; 0\rangle$, the interaction element between the two weak-field terms is

$$\langle {}^3T_1 \; 1 \; z({}^3F)|\mathbf{V}'|{}^3T_1 \; 1 \; z({}^3P)\rangle = \tfrac{2}{5}(10Dq) \tag{9.15}$$

The secular equation for the interaction of the two states is thus

$$\begin{vmatrix} 15B - E & \tfrac{2}{5}(10Dq) \\ \tfrac{2}{5}(10Dq) & -\tfrac{3}{5}(10Dq) - E \end{vmatrix} = 0 \tag{9.16}$$

where we have taken the energy of the ${}^3F$ state as zero. The energies of the two states are thus

$$E = -\frac{3}{10}(10Dq) + \frac{15B}{2} \pm \frac{1}{2}\sqrt{(10Dq)^2 + 18B(10Dq) + 225B^2} \tag{9.17}$$

In the limit that $15B \gg 10Dq$ (the weak-field limit) we may expand the square root‡ to give

$$E = -\frac{3}{10}(10Dq) + \frac{15B}{2} \pm \left[\frac{15B}{2} + \frac{3}{10}(10Dq) \right.$$
$$\left. + \frac{4(10Dq)^2}{25 \cdot 15B} + \cdots \right] \tag{9.18}$$

This is exactly the same result one obtains from the use of perturbation theory. At the opposite extreme, when $10Dq \gg 15B$, we may again expand the square root to obtain

$$E = -\frac{3}{10}(10Dq) + \frac{15B}{2} \pm \left[\frac{1}{2}(10Dq) + \frac{9B}{2} + \frac{36B^2}{10Dq} + \cdots \right] \tag{9.19}$$

In the *strong-field limit*, the two energies are $E_1 = -\tfrac{4}{5}(10Dq) + 3B$ and $E_2 = \tfrac{1}{5}(10Dq) + 12B$. The term $36B^2/10Dq$ acts as a perturbation which further splits the two levels. In the next section we shall show an easier path to the strong-field limit, and we shall also derive the perturbation matrix element which mixes the two levels. Both approaches, through the weak-field limit and through the strong-field limit, give identical results when the secular equation [such as Eq. (9.16)] is accurately solved. Which method one uses is therefore largely a matter of convenience.

In the weak-field limit, the ground-state wavefunction belongs to the ${}^3F$ term of the free ion. In the strong-field limit the ground state arises both from the ${}^3F$ and the ${}^3P$ and is therefore rather complicated to derive. One advantage which we will discover in the strong-field method

‡ $\sqrt{1+x} = 1 + \tfrac{1}{2}x - \tfrac{1}{8}x^2 + \cdots$

developed in the next section is that the strong-field-limit wavefunctions belong to distinct configurations. For example, we will be able to assign the ground state $^3T_1$ of d^2 to the strong-field configuration t_{2g}^2, whereas the excited $^3T_1$ will be assigned to the configuration $t_{2g}e_g$. This will make it much easier to determine the proper wavefunctions.

PROBLEMS

9.2 Use perturbation theory to verify (9.18).

9.3 Using Table 9.2, confirm the results of (9.13).

9.4 Prove that the ground state of a d^3 complex is $^4A_2$.

9.4 CALCULATIONS IN THE STRONG-FIELD LIMIT‡

The phrases "weak field" and "strong field" are used not only to indicate the kind of calculation performed, but also to specify roughly the actual size of the field. Consider the configurations of transition metal complexes with from one to nine d electrons. In the limit that $10Dq$ is very small, the electrons will fill the five spatial orbitals according to Hund's rule; i.e., the spin will be at a maximum in all cases. On the other hand, when $10Dq$ is very large, we expect that the t_{2g} orbitals will be filled first, before any electrons go into the e_g shell. Of course, within the t_{2g} shell, Hund's rule will be obeyed.

These observations are summarized in Fig. 9.3. The size of the ligand field affects the electronic configuration only in the d^4, d^5, d^6, and d^7 cases. For obvious reasons, it is therefore customary to distinguish *high-spin* and *low-spin* configurations. The ground state of each configuration can be derived from fairly elementary considerations of spin and spatial multiplicity. Consider first the low-spin (strong-field) configurations. In the d^1 and d^9 cases, the ground states are $^2T_{2g}$ and ^2E_g, respectively, and, in reverse order, these are the only excited states. The reader will recall that the closed t_{2g} shell in the d^9 case can be neglected in working out the ground state and that because of the hole-particle relationship, a single hole in a closed shell gives rise to the same state as a single particle in that shell. Thus, in addition, the low-spin state in the d^6 case is $^1A_{1g}$, in the d^7 case it is ^2E_g, and in the d^5 case it is $^2T_{2g}$. The ground state of d^3 is spatially nondegenerate, since there is only one way of placing three electrons with the same spin in three spatial orbitals. Therefore, the state is either $^4A_{1g}$ or $^4A_{2g}$. From Chap. 6, the ground state of the free ion is ^4F, which, in an octahedral field, splits into $^4A_2$, $^4T_1$, and $^4T_2$. Furthermore,

‡ This section may also be skimmed if the reader is primarily interested in the results and not the derivation of ligand field theory. However, close attention should be paid to Fig. 9.3, Table 9.6, Table 9.8, and Figs. 9.4 to 9.11.

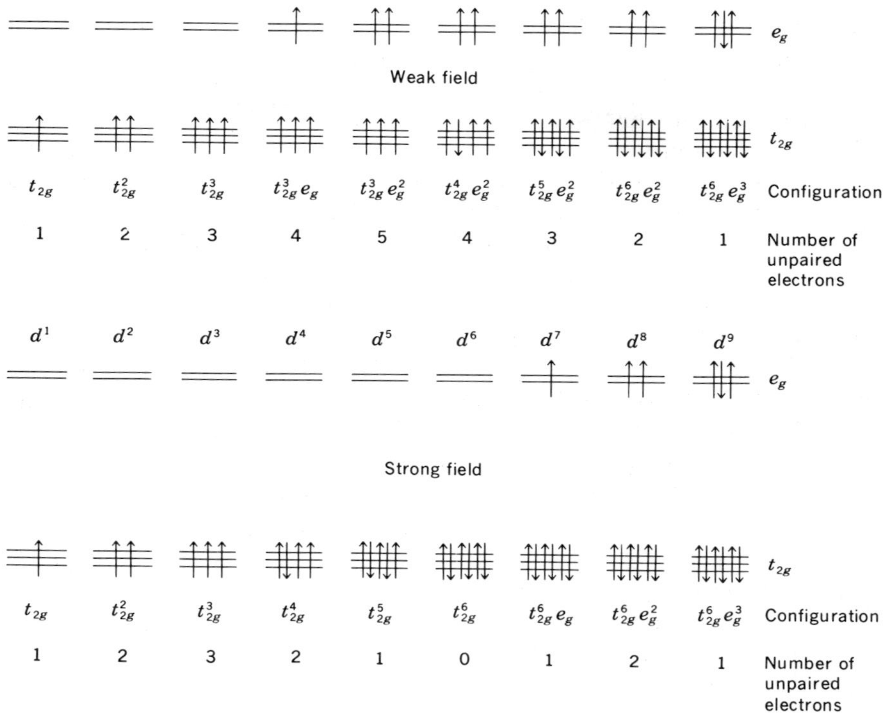

Fig. 9.3 Weak-field and strong-field configurations for d^n systems.

the only other free-ion quartet is the ^4P, which becomes $^4T_1$ in an octahedral environment. Thus there does not exist a $^4A_{1g}$, and the ground state of the t_{2g}^3 configuration must be the $^4A_{2g}$. There are only three remaining problems in the strong-field limit. In the t_{2g}^2 configuration, there are three ways of placing two up-spin electrons in the three orbitals, and so the ground state must have three-fold spatial degeneracy. The state is therefore either $^3T_{1g}$ or $^3T_{2g}$; we showed in the previous section that the former is correct. In the e_g^2 (that is, $t_{2g}^6 e_g^2$) configuration, the ground state must again be spatially nondegenerate in order to accommodate two parallel spins and obey Hund's rule. The free-ion ground state is ^3F which splits into $^3T_1$, $^3T_2$, and $^3A_2$, the latter state being the only one meeting the requirement of spatial nondegeneracy. The ground state of the t_{2g}^4 configuration may be determined by considering the addition of a t_{2g} electron to the $^4A_{2g}$ state of t_{2g}^3. The resulting state has $A_2 \times T_2$ symmetry, and is therefore a $^3T_{1g}$.

We may consider the high-spin configurations in the same way. The only free-ion state with five unpaired electrons is the ^6S, which becomes $^6A_{1g}$ in the octahedral field. The ground states of the two configurations $t_{2g}^3 e_g$ and $t_{2g}^4 e_g^2$ are, respectively, ^5E_g and $^5T_{2g}$ since these two states are produced by adding a hole or a particle to the totally symmetric $^6A_{1g}$. Alternatively, the ground state of the $t_{2g}^3 e_g$ configuration may be constructed from the ground state of the t_{2g}^3 configuration by adding a single e_g electron; the result is $A_2 \times E = E$, where A_2 describes the spatial symmetry of the ground state of the t_{2g}^3 configuration. There is no ambiguity here since $A_2 \times E$ is irreducible. Finally, the ground state of the configuration $t_{2g}^5 e_g^2$ may be derived by considering that this state can be produced by removing a single electron from the ground state of $t_{2g}^6 e_g^2$, $^3A_{2g}$. The hole is in the t_{2g} shell, so the state has $T_2 \times A_2 = T_1$ symmetry and is therefore $^4T_{1g}$.

Wavefunctions for all states may be derived by means of the coupling coefficients in Table 9.4. For example, consider the states of the t_2^2 configuration. According to Table 9.3, $T_2 \times T_2 = A_1 + E + T_1 + T_2$, and the coupling coefficients determine the actual wavefunctions. Thus, the linear combination of Slater determinants $\sqrt{1/6}\,(xx + yy + zz)$ $(\alpha_1\beta_2 - \beta_1\alpha_2)$ forms a $^1A_{1g}$ state. Similarly, the functions $-(1/\sqrt{12})(2zz - xx - yy)(\alpha_1\beta_2 - \beta_1\alpha_2)$ and $\frac{1}{2}(yy - xx)(\alpha_1\beta_2 - \beta_1\alpha_2)$ form the θ and ϵ bases, respectively, of a ^1E_g state.

Finally, we note that the coefficients giving rise to T_1 provide an antisymmetric basis, whereas the T_2 basis is symmetric. Therefore the t_{2g}^2 configuration produces $^3T_{1g}$, $^1T_{2g}$, ^1E_g, and $^1A_{1g}$ states, with the $M_S = 1$, $\gamma = z$ basis for the triplet equal to $-(1/\sqrt{2})(xy - yx)\alpha_1\alpha_2$.

In a similar way, basis functions may be built up for states arising from configurations with more than two electrons. For example, we can form states of the t_{2g}^3 configuration by coupling the four states of t_{2g}^2 with an extra t_{2g} electron. Clearly, if there are any quartet states in the t_{2g}^3 configuration, they originate in the $^3T_{1g}$ state of t_{2g}^2. Consider first the A_2 representation occuring in the product $T_1 \times T_2$. Its basis is

$$\frac{1}{\sqrt{3}}\left[\frac{1}{\sqrt{2}}\,(zy - yz)x + \frac{1}{\sqrt{2}}\,(xz - zx)y - \frac{1}{\sqrt{2}}\,(xy - yx)z\right]$$

$$= \frac{1}{\sqrt{6}}\,(zyx + xzy + yxz - yzx - zxy - xyz)$$

which is suitable as the spatial part of a $^4A_{2g}$ term. An E state also arises from t_{2g}^3. According to Table 9.4, both the $^3T_{1g}$ and $^1T_{2g}$ of t_{2g}^2 can participate in this state. The coupling of the $^3T_{1g}$ state with the extra t_{2g} electron cannot produce a ^4E_g state since the basis formed by the coupling cannot be antisymmetric. This term must therefore be a doublet. The derivation of the actual wavefunctions is complicated somewhat by spin problems, but fortunately Griffith‡ has worked out the complete states

‡ J. S. Griffith, 1961, p. 406.

of all t_{2g}^m and e_g^m configurations. States arising from configurations with two open shells $(t_{2g}^m e_g^n)$ can be obtained readily from Griffith's Table A 24 by coupling the two open shells. This coupling is completely trivial since the two shells involve inequivalent electrons. A list of all such states is given in Table 9.6.

We are now very nearly in a position to evaluate the matrix elements between the two $^3T_{1g}$ terms of d^2. Before we can do so, however, we need to be able to determine the coulomb and exchange integrals between the real forms of the d orbitals. A list of such integrals is given for the complex forms in Table 6.6, and it is straightforward to convert them to real orbitals using Eq. (9.5). The results are given in Table 9.7. There are a few nonzero integrals which are properly neither coulomb nor exchange, and we have listed these as well.

Both the t_{2g}^2 and $t_{2g}e_g$ configurations give rise to $^3T_{1g}$ terms. We have already shown that $|{}^3T_{1g}\ 1\ z\rangle = |y^+ x^+\rangle$ for t_{2g}^2. From Table 9.4, $|{}^3T_{1g}\ 1\ z\rangle = |z^+ \epsilon^+\rangle$ for $t_{2g}e_g$. The energy of the first state is

$$-\tfrac{4}{5}(10Dq) + \langle xy\|xy\rangle - \langle xy\|yx\rangle = A - 5B - \tfrac{4}{5}(10Dq)$$

Similarly, the energy of the second state is

$$\tfrac{1}{5}(10Dq) + \langle z\epsilon\|z\epsilon\rangle - \langle \epsilon z\|z\epsilon\rangle = A + 4B + \tfrac{1}{5}(10Dq)$$

The off-diagonal matrix element between the two states is

$$-\langle xy\|z\epsilon\rangle + \langle xy\|\epsilon z\rangle = -6B$$

These results are exactly the same as those obtained in the previous section for the strong-field limit except for a shift in the zero of energy. (In the previous discussion all results were referred to the undisturbed atomic ^3F term. If we add that energy, $A - 8B$, to the results of the previous section, the results will be identical.)

Table 9.6   States of the strong-field configurations $t_{2g}^m e_g^n$

$t_{2g},\ t_{2g}^5$	$^2T_{2g}$
$t_{2g}^2,\ t_{2g}^4$	$^3T_{1g},\ ^1A_{1g},\ ^1E_g,\ ^1T_{2g}$
t_{2g}^3	$^4A_{2g},\ ^2E_g,\ ^2T_{2g},\ ^2T_{1g}$
$e_g,\ e_g^3$	^2E_g
e_g^2	$^3A_{2g},\ ^1A_{1g},\ ^1E_g$
$t_{2g}e_g,\ t_{2g}e_g^3,$ $t_{2g}^5 e_g,\ t_{2g}^5 e_g^3$	$^3T_{2g},\ ^3T_{1g},\ ^1T_{2g},\ ^1T_{1g}$
$t_{2g}^2 e_g,\ t_{2g}^4 e_g,$ $t_{2g}^2 e_g^3,\ t_{2g}^4 e_g^3$	$^4T_{2g},\ ^4T_{1g},\ ^2A_{1g},\ ^2A_{2g},\ ^2E_g(2),\ ^2T_{2g}(2),\ ^2T_{1g}(2)$
$t_{2g}^3 e_g,\ t_{2g}^3 e_g^3$	$^5E_g,\ ^3A_{1g},\ ^3A_{2g},\ ^3E_g(2),\ ^3T_{2g}(2),\ ^3T_{1g}(2),\ ^1A_{1g},\ ^1A_{2g},\ ^1E_g,\ ^1T_{2g}(2),\ ^1T_{1g}(2)$
$t_{2g}e_g^2,\ t_{2g}^5 e_g^2$	$^4T_{1g},\ ^2T_{2g}(2),\ ^2T_{1g}(2)$
$t_{2g}^2 e_g^2,\ t_{2g}^4 e_g^2$	$^5T_{2g},\ ^3A_{2g},\ ^3E_g,\ ^3T_{2g}(2),\ ^3T_{1g}(3),\ ^1A_{1g}(2),\ ^1A_{2g},\ ^1E_g(3),\ ^1T_{2g}(3),\ ^1T_{1g}$
$t_{2g}^3 e_g^2$	$^6A_{1g},\ ^4A_{1g},\ ^4A_{2g},\ ^4E(2),\ ^4T_{2g},\ ^4T_{1g},\ ^2A_{1g}(2),\ ^2A_{2g},\ ^2E_g(3),\ ^2T_{2g}(4),\ ^2T_{1g}(4)$

Table 9.7 Two-electron integrals for real d orbitals in terms of Racah parameters. All integrals not shown are zero

		Coulomb integrals $\langle ab\|ab\rangle$	Exchange integrals $\langle ab\|ba\rangle$	Other integrals	
x	x	$A + 4B + 3C$	$A + 4B + 3C$	$\langle x\theta\|x\epsilon\rangle$	$2\sqrt{3}\,B$
x	y	$A - 2B + C$	$3B + C$	$\langle y\theta\|y\epsilon\rangle$	$-2\sqrt{3}\,B$
x	z	$A - 2B + C$	$3B + C$	$\langle yz\|x\theta\rangle$	$-2\sqrt{3}\,B$
x	θ	$A + 2B + C$	$B + C$	$\langle xy\|z\theta\rangle$	$\sqrt{3}\,B$
x	ϵ	$A - 2B + C$	$3B + C$	$\langle xy\|\theta z\rangle$	$\sqrt{3}\,B$
y	y	$A + 4B + 3C$	$A + 4B + 3C$	$\langle xx\|\theta\epsilon\rangle$	$-\sqrt{3}\,B$
y	z	$A - 2B + C$	$3B + C$	$\langle yy\|\theta\epsilon\rangle$	$\sqrt{3}\,B$
y	θ	$A + 2B + C$	$B + C$	$\langle xy\|\epsilon z\rangle$	$-3B$
y	ϵ	$A - 2B + C$	$3B + C$	$\langle xy\|z\epsilon\rangle$	$3B$
z	z	$A + 4B + 3C$	$A + 4B + 3C$		
z	θ	$A - 4B + C$	$4B + C$		
z	ϵ	$A + 4B + C$	C		
θ	θ	$A + 4B + 3C$	$A + 4B + 3C$		
θ	ϵ	$A - 4B + C$	$4B + C$		
ϵ	ϵ	$A + 4B + 3C$	$A + 4B + 3C$		

Notation: $x = d_{yz}$, $y = d_{xz}$, $z = d_{xy}$, $\theta = d_{z^2}$, $\epsilon = d_{x^2-y^2}$.

$$\langle ab\|cd\rangle = \langle a(1)b(2)| \frac{1}{r_{12}} |c(1)d(2)\rangle$$

As further examples of calculations in the strong-field coupling scheme, we consider the energies of the ground states of all systems which have different high-spin and low-spin configurations, that is, d^4, d^5, d^6, and d^7. We state without proof that the low-spin ground states of these systems are $|^3T_1\,1\,z\rangle = |x^+y^+z^2\rangle$, $|^2T_2\,\tfrac{1}{2}\,z\rangle = |x^2y^2z^+\rangle$, $|^1A_1\rangle = |x^2y^2z^2\rangle$, $|^2E\,\tfrac{1}{2}\,\theta\rangle = |x^2y^2z^2\,\theta^+\rangle$ and $|^3A_2\,1\,a_2\rangle = |x^2y^2z^2\theta^+\epsilon^+\rangle$. It is then straightforward to derive the energies of Table 9.8 by using the results of Table 9.7. Similarly the high-spin ground states are $|^5E\,2\,\theta\rangle = |x^+y^+z^+\epsilon^+\rangle$, $|^6A_1\,\tfrac{5}{2}\,a_1\rangle = |x^+y^+z^+\theta^+\epsilon^+\rangle$, $|^5T_2\,2\,z\rangle = |x^+y^+z^2\theta^+\epsilon^+\rangle$, and $|^4T_1\,\tfrac{3}{2}\,z\rangle = |x^2y^2z^+\theta^+\epsilon^+\rangle$ and the energies of these states are also given in Table 9.8. It is clear from this table that for each system there is a critical value of $10Dq$ for which the two possible ground states have the same energy. Above this critical value, the systems will be in the low-spin state, but for smaller values of $10Dq$ the systems will have the same spin as the free ion. Writing this value as $(10Dq)_c$, we have

$$\begin{aligned}
(10Dq)_c &= 6B + 5C \quad (d^4) \\
&= 7.5B + 5C \quad (d^5) \\
&= 2.5B + 4C \quad (d^6) \\
&= 4B + 4C \quad (d^7)
\end{aligned}$$

▶ (9.20)

Table 9.8   Energies of the ground states of the systems d^4, d^5, d^6, and d^7 calculated in the strong-field limit. The energies of all but the high-spin states of d^4, d^5, and d^6 are subject to lowering by configuration interaction with other d^n states

Configuration	Low-spin term	Energy	Configuration	High-spin term	Energy
d^4 t_{2g}^4	$^3T_{1g}$	$6A - 15B + 5C - \frac{3}{5}(10Dq)$	$t_{2g}^3 e_g$	^5E_g	$6A - 21B - \frac{3}{5}(10Dq)$
d^5 t_{2g}^5	$^2T_{2g}$	$10A - 20B + 10C - 2(10Dq)$	$t_{2g}^3 e_g^2$	$^6A_{1g}$	$10A - 35B$
d^6 t_{2g}^6	$^1A_{1g}$	$15A - 30B + 15C - 12\frac{2}{5}(10Dq)$	$t_{2g}^4 e_g^2$	$^5T_{2g}$	$15A - 35B + 7C - \frac{2}{5}(10Dq)$
d^7 $t_{2g}^6 e_g$	^2E_g	$21A - 36B + 18C - \frac{9}{5}(10Dq)$	$t_{2g}^5 e_g^2$	$^4T_{1g}$	$21A - 40B + 14C - \frac{4}{5}(10Dq)$

Therefore, assuming B and C do not depend too much on the specific nature of the complex, the d^6 systems should be the most likely to take up a low-spin configuration, and the d^5 systems the least likely. This is in fact experimentally observed; $[Fe(H_2O)_6]^{3+}$ has a $^6A_{1g}$ ground term, whereas $[Co(H_2O)_6]^{3+}$ belongs to $^1A_{1g}$.

The ground state of a complex has very important consequences for its general physical properties. Spectra and magnetic moment are the most obviously affected. In the next two sections we will consider how these properties may be calculated using ligand field theory. The calculations are surprisingly straightforward, and even a novice in the field can perform them provided that he makes use of the appropriate tables of wavefunctions and matrix elements given by Griffith and others. For example, complete tables of matrix elements in the strong-field-coupling scheme have been tabulated by Griffith‡ and by Stevenson.§ These tables include *all* off-diagonal matrix elements *between* configurations and so can be used equally well in the weak-field limit or in any intermediate case. Given values of $10Dq$, B, and C, one can solve for the energies of all ligand field states. Even more useful are the so-called Orgel diagrams and Tanabe-Sugano diagrams in which the splittings of the levels are plotted as functions of $10Dq$ (or $10Dq/B$) for fixed B and C (or fixed B/C). In the Orgel diagrams it is conventional to take the ground state of the free ion as the zero of energy; this shows how the energy of the ground state is lowered by the splitting but is somewhat misleading because the actual zero of energy is very strongly affected by the operator $\mathbf{V}_0$ (see Sec. 9.1). In the Tanabe-Sugano diagrams the *actual* ground state is always taken as the zero of energy, and this leads to peculiar results when the state of the system changes from one spin to another. Tanabe-Sugano diagrams for d^2 through d^8 are given in Figs. 9.4 to 9.10; for comparison, the Orgel diagram for d^8 is given in Fig. 9.11. As will be seen in the next section, these diagrams are very useful in discussing spectra.

‡ Griffith, 1961, pp. 410 *ff.*
§ R. Stevenson, 1965, pp. 164 *ff.*

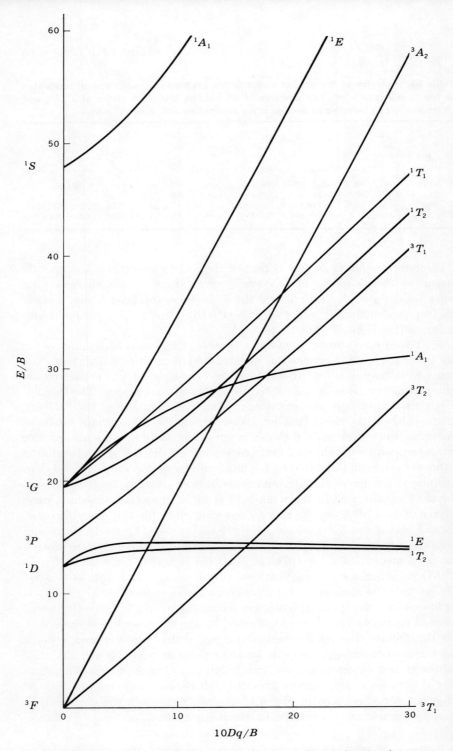

Fig. 9.4 Tanabe-Sugano-type diagram for the d^2 system. In this figure and the ones which follow, $\gamma = C/B = 3.70$, a value close to that calculated theoretically by Watson (1959) for the M^{+2} and M^{+3} ions. Experimental values vary erratically in the range 3.5 to 4.5.

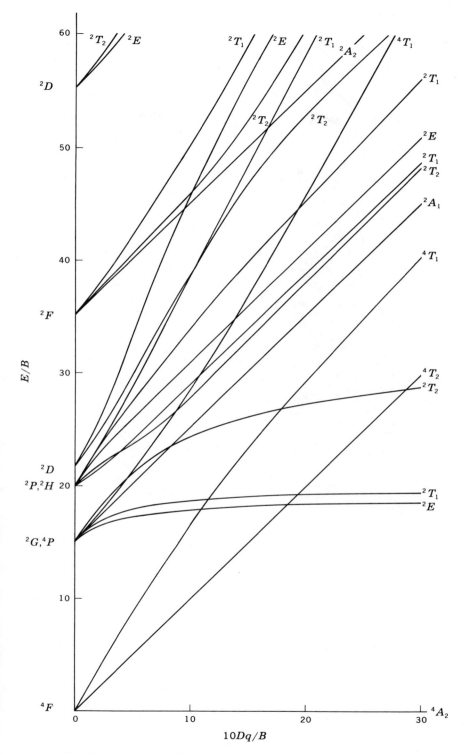

Fig. 9.5 Tanabe-Sugano-type diagram for the d^3 system. See Fig. 9.4.

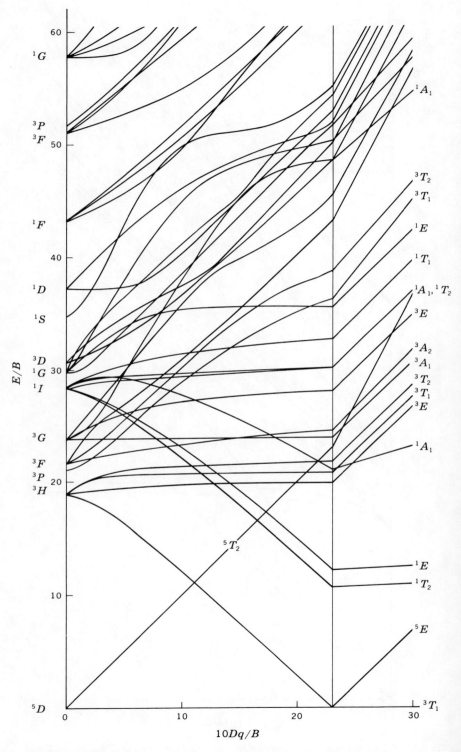

Fig. 9.6 Tanabe-Sugano-type diagram for the d^4 system. See Fig. 9.4.

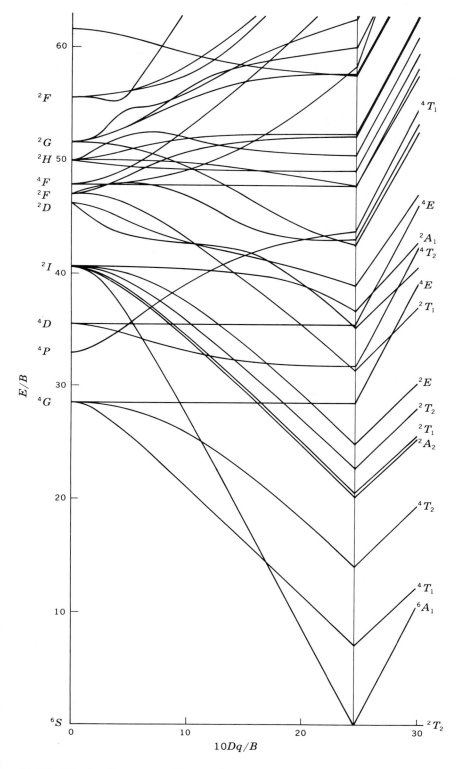

Fig. 9.7 Tanabe-Sugano-type diagram for the d^5 system. See Fig. 9.4.

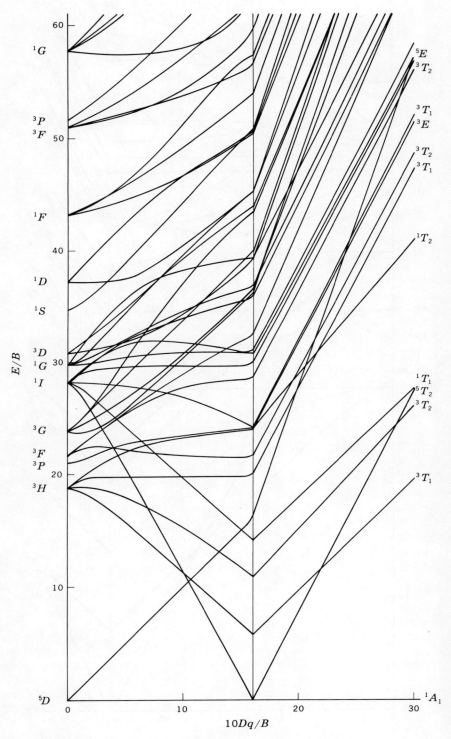

Fig. 9.8 Tanabe-Sugano-type diagram for the d^6 system. See Fig. 9.4.

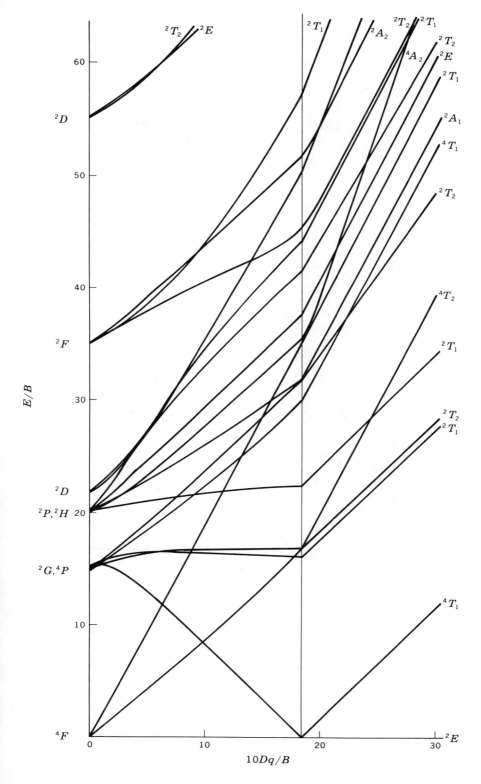

Fig. 9.9 Tanabe-Sugano-type diagram for the d^7 system. See Fig. 9.4.

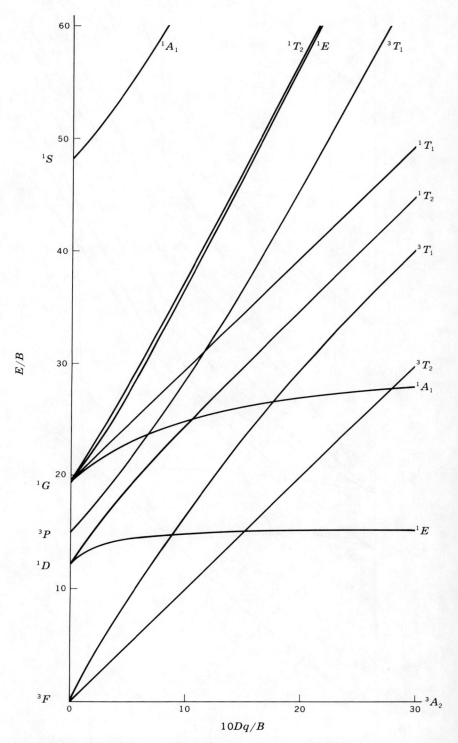

Fig. 9.10 Tanabe-Sugano-type diagram for the d^8 system. See Fig. 9.4.

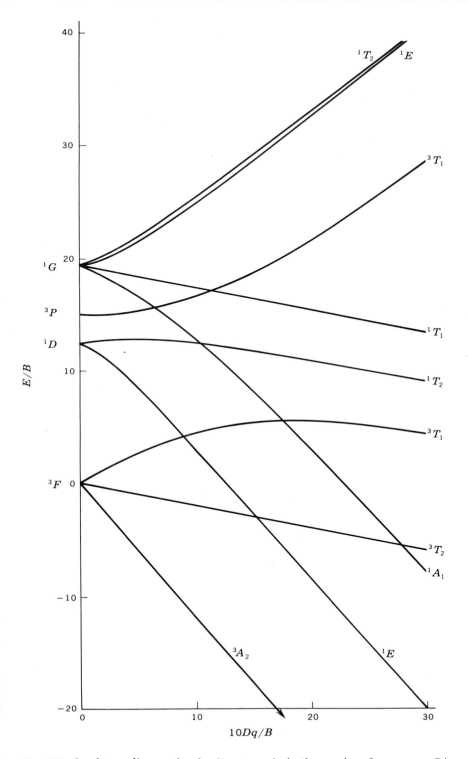

Fig. 9.11 Orgel-type diagram for the d^8 system. As in the previous figures, $\gamma = C/B = 3.70$. Note that in the Orgel-type diagram, the zero of energy is taken as the ground-state energy of the uncomplexed ion.

PROBLEMS

9.5 Work out the wavefunctions for all states of the e_g^2 and $t_{2g}e_g$ configurations.

9.6 Solve for the energies of the other two triplet states in the d^2 system.

9.7 Verify that the result of solving the secular equation for the two $^3T_{1g}$ states of d^2 leads to the same result as (9.17).

9.5 OPTICAL SPECTRA OF TRANSITION METAL COMPLEXES

As noted previously, all transitions within the d^n configurational framework are forbidden since all such states have even parity. This assumes the truth of the Born-Oppenheimer approximation, but in fact, due to vibrations, the environment around the metal will not always be strictly octahedral. In slightly asymmetric surroundings, there will be configuration interaction between d^n states and excited states (such as those arising from the $3d^{n-1}4p$ configuration) of odd parity. The intensities of the resulting d-d transitions are rather weak, with oscillator strengths f on the order of 10^{-4} (for a fully allowed transition, $f = 1$). Typically, extinction coefficients are on the order of 10 to 100. Because of the vibronic coupling, the ordinary symmetry selection rules break down, and all spin-allowed bands appear with much the same intensity.

We can use either tables of strong-field matrix elements or the Orgel diagrams to obtain the spectrum of a given complex. For example, consider once again the d^2 system. If the reader did the problems of the previous section, he will have found that the energy of the $^3A_{2g}$ term is $A - 8B + \frac{6}{5}(10Dq)$, whereas the energy of the $^3T_{2g}$ is $A - 8B + \frac{1}{5}(10Dq)$. The energies of the two $^3T_{1g}$ terms are $-\frac{3}{10}(10Dq) + A - \frac{1}{2}B \pm K$ where $K = \frac{1}{2}[(10Dq)^2 + 18B \cdot 10Dq + 225B^2]^{1/2}$. From this we can derive the following energy differences with respect to the ground state:

$$E(^3T_{1g}) = 2K$$

$$E(^3A_{2g}) = K - \frac{15B}{2} + \frac{3}{2}(10Dq)$$

$$E(^3T_{2g}) = K - \frac{15B}{2} + \frac{1}{2}(10Dq)$$

(9.21)

We therefore predict the existence of three bands, with energies related according to (9.21). Only one parameter, $10Dq$, should be necessary; the value of B can be obtained from atomic spectra. Actually, the experimental bands in $[V(H_2O)_6]^{3+}$ can be fit rather more accurately by reducing the value of B by about 10 percent, but even with the atomic B the fit is quite good. Unfortunately, the d^2 case is not entirely ideal as an introduction to spectra since, experimentally, the energy of the $^3A_{2g}$ states lies close to a charge transfer band (*vide infra*) in $[V(H_2O)_6]^{3+}$, and thus is difficult to locate with any precision.

Because of the hole-particle isomorphism, the states of d^8 systems are identical to those of d^2. Furthermore, the matrix elements between states are also identical except for possible choices of phase. However, the sign of $10Dq$ must be changed to account for the difference between particles and holes. The formulas of (9.21) therefore apply to complexes of Ni^{2+} with the sign of $10Dq$ reversed throughout. However, the ground state is now $^3A_{2g}$, and with respect to *this* state we have

$$E(^3T_{2g}) = 10Dq$$

$$E(^3T_{1g}a) = -K + \frac{15B}{2} + \frac{3}{2}(10Dq)$$

$$E(^3T_{1g}b) = K + \frac{15B}{2} + \frac{3}{2}(10Dq)$$

(9.22)

The experimental spectrum of $[Ni(H_2O)_6]^{2+}$ is shown in Fig. 9.12; three bands are indeed observed. From the first band we have $10Dq \sim 8,500$ cm^{-1}, and together with the free-ion $B = 1,056$ cm^{-1}, we predict bands at 14,300 and 27,000 cm^{-1}. The agreement with experiment can be improved if we reduce B to 887 cm^{-1}, in which case the two bands are predicted to lie 14,100 and 24,700 cm^{-1}.

The reader will note that the first $^3T_{1g}$ band in $[Ni(H_2O)_6]^{2+}$ is observed to be split. Correspondingly, on the d^8 Orgel diagram in Fig. 9.11, a ^1E_g state crosses this $^3T_{1g}$ close to the experimental $10Dq$. As we shall show in the next section, the two states are so close together that they are strongly mixed due to spin-orbit coupling, and the splitting is

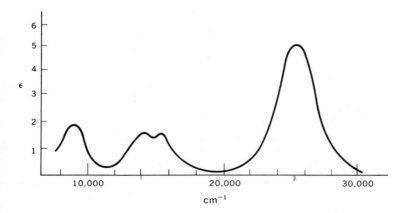

Fig. 9.12   The spectrum of $[Ni(H_2O)_6]^{2+}$. (*Data from Holmes and McClure, 1957.*)

due to this mixing. Spin-orbit coupling can be observed in most nickel
+2 complexes, but in none is it so clear as in the hexaquo complex, for
here the value of $10Dq$ is exactly right. For complexes with much larger
$10Dq$, the ^1E_g can again be observed clearly as it crosses the $^3T_{2g}$; for
intermediate $10Dq$, the singlet is weak in intensity since it is not close
enough to any triplet to interact strongly.

In d^1 and d^9 systems there is only one possible excited state. High-
spin d^4 and d^6 cases are equally elementary as there is only one spin-
allowed band. We have therefore discussed the spectra of all ions with
high-spin ground states except those having d^5 configurations. The ground
state in this case is a $^6A_{1g}$, essentially identical to the free-ion ^6S. There
are no other states of the same spin multiplicity, and thus all d-d transitions are spin-forbidden. Therefore, high-spin complexes of Mn^{2+} and
Fe^{3+} are only very weakly colored. Nevertheless, spin-forbidden transitions do occur from the ground sextet to excited quartets. Transitions
to doublets are very strictly forbidden, since the spin-orbit coupling
operator only couples states with $\Delta S = 1$. Shown in Fig. 9.13 is a classic
example of a spin-forbidden spectrum. The bands are easily identified
using Fig. 9.7. The oscillator strengths here are on the order of 10^{-7}, and
the extinction coefficients are consequently very small ($\sim 10^{-2}$). The
narrowness of the transition to the pair of bands $^4A_{1g}$ and ^4E_g is especially
remarkable, and the reason for this can be seen at once from Fig. 9.7.
The position of these bands with respect to the ground state is entirely
insensitive to the value of $10Dq$, since ground and excited states both

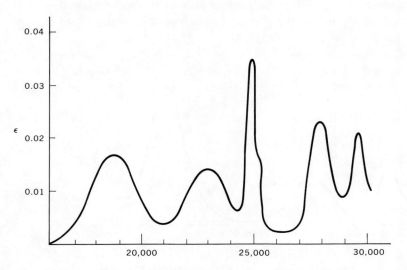

Fig. 9.13 The spin-forbidden absorption spectrum of $[Mn(H_2O)_6]^{2+}$. (*Data
from C. K. Jørgenson, 1954.*)

arise from the configuration $t_{2g}^3 e_g^2$. Since the source of the bandwidths is the change of $10Dq$ with internuclear distance in the vibronic coupling, we do expect this band to be quite narrow. Another classic example occurs in the d^3 case. As can be seen from Fig. 9.5, the energy of the spin-forbidden transition to the lowest doublets is nearly independent of the value of $10Dq$. In Cr^{3+}, these transitions are responsible for the exquisite color of natural ruby, and phosphorescence from the doublets in ruby is used in the design of the ruby laser.

The spectra of the low-spin d^4, d^5, d^6, and d^7 complexes are rather more complicated, but the reader should have no difficulty interpreting a given spectrum using Figs. 9.4 to 9.11. Two difficulties, one major and one minor, should be kept in mind. Complexes having a ground-state configuration containing an odd number of e_g electrons (d^9, high-spin d^4, and low-spin d^7) do not fit the theory at all well. Bands are not found in the predicted positions, and extra splittings are common. These observations can be explained by assuming that the ground state of the complex is not in fact octahedrally symmetric, but is tetragonally distorted. The distortion is predicted by the Jahn-Teller theorem (1937), which states that if the ground state of *any* nonlinear compound is spatially degenerate, the compound *will distort* to remove the degeneracy. The theorem does not say how large the distortion must be, and in fact complexes with spatially degenerate filling of the t_{2g} shell have negligible distortions.

The physical basis for the distortion is easy enough to understand. Consider the configuration $t_{2g}^6 e_g$. The single e_g electron can be placed in either the $d_{x^2-y^2}$ or d_{z^2} orbital. In the former case, the electronic density points entirely toward the four ligands in the xy plane. The d_{z^2} orbital is empty and can be used for binding to the *trans*-ligands. We would then expect the x and y ligands to move out and the z ligands to move in closer to the metal. In the latter case, the electronic density points primarily in the z directions, and so the $d_{x^2-y^2}$ orbital can be used for binding, and the x and y ligands move in while the z ligands move out. This latter pattern seems to occur experimentally, and only two ligands move out.

The Jahn-Teller effect is responsible not only for splittings but also for anomalously large observed values for $10Dq$. For example, in complexes of Cu^{2+}, the observed $10Dq$ is about twice that expected. The mechanism involved is somewhat complicated, and so we do not discuss it. However, it is possible to discuss what we mean when we say that a certain value for $10Dq$ is "expected." Jørgenson (1958) has shown that the experimental values fit the general formula

$$10Dq = f \cdot g$$

where f depends only on the ligands and g only on the metal. A list of typical f and g values is given in Table 9.9. The metal-dependent part decreases somewhat irregularly across the periodic table for ions of the

Table 9.9 Parameters for the estimation of $10Dq$ in ML_6 complexes, $10Dq = fg$ and $1 - \beta = hk$

Ligand	f	h	Metal ion	g, cm^{-1}	k
H_2O	1.0	1.0	V^{2+}	12,300	0.08
NH_3	1.25	1.4	Mn^{2+}	7,600	0.07
F^-	0.9	0.8	Ni^{2+}	8,900	0.12
Cl^-	0.8	2.0	Cr^{3+}	17,400	0.21
Br^-	0.76	2.3	Fe^{3+}	14,400	0.24
CN^-.	1.7	2.0	Co^{3+}	19,000	0.30

same charge. Thus one "expects" complexes of copper to have $10Dq$ values close to those of nickel, whereas in fact they are about twice as large because of the Jahn-Teller effect. The reader should note that the tripositive ions have $10Dq$ about half again as large as the dipositive ions. Furthermore, the ligands (excepting CN^- and other groups with empty low-lying π orbitals) have similar f values, independent of charge or binding strength to the metal. This clearly shows the weakness in any electrostatic theory of the origin of ligand field splitting.

The minor difficulty in fitting spectra, as mentioned above, concerns the reductions in free-ion B and C values observed in all complexes. According to the theory outlined so far, the B and C values should be unchanged by complex formation. They are a measure of the strength with which the d electrons interact, and in the simple theory this would be unchanged. Actually, there are interactions with t_{2g} and e_g symmetry orbitals on the ligands, and in place of d orbitals we have molecular orbitals belonging to the same irreducible representations.[‡] Furthermore, the charge distribution on the metal may be altered by complex formation. Thus B and C may be greatly changed; in fact in the exact theory B and C lose all meaning, and instead we have 10 independent parameters which express all possible interactions within the open shells. Experimentally, however, B and C are only slightly reduced from the free-ion values, and in fact a single reduction parameter β would appear to suffice, with the complex ion values given by βB and βC. The reduction has been called by Jørgenson the *nephalauxetic effect*, using a Greek-derived word meaning *cloud-expanding*. Any tendency to expand the d orbitals should in fact result in reduced B and C values.

Fortunately, the pattern for the nephalauxetic effect also fits a simple ligand and metal dependence. As noted by Jørgenson,[§] $(1 - \beta) = hk$, where k is a function only of the metal and h is a function only of the ligands. Values of k roughly parallel to those of g, but there is no relation

[‡] See Sec. 9.7.
[§] C. K. Jørgenson, 1962, p. 138.

between h and f. Values of k and h are given in Table 9.9. The theory behind these parameters is not understood.

Using Figs. 9.4 to 9.11 and Table 9.9, we can predict the spectrum of nearly any complex. Complexes with mixed ligands, such as $[\mathrm{Co}(\mathrm{NH}_3)_4\mathrm{X}_2]^+$ will have values of $10Dq$ intermediate between those of the hexahalide and hexamine complexes. Spectral intensities may also increase if the mixed complex no longer has a center of symmetry. Some of the bands may be split. Generally speaking, all such difficulties are minor, and the bands are easily assigned. Conversely, optical information may be used to confirm the solution structure of a complex.

The six-coordinate octahedron is not the only form for a complex. Tetrahedral bonding is also found. The theory here is very similar to that for octahedral complexes since in T_d symmetry the d orbitals transform as $T_2 + E$. In fact, it is possible to show that nearly the entire theory for the octahedral field can be taken over with a change only in the sign of $10Dq$. For this reason Orgel diagrams for the octahedral field often give the splittings for negative, as well as for positive, values of $10Dq$.

Tetragonal coordination is also common, primarily because of the Jahn-Teller effect. A related phenomena occurs in d^8 complexes, where the e_g shell is ordinarily occupied by two electrons, one in each orbital. Although this configuration is Jahn-Teller stable (the ground state is $^3A_{2g}$, which is spatially nondegenerate), an even more stable state can be found for many complexes by moving the two z ligands out. The d_{z^2} orbital is thereby lowered in energy, and the configuration *changes spin* and becomes $t_{2g}^6 d_{z^2}^2$. The $d_{x^2-y^2}$ is then used for bonding. Thus complexes of Ni^{2+} are often "square planar."

Another spectral effect common in many transition metal complexes is the occurrence of so-called charge-transfer bands. Electrons can be transferred from the ligands to the metal, or vice-versa. Of course, such a classification is necessarily approximate since the orbitals of the metal and ligand may be thoroughly mixed. The charge-transfer bands are distinguished by their great intensity, since they may be fully allowed. Fortunately, they usually occur at high enough energies and at such high intensities that they are not confused with d-d bands. (However, a partially forbidden charge-transfer band could cause trouble.) It is possible to work out a partial theory for charge-transfer spectra if something is known about the symmetry of the donor (or acceptor) orbitals of the ligand. For example, consider a simple system in which there is but a single ligand acceptor orbital of symmetry a_{1u}, and assume the metal has the t_{2g}^2 configuration. If the ground state of the complex is $^3T_{1g}$, a charge-transfer excited state can occur when a t_{2g} electron is transferred to the a_{1u} ligand orbital. The resulting state will be $^3T_{2u}(t_{2g} \times a_{1u})$, and the transition is fully allowed ($T_{1g} \times T_{1u} \times T_{2u}$ contains A_{1g}). In systems with a greater number of d electrons and in systems with ligand-to-metal charge transfer, there will be more than one charge-transfer state, and the *relative* energies of these states can be worked out using ligand field theory.

A third kind of transition, occurring within the ligands, need not concern us in this section of the text. Usually such transitions are only observed when the free ligand itself has internal transitions. Binding to a metal usually shifts the position of the bands without greatly affecting their shape.

PROBLEMS

9.8 Show that the symmetries of the high-spin terms of d^3 are the same as those of d^2.

9.9 Assign the bands in the spectrum of $[Mn(H_2O)_6]^{2+}$ (see Fig. 9.13). What problems occur in the assignment? How might these be resolved?

9.6 MAGNETIC EFFECTS: SUSCEPTIBILITIES, SPIN RESONANCE, AND SPIN–ORBIT COUPLING

A measurement of the magnetic susceptibility of a compound is a most useful technique for studying the ground state of the complex. Substances with singlet ground states are invariably diamagnetic, that is, they tend to move out of a magnetic field. On the other hand, paramagnetic substances, having unpaired electrons, are attracted by a magnetic field; the field induces a dipole moment in the substance in the direction of the field. As we shall see, the size of the dipole moment depends primarily on the number of unpaired electrons but also on orbital factors.

The molar susceptibility of a substance is defined as

$$\chi = -\frac{N_0}{H} \frac{\sum_i (\partial E_i/\partial H)e^{-E_i/kT}}{\sum_i e^{-E_i/kT}} \qquad \blacktriangleright (9.23)$$

where N_0 is Avogadro's number and H is the magnitude of the field. The sum extends over all *states*. Essentially χ is the weighted magnetic moment of each state ($\mu_i = -\partial E_i/\partial H$) divided by the field strength. We also define the total magnetic moment

$$\mu = \sqrt{\frac{3\chi kT}{N_0\beta^2}} \qquad \blacktriangleright (9.24)$$

where $\beta = e\hbar/2mc$ is the electronic Bohr magneton. The quantity μ is dimensionless and typically has values between 0 and 6.

Each term will be split by the interaction with the magnetic field, as discussed in Chap. 6. The degeneracies of the M levels belonging to L

and S will be removed. Let us assume that each energy level can be expanded as a power series in H. Writing

$$E_i = W_t^{(0)} + W_i^{(1)}H + W_i^{(2)}H^2 + \cdots \qquad (9.25)$$

we obtain

$$\frac{\partial E_i}{\partial H} = W_i^{(1)} + 2HW_i^{(2)} + \cdots \qquad (9.26)$$

$$e^{-E_i/kT} = e^{-W_t^{(0)}/kT}\left(1 - \frac{HW_i^{(1)}}{kT} + \cdots\right) \qquad (9.27)$$

All states of the same *term* have the same $W_t^{(0)}$, hence the subscript t. Substituting in (9.23), we find

$$\chi = -\frac{N_0}{H}\frac{\displaystyle\sum_t\sum_i [W_i^{(1)} - (W_i^{(1)})^2(H/kT) + 2HW_i^{(2)} + \cdots]e^{-W_t^{(0)}/kT}}{\displaystyle\sum_t\sum_i e^{-W_t^{(0)}/kT}\left(1 - HW_i^{(1)}\frac{1}{kT} + \cdots\right)}$$

$$(9.28)$$

The sum over t is over terms; the sum over i is over states of a given term. If we assume that the spacing between terms is much *greater* than kT and that the splitting caused by the field is much *less* than kT, we can drop all elements in the sum except the ground term. Furthermore, the magnetic moment μ is zero at zero field, requiring that

$$\sum_t\sum_i W_i^{(1)}e^{-W_t^{(0)}/kT} = 0 \qquad (9.29)$$

We therefore obtain

$$\chi = \frac{N_0}{g_0}\sum_i\left[\frac{(W_i^{(1)})^2}{kT} - 2W_i^{(2)}\right] \qquad \blacktriangleright(9.30)$$

where g_0 is the degeneracy of the ground term, and the sum extends only over states of the ground term. (Note that it is possible to define a susceptibility for each *term*. For high accuracy, one can include the susceptibility of excited terms weighted by their Boltzman factors.) For an isolated atom, the sum over i would extend only over the $2J + 1$ states of a given level $^{2S+1}L_J$ if the splitting due to spin-orbit coupling were much greater than kT, and over *all* states of the term if the spin-orbit splittings were much less than kT.

We must now obtain the coefficients in the energy expression (9.25). According to Sec. 6.7, the Hamiltonian for the interaction between the external field and the molecular system is

$$\mathcal{K}_H = \frac{\beta}{\hbar}\mathbf{H}\cdot(\mathbf{L} + 2\mathbf{S}) \qquad \blacktriangleright(9.31)$$

It is convenient to take the direction of H along the z axis and then

$$\mathfrak{IC}_H = \frac{\beta}{\hbar} H (\mathbf{L}_z + 2\mathbf{S}_z) \tag{9.32}$$

Using second-order perturbation theory, we can immediately write

$$W_i^{(1)} = \langle \psi_i | \beta (\mathbf{L}_z + 2\mathbf{S}_z) | \psi_i \rangle$$
$$W_i^{(2)} = \sum_{j \neq i} \frac{|\langle \psi_i | \beta (\mathbf{L}_z + 2\mathbf{S}_z) | \psi_j \rangle|^2}{E_i - E_j} \qquad \blacktriangleright (9.33)$$

Therefore

$$\chi = \frac{N_0 \beta^2}{g_0} \sum_i \left(\frac{|\langle \psi_i | \mathbf{L}_z + 2\mathbf{S}_z | \psi_i \rangle|^2}{kT} + 2 \sum_{j \neq i} \frac{|\langle \psi_i | \mathbf{L}_z + 2\mathbf{S}_z | \psi_j \rangle|^2}{E_j - E_i} \right) \qquad \blacktriangleright (9.34)$$

The spin part of the above formula is trivial, provided the functions ψ_i are eigenstates of $\mathbf{S}_z$ (that is, neglecting the spin-orbit coupling between states of different spin). The difficult part comes from $\mathbf{L}_z$ since, in an octahedral field, the wavefunctions are no longer eigenfunctions of $\mathbf{L}^2$ or $\mathbf{L}_z$. However, there is one special case of importance. If the ground term is spatially non-degenerate ($^{2S+1}A_1$ or $^{2S+1}A_2$), its orbital angular momentum is necessarily zero. The $\mathbf{L}_z$ part of the first sum is thus zero. *Within* the states of the term the second sum is also zero, and if we neglect matrix elements to the excited terms, we have

$$\chi = \frac{4N_0 \beta^2}{kT(2S + 1)} \sum_{i=-S}^{S} |\langle \psi_i | \mathbf{S}_z | \psi_i \rangle|^2 = \frac{4N_0 \beta^2}{(2S + 1)kT} \sum_{M=-S}^{S} M^2$$
$$= \frac{4N_0 \beta^2 S(S + 1)}{3kT} \tag{9.35}$$

Expressed in terms of the total magnetic moment,

$$\mu = 2\sqrt{S(S + 1)} \qquad \blacktriangleright (9.36)$$

This is known as the "spin-only" formula.

It is possible to show that $\mathbf{L}$ transforms as T_1 in octahedral symmetry (Prob. 9.10). Therefore, the matrix elements of $\mathbf{L}$ *within* a term such as $^{2S+1}A_1$, $^{2S+1}A_2$, or ^{2S+1}E are all zero. The spin-only formula therefore applies to ^{2S+1}E terms as well. Furthermore (*vide infra*), these three types of terms are not split to first order by the spin-orbit coupling, and thus there is no difficulty regarding the relative magnitude of kT and the spin-orbit coupling constant.

The configurations d^1, d^2, low-spin d^4 and d^5, and high-spin d^6 and d^7 all have ground terms of the form $^{2S+1}T_1$ or $^{2S+1}T_2$. For these we cannot expect the spin-only formula to work with much accuracy. Evaluating

Table 9.10   Experimental and calculated magnetic susceptibilities for typical transition metal complexes. Complexes with $^{2S+1}T_1$ and $^{2S+1}T_2$ ground states have their calculated μ values in parentheses, since it is not expected that the spin-only formula should give accurate predictions in these cases. Data from Griffith (1961, p. 274)

	Ion	μ [spin-only, Eq. (9.36)]	μ (experimental)
d^1	V^{4+}, Ti^{3+}	(1.73)	1.7–1.9
d^2	V^{3+}	(2.83)	2.7–2.9
d^3	V^{2+}, Cr^{3+}	3.87	3.8–3.9
d^4	Cr^{2+}, Mn^{3+}	4.90 (h.s.)	4.8–4.9
d^5	Mn^{2+}, Fe^{3+}	5.92 (h.s.)	5.8–5.9
d^6	Fe^{2+}	(4.90) (h.s.)	5.2–5.5
d^7	Co^{2+}	(3.87) (h.s.)	4.8–5.1
d^8	Ni^{2+}	2.83	2.8–3.3
d^9	Cu^{2+}	1.73	1.8–2.0

the matrix elements of **L** is not simple, so one often uses the spin-only formula as a rough approximation nevertheless. Table 9.10 illustrates the agreement with experiment. The reader will note that even in cases in which the spin-only formula should give the correct answer, there are discrepancies. Presumably these are due to contributions to the susceptibility from excited terms. In cases in which the spin-only formula should not work, Griffith‡ has shown how the correct susceptibility may be calculated. The corrected results are in good agreement with experiment.

The splitting between the states of a term in a magnetic field can be detected directly by electron spin resonance. According to Eqs. (9.25) and (9.33), a spatially nondegenerate state is split according to the relation

$$E_{M_S} = W + 2\beta H M_S \qquad \blacktriangleright(9.37)$$

For $H \sim 10,000$ gauss, two states with $M_S = \frac{1}{2}$ and $M_S = -\frac{1}{2}$ are split by about 1 cm^{-1}, and transitions between them may be induced by radio-frequency radiation. Only transitions with $M_S = \pm 1$ are allowed, and so (9.37) may be rewritten as

$$\Delta E = 2\beta H \qquad (9.38)$$

This relation is actually insufficiently general for most purposes, and it is customary to write

$$\Delta E = g\beta H \qquad \blacktriangleright(9.39)$$

The parameter g may even be field-dependent for complete generality.

‡ Griffith, 1961, pp. 271 and 281.

The theory for the calculation of g is moderately complicated, but one can determine a great deal about the electronic structure of a complex by comparing experimental values with calculations on model structures. The subject has been treated with great clarity and precision in the recent text by Carrington and McLachlan (1967), so we do not consider it further in this text. We note, however, that spin-resonance experimental methods are of the highest importance in spectroscopy and are capable of yielding an astonishing amount of information about the electronic structures of all kinds of complexes with unpaired electrons.

We next discuss the spin-orbit coupling. As noted in Chap. 6, the full Hamiltonian for a system contains a one-electron operator of the form

$$\sum_i \xi(r_i)\mathbf{l}_i \cdot \mathbf{s}_i \tag{9.40}$$

where the sum extends over all electrons in the system. In free atoms this operator does not commute with $\mathbf{L}^2$, $\mathbf{S}^2$, $\mathbf{L}_z$, or $\mathbf{S}_z$, although it "nearly" commutes with the first two. We therefore reclassified our terms from the first Russell-Saunders scheme $|{}^{2S+1}LM_SM_L\rangle$ to the second scheme $|{}^{2S+1}LJM_J\rangle$. This involved coupling $\mathbf{S}$ and $\mathbf{L}$ to form $\mathbf{J}$. $\mathbf{J}^2$ and $\mathbf{J}_z$ commute with the spin-orbit operator, and in the new coupling scheme the spin-orbit operator split terms $({}^{2S+1}L)$ into levels $({}^{2S+1}L_J)$ and mixed together levels with the same J belonging to different S and L.

In an octahedral field, we classify states according to the $|{}^{2S+1}\Gamma M_S\gamma\rangle$ scheme. In order to discuss spin-orbit coupling properly, we need to reclassify our states in a new scheme in which M_S and γ are replaced by a J-type quantum number. This involves coupling $\mathbf{S}$ to Γ, and this can be done, for integral‡ S, by considering the direct product of the representations spanned by S (Table 9.5) with Γ. Thus, for example, a ${}^3A_2$ term is classified as $|{}^3A_2T_2\gamma\rangle$ since $S = 1$ spans the T_1 representation and $A_2 \times T_1 = T_2$. Similarly, a ${}^3T_1$ term breaks up into $|{}^3T_1A_1\rangle$, $|{}^3T_1T_1\rangle$, $|{}^3T_1T_2\rangle$, and $|{}^3T_1E\rangle$ levels. Then, in the $|{}^{2S+1}\Gamma\Gamma'\gamma'\rangle$ scheme, levels of different S and Γ but the same Γ' may mix, and within a term the different levels will be split apart. For example, the T_2 level of the ${}^3T_1$ term can mix with the ${}^3A_2$ term. Note that a ${}^1T_2$ term can mix with both triplets, whereas a ${}^1T_1$ can only mix with the ${}^3T_1$.

The spin-orbit coupling operator can be shown to transform as T_1. Therefore, as noted above, the terms ${}^{2S+1}A_1$, ${}^{2S+1}A_2$, and ${}^{2S+1}E$ are not split to first order. It can further be shown that the terms ${}^{2S+1}T_1$ and ${}^{2S+1}T_2$ obey a Landé interval rule exactly the same as that followed by a ${}^{2S+1}P$ atomic term. Values for the spin-orbit coupling constant ζ for transition metal complexes are in the range of 100 to 1,000 cm^{-1}, so the

‡ For half-integral S it is necessary to define a special octahedral "double group" and determine how half-integral S transforms. The theory is not especially complicated, but to develop it would take us outside the scope of this text.

different levels of the $^{2S+1}T_1$ and $^{2S+1}T_2$ ground terms are partially populated at room temperature ($kT \sim 200$ cm^{-1}). As noted above, this complicates the calculation of the magnetic susceptibility of these complexes.

Spin-orbit coupling is often responsible for deviations of the observed g value from 2.0. For example, consider the $^3A_2$ ground state of Ni^{2+} complexes with configuration d^8. Since this state is spatially nondegenerate, the matrix elements of $\mathbf{L}_z$ are zero, and [according to (9.33)], the state is split into three components with $M_S = 1, 0,$ and -1. Applying (9.25), we have energy levels (Fig. 9.14)

$$
\begin{aligned}
E_1 &= W + 2\beta H \\
E_0 &= W \\
E_{-1} &= W - 2\beta H
\end{aligned}
\tag{9.41}
$$

Transitions can only occur with $\Delta M_S = \pm 1$, so we verify Eq. (9.38). However, spin-orbit coupling mixes a small amount of the $^3T_2$ and $^1T_2$ terms into the ground term. This mixing alters the magnetic splitting of the levels, and it can be shown that‡

$$
g = 2 + \frac{4\zeta}{10Dq}
\tag{9.42}
$$

‡ Griffith, 1961, p. 347.

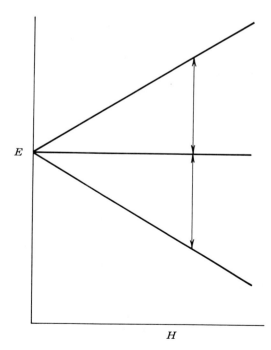

Fig. 9.14 Splitting of a spatially nondegenerate triplet in a magnetic field. The vertical lines indicate allowed transitions induced by radio-frequency radiation.

H

For example, in $[Ni(H_2O)_6]^{2+}$, with $10Dq = 8,500$ cm^{-1}, $g = 2.25$. This implies that $\zeta = 530$ cm^{-1}, somewhat reduced from the free-ion value $\zeta = 650$ cm^{-1}. The spin-orbit coupling constant usually shows a reduction in complexes, similar in magnitude to the corresponding reduction for B and C.

PROBLEMS

9.10 Show that $\mathbf{L}$ transforms as T_1 in octahedral symmetry.

9.11 What physical mechanisms could be responsible for the reduction in the spin-orbit coupling constant?

9.7 THE ORIGIN OF $10Dq$

It is often stated that $10Dq$ is the difference between the energies of the e_g and t_{2g} molecular orbitals having predominately metal d character. This is not at all true, given the usual definition of an orbital energy. It *is* true (in the approximation that the molecular orbitals are d orbitals) that *in the absence of all d-d interactions*, $10Dq$ is the difference between the t_{2g} and e_g "orbital energies," but this is a rather restricted definition.

In order to define the exact relationship between orbital energy and $10Dq$ and to examine the factors responsible for the ligand field splitting, we now proceed to a careful discussion of the molecular orbital framework of ligand field theory. First we consider the orbital energies. As is shown in the next chapter, the Fock operator for a system with one or more open shells is quite complicated, and so, for the first discussion, we choose the one ligand field system of interest with a closed-shell ground state, that with ground configuration t_{2g}^6. According to conventional Hartree-Fock theory, the orbital energies are given by the expression

$$\epsilon_i = H_i + \sum_j (2J_{ij} - K_{ij}) \qquad \blacktriangleright (9.43)$$

H_i is the one-electron energy, consisting of the kinetic energy and the sum of all nuclear potential energies, whereas J_{ij} and K_{ij} are the coulomb and exchange integrals, respectively,

$$H_i = \langle \psi_i | - \frac{\nabla^2}{2} | \psi_i \rangle - \sum_n \langle \psi_i | \frac{Z_n}{R_n} | \psi_i \rangle$$

$$J_{ij} = \langle \psi_i(1)\psi_j(2) | \frac{1}{r_{12}} | \psi_i(1)\psi_j(2) \rangle \qquad (9.44)$$

$$K_{ij} = \langle \psi_i(1)\psi_j(2) | \frac{1}{r_{12}} | \psi_j(1)\psi_i(2) \rangle$$

The sum extends over all (doubly occupied) molecular orbitals in the system.

In crystal field theory, we treat the ligands as *external* to the orbital system, and then the sum extends only over the (atomic) orbitals on the metal. We then write

$$\epsilon_i = H_i + \sum_{j(\text{metal})} (2J_{ij} - K_{ij}) + V_i \qquad (9.45)$$

where V_i is the potential due to the ligands. This potential could include the coulombic, nuclear, and exchange potentials of each ligand; ordinarily, it is approximated by point charges or dipoles placed at the ligand nuclei. Note that (9.45) is a natural extension of (9.43), in the limit of no mixing of metal and ligand orbitals; we have simply separated out that part of the sum due to the ligands and labeled it **V**.

The sum in (9.45) may again be separated into a part originating in the inner shells and a part due to the d-d interactions. We may thus write‡

$$\epsilon_i = U + \sum_{j(d)} (2J_{ij} - K_{ij}) + V_i \qquad \blacktriangleright (9.46)$$

where U is the same for t_{2g} and e_g electrons. (We are imposing the restriction that the t_{2g} and e_g d orbitals are not distorted by the field, and so have the same value of H_i, etc.) The sum over d-electron interactions is easily evaluated using Table 9.7, and we obtain

$$\begin{aligned} \epsilon_t &= U + V_t + 5A - 10B + 5C \\ \epsilon_e &= U + V_e + 6A - 6B + 3C \end{aligned} \qquad \blacktriangleright (9.47)$$

where

$$\begin{aligned} V_t &= \langle \varphi_{t_{2g}} | \mathbf{V} | \varphi_{t_{2g}} \rangle \\ V_e &= \langle \varphi_{e_g} | \mathbf{V} | \varphi_{e_g} \rangle \end{aligned} \qquad \blacktriangleright (9.48)$$

The entire discussion in the chapter thus far has assumed (for the crystal field limit) that $10Dq = V_e - V_t$, but from (9.47) we see that

$$\epsilon_e - \epsilon_t = V_e - V_t + A + 4B - 2C$$

The difference lies in the fact that an e_g orbital "sees" six d electrons, whereas a t_{2g} orbital "sees" only five. The energy of the (empty) e_g orbital therefore lies far above the occupied t_{2g} shell. Clearly, $10Dq$ is not the simple difference of two orbital energies. It is, however, the difference in e_g and t_{2g} energies *in the absence of all d-d interactions*.

The above argument applies only to crystal field theory, in which the metal and ligands can be separated. In molecular orbital theory, however, these orbitals can mix. In particular, the metal d orbitals, which are of e_g and t_{2g} symmetry, mix with linear combinations of ligand orbitals of

‡ In the weak-field coupling scheme, V is treated as a perturbation, whereas in the strong-field scheme, the d-d interactions are treated as perturbations. In the present theory both are simultaneously diagonalized.

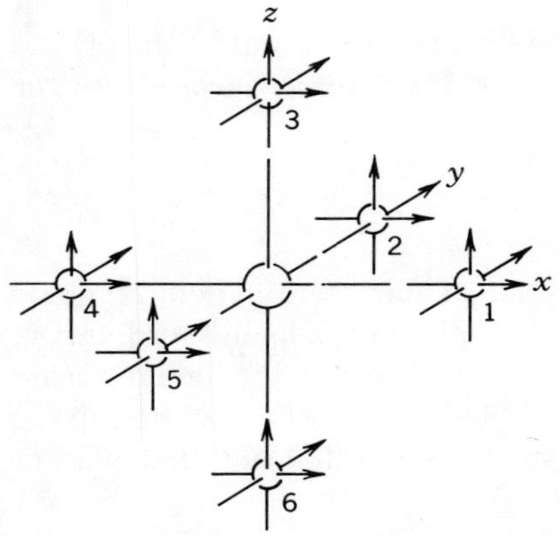

Fig. 9.15 Numbering of the ligands in an octahedrally coordinate complex.

the same symmetry. For example, consider either of the complexes MX_6 or $M(XY)_6$, where XY is a ligand which binds linearly to the metal. CrF_6^{3-} and $Cr(CO)_6^{3+}$ are typical examples. Then on each ligand *atom* we define σ and π orbitals according to Fig. 9.15. The σ orbitals are either spherically symmetric or point toward the metal, whereas the π orbitals are perpendicular to the metal-ligand axes. As in Sec. 8.5, we let the σ and π orbitals separately form bases for *reducible* representations. Then the characters of the two bases are easily shown to be

$$
\begin{array}{cccccc}
 & E & 8C_3 & 3C_2 & 6C_4 & 6C_2' \\
\chi_\sigma & 6 & 0 & 2 & 2 & 0 \\
\chi_\pi & 12 & 0 & -4 & 0 & 0
\end{array}
\qquad (9.49)
$$

from which, using Table 8.1, we determine that $\Gamma_\sigma = A_{1g} + E_g + T_{1u}$ and $\Gamma_\pi = 2T_{1u} + 2T_{2g}$. It is not difficult to find linear combinations of ligand s and p orbitals having the proper transformation properties. For example, the combination‡

$$
\chi_{a_{1g}} = \frac{N_{a_{1g}}}{\sqrt{6}} (\sigma^1 + \sigma^2 + \sigma^3 + \sigma^4 + \sigma^5 + \sigma^6) \qquad (9.50)
$$

transforms as a_{1g}, where σ^i is an s or p_σ orbital on center i. Similarly, the combinations

$$
\chi_\theta = \frac{N_{e_g}}{\sqrt{12}} (2\sigma^3 + 2\sigma^6 - \sigma^1 - \sigma^2 - \sigma^5 - \sigma^4)
$$

$$
\chi_\epsilon = \frac{N_{e_g}}{2} (\sigma^1 + \sigma^4 - \sigma^2 - \sigma^5) \qquad (9.51)
$$

‡ The symbol χ is used in three different ways in this chapter: for the character of a representation, for the magnetic susceptibility, and for symmetry orbitals. All three uses are conventional. Hopefully, the uses are different enough so that there is little possibility of confusion.

transform as the two indicated components of e_g. The only other set of components of immediate importance are those transforming as t_{2g}, and we have

$$
\begin{aligned}
\chi_x &= \tfrac{1}{2} N_{t_{2g}} (\pi_y^3 - \pi_y^6 + \pi_z^2 - \pi_z^5) \\
\chi_y &= \tfrac{1}{2} N_{t_{2g}} (\pi_z^1 - \pi_z^4 + \pi_x^3 - \pi_x^6) \\
\chi_z &= \tfrac{1}{2} N_{t_{2g}} (\pi_y^1 - \pi_y^4 + \pi_x^2 - \pi_x^5)
\end{aligned}
\tag{9.52}
$$

The normalization factors have been chosen so that they are all equal to unity in the limit of infinite ligand-ligand separation (zero ligand-ligand overlap) and so that all components of a given representation have the same normalizing factor. For example, when σ^i is a ligand s orbital, $N_{e_g}^{-2} = 1 - 2S_{12} + S_{14}$, where $S_{12} = \langle \sigma^1 | \sigma^2 \rangle$ and $S_{14} = \langle \sigma^1 | \sigma^4 \rangle$.

The molecular orbitals of e_g and t_{2g} symmetry may be approximated as finite linear combinations of atomic orbitals. Thus

$$
\psi_\theta = c_1 \varphi_{d_{z^2}} + c_2 \chi_\theta
\tag{9.53}
$$

or, more generally,

$$
\psi_\theta = c_1 \varphi_{d_{z^2}} + c_2 \chi_\theta + c_3 \chi_\theta' + \cdots
\qquad \blacktriangleright (9.54)
$$

where χ_θ and χ_θ' are linear combinations of *different* σ orbitals, such as the $1s$, $2s$, $2p_\sigma$, . . . orbitals of the ligands. For that matter, the sum may be extended to include other e_g orbitals of the metal, such as $4d$ or $5d$, or simply nonorthogonal Slater-type d orbitals.

$10Dq$ can still be defined as the difference between the e_g and t_{2g} orbital energies *in the absence of interactions among these orbitals*. Thus, in the t_{2g}^6 configuration,

$$
10Dq = \epsilon_{e_g} - \epsilon_{t_{2g}} - \sum_{i=1}^{3} (2J_{\theta i} - K_{\theta i} - 2J_{xi} + K_{xi})
\qquad \blacktriangleright (9.55)
$$

where the sum extends over the three doubly occupied components of the t_{2g} shell ($i = x$, y, and z) and the J's and K's have the usual meaning. Note that $J_{xx} = K_{xx}$.

Nonempirical SCF calculations have not been carried out on any system with t_{2g}^6 configuration, but several open-shell systems have been treated.[‡] The calculations are complicated, and the results are somewhat inconclusive. Owing to the extremely large number of many-centered integrals involved in any molecular orbital treatment with more than two or three atoms, the calculations have involved numerous approximations and simplications. NiF_6^{4-}, MnF_6^{4-}, CrF_6^{3-}, and FeF_6^{3-} are the only systems to have been treated. The fluoride ligand was chosen because it has only 10 electrons and because fluorine has a high electron affinity, implying

[‡] R. E. Watson and A. J. Freeman, 1964; S. Sugano and Y. Tanabe, 1965; J. Hubbard, D. E. Rimmer, and F. R. A. Hopgood, 1966; P. O'D. Offenhartz, 1967a.

that the metal-fluoride binding should be quite ionic. The basis sets have been restricted to the atomic orbitals of the free ions (F^- and M^{2+} or M^{3+}). All calculations (except a valence-bond calculation of MnF_6^{4-}) have predicted values of $10Dq$ which are less than half that experimentally observed,‡ which is certainly disappointing. Nevertheless, some definite conclusions can be drawn. The crystal field approach, when taken to its logical limit, gives completely erroneous results; the calculated differences $V_e - V_t$ of (9.48) are $-3,580$, $-4,840$, and $-7,180$ cm^{-1} for NiF_6^{4-}, FeF_6^{3-}, and CrF_6^{3-}, respectively, compared to 7,250, 14,000, 15,200 cm^{-1} for the experimental values of $10Dq$. In these calculations, V includes all the nuclear, coulombic, and exchange potentials of the ligands, but the metal and ligand orbitals are not orthogonalized. As noted by Freeman and Watson,§ this approach is logically somewhat inconsistent, but it is, nevertheless, the approach of classical crystal field theory.

Since crystal field theory give results of the wrong sign, it is clear that $10Dq$ directly results from either the mixing of the metal and ligand orbitals or a distortion of the metal orbitals. In either case, $10Dq$ depends on the off-diagonal matrix elements between the metal and the ligands. This explains why $10Dq$ is so insensitive to ligand charge (Table 9.9). It also helps explain why $10Dq$ is so large for systems with carbonyl and cyanide ligands; the $\psi_{t_{2g}}$ molecular orbitals are greatly lowered in energy through mixing ("back-bonding") with the empty excited π orbitals of the ligands.

There is also considerable direct experimental evidence for the importance of metal-ligand orbital mixing in explaining the origin of $10Dq$. We have already noted the reduction in B, C, and ζ. In addition, through nuclear magnetic and electronic spin resonance, the presence of unpaired electron density (spin density) has been detected at the ligand nuclei. This is a direct result of the metal-ligand binding, and in fact the experimental results have been used to estimate the degree of mixing.

It should be emphasized that the usefulness of ligand field theory is in no way imperiled by our present inability to calculate $10Dq$. The fundamental machinery of the theory depends on the high accuracy with which the relative energies of the states of the d^n configurations can be calculated within an orbital framework. It is quite possible that molecular orbital theory will not prove capable of providing an accurate calculation of $10Dq$, just as, in Chap. 6, we noted that Hartree-Fock theory gives values of B and C which are quite a bit higher than the experimental. In both cases it is still true that an effective orbital (one-electron) picture provides an accurate description of the experimental results.

‡ However, see J. W. Richardson, D. M. Vaught, T. F. Soules, and R. R. Powell, *J. Chem. Phys.*, **50**:3633 (1969) and P. O'D. Offenhartz, *J. Am. Chem. Soc.*, **91**:5699 (1969).

§ A. J. Freeman and R. E. Watson, 1960.

Both the atomic theory of Chap. 6 and the present theory of ligand fields predict relative state energies arising from one or two open shells. In the following chapters we shall attempt to calculate state energies in systems with closed-shell ground states. This implies that the *kind* of information developed in the present chapter and in Chap. 6 is no longer relevant; closed-shell molecules have these properties, but they are no longer so easily measured, and their calculation is often much more complicated than for free atoms or atoms in a field. We will therefore be turning our attention to *different* ground-state properties, such as bond lengths and angles, dipole moments, and reactivities. Our predictions of spectra will involve many different configurations instead of just one or two. We will therefore obtain less ready information than in ligand field theory, yet a far greater investment in *ab initio* computational labor, usually via computers, will be required. Furthermore, as noted in Chap. 7, such calculations are tedious even on diatomic molecules, and high accuracy is virtually out of the question for larger systems. Our theories will therefore be more and more semiempirical, with a progressively weaker theoretical foundation. Furthermore, connections between theory and experiment will become more difficult to make. Nevertheless, the situation is far from hopeless, and a great deal of progress has been made in finding approximate methods of calculation which simulate either the experimental results or the Hartree-Fock predictions. (As in the case of the atomic B and C values, it is doubtful if both can be simulated at once.) In the next two chapters we first lay down the general theoretical background and methods of approach and secondly examine some of the approximate techniques which have been proposed.

PROBLEMS

9.12 In the crystal field approach, show in detail that $\epsilon_e - \epsilon_t = V_e - V_t + A + 4B - 2C$ for the t_{2g}^6 configuration.

9.13 Verify the characters of the reducible bases given in (9.49), and include information on the transformation properties in O_h to be able to verify the given decompositions, including parity.

9.14 Determine the normalizing factors in (9.51) and (9.52) in terms of the overlap integrals between ligands. Let the σ orbitals represent $2p_\sigma$ functions rather than $2s$ functions.

9.15 Carbon monoxide and the cyanide ion both combine with metal ions to produce complexes with exceptionally large values of $10Dq$. Which diatomic molecules or ions could combine to give small or negative values of $10Dq$?

SUGGESTIONS FOR FURTHER READING

The literature in ligand field theory is vast, but fortunately there are now several excellent texts.

1. Griffith, J. S.: "The Theory of Transition Metal Ions," Cambridge University Press, London, 1961.

   This is the best source of theoretical material. The book is both lucid and comprehensive, and is to be very highly recommended.

2. Ballhausen, C. J.: "Introduction to Ligand Field Theory," McGraw-Hill Book Company, New York, 1962.

   This book is not so clearly written, but it contains more references to the experimental literature.

3. Jørgenson, C. K.: "Absorption Spectra and Chemical Bonding in Complexes," Addison-Wesley Publishing Co., Inc., Reading, Mass., 1962.

   This book has the best summary of experimental work.

4. Orgel, Leslie E.: "An Introduction to Transition Metal Chemistry," Methuen & Co., Ltd. London, 1960.

   The monograph is short, clear, and almost entirely nonmathematical, and the author has a true genius for making complicated points seem perfectly obvious. It certainly provides the ideal introduction to the subject as a whole, and is by far the best short summary of both the theoretical and experimental aspects of ligand field theory.

5. Murrell, J. N., S. F. A. Kettle, and J. M. Tedder: "Valence Theory," John Wiley & Sons, Inc., New York, 1965.

   A good short discussion of ligand field theory is also to be found in this book, and elementary discussions will be found in most basic books on inorganic chemistry.

6. Stevenson, Richard: "Multiplet Structure of Atoms and Molecules," W. B. Saunders Co., Philadelphia, 1965.

   This is a terse but readable account of some of the more sophisticated means of handling open-shell calculations. The chapter on ligand field theory contains many references to the recent theoretical literature.

10

GENERAL ORBITAL THEORY

The theory yields much, but it
hardly brings us nearer to the
secret of the Old One.
A. Einstein

10.1 DERIVATION OF THE ROOTHAAN–HARTREE–FOCK EQUATIONS FOR CLOSED SHELLS‡

In Sec. 6.8 we developed the method of the self-consistent field, and stated that for sufficiently large and well-chosen basis sets, the energy obtained in the SCF method converged to the Hartree-Fock limit. It presumably then gives the lowest possible energy for a closed-shell, single-configuration orbital (one-electron) wavefunction. This assumption has been implicitly used in discussing molecules as well since, in the LCAO expansion, one supposes that a consistent set of orbital coefficients will optimize the energy of the molecule. Furthermore, almost all modern approximate

‡ This section and the one following are based on two papers by C. C. J. Roothaan (1951, 1960). Neither paper is easy reading, and I have benefited much from attending lectures by Professor J. de Heer, who has gone to great lengths to clarify some of the more difficult points in these papers.

orbital treatments of molecules rely on the concept of the self-consistent field. It is therefore worthwhile examining the exact relationship between the SCF method and the variation theorem, and thus we explicitly derive the Roothaan-Hartree-Fock equations.

We begin by listing the crucial assumptions and quoting the derived results. First, each orbital‡ ψ_i is expressed as a linear combination of basis functions φ_p; thus

$$\psi_i = \sum_{p=1}^{m} c_{ip}\varphi_p \qquad \blacktriangleright (10.1)$$

Secondly, the ground state of the system is to be written as a single Slater determinant,

$$\Psi = |\psi_1(1)\bar{\psi}_1(2)\psi_2(3)\bar{\psi}_2(4) \cdots \psi_n(2n-1)\bar{\psi}_n(2n)| \qquad \blacktriangleright (10.2)$$

Thirdly, using the system Hamiltonian

$$\mathcal{K} = \sum_{i=1}^{2n} \mathbf{H}(i) + \sum_{\substack{i=1 \\ i<j}}^{2n-1} \sum_{j=i+1}^{2n} \frac{e^2}{r_{ij}} \qquad \blacktriangleright (10.3)$$

we can derive the fundamental expression for the total energy of the system,

$$E = 2\sum_{i=1}^{n} H_i + \sum_{i=1}^{n} \sum_{j=1}^{n} (2J_{ij} - K_{ij}) \qquad \blacktriangleright (10.4)$$

$\mathbf{H}(i)$ is the one-electron operator,

$$\mathbf{H}(i) = -\frac{\nabla_i^2}{2} - \sum_{k=1}^{K} \frac{e^2}{R_{ki}} \qquad (K \text{ nuclei in the system}) \qquad (10.5)$$

The matrix elements H_i, J_{ij}, and K_{ij} are defined as

$$\begin{aligned}
H_i &= \langle \psi_i(1)|\mathbf{H}(1)|\psi_i(1)\rangle \\
J_{ij} &= \langle \psi_i(1)\psi_j(2)| \frac{e^2}{r_{12}} |\psi_i(1)\psi_j(2)\rangle \\
K_{ij} &= \langle \psi_i(1)\psi_j(2)| \frac{e^2}{r_{12}} |\psi_j(1)\psi_i(2)\rangle
\end{aligned} \qquad (10.6)$$

‡ Compare Eq. (6.96). In Chap. 6 it was convenient to call the basis functions χ_p and define the atomic orbitals as φ_i. In this and subsequent chapters, ψ_i is a molecular orbital, φ_p is a basis function (often an atomic orbital), and χ is a special linear combination of atomic orbitals referred to as a symmetry orbital; see Chap. 9, especially Sec. 9.7 and Eqs. (9.50) to (9.54).

Then, according to the Roothaan theory, the ψ_i are solutions to the pseudo-eigenvalue equations

$$\mathbf{F}\psi_i = \epsilon_i\psi_i \qquad\qquad\blacktriangleright (10.7)$$

The operator $\mathbf{F}$ is given by the expression

$$\mathbf{F} = \mathbf{H} + \sum_{j=1}^{n} (2\mathbf{J}_j - \mathbf{K}_j) \qquad\qquad\blacktriangleright (10.8)$$

where $\mathbf{J}_j$ and $\mathbf{K}_j$ are defined by the equations

$$\begin{aligned}\langle \psi_i|\mathbf{J}_j|\psi_i\rangle &= J_{ij}\\ \langle \psi_i|\mathbf{K}_j|\psi_i\rangle &= K_{ij}\end{aligned} \qquad\qquad (10.9)$$

It is easily shown that the orbital energies are given by the formula

$$\epsilon_i = H_i + \sum_{j=1}^{n} (2J_{ij} - K_{ij}) \qquad\qquad\blacktriangleright (10.10)$$

so that

$$E = \sum_{i=1}^{n} (H_i + \epsilon_i) \qquad\qquad\blacktriangleright (10.11)$$

Equation (10.7) is a pseudo-eigenvalue equation in the sense that both $\mathbf{F}$ and the ψ_i are determined by the coefficients c_{ip}. The operator satisfying (10.7) can only be found by trial and error, hence the phrase "self-consistent" field.

Now let us consider the derivation of these equations from the variation theorem in order to prove that the SCF method actually minimizes the energy of a given configurational wavefunction. We seek to minimize‡ the energy of Ψ [Eq. (10.2)] with respect to the coefficients c_{ip}, subject to the condition that the resulting orbitals ψ_i are all orthogonal, or

$$\langle \psi_i|\psi_j\rangle = \delta_{ij} \qquad\qquad\blacktriangleright (10.12)$$

The energy of the system is

$$E = \langle \Psi|\mathfrak{IC}|\Psi\rangle$$

In Secs. 6.3 and 6.8 we showed that this expression can be rewritten as in (10.4), where the sums extend over the doubly occupied orbitals of the system (there are $2n$ electrons). Defining the operators $\mathbf{J}_j$ and $\mathbf{K}_j$ as in

‡ Slater has shown that it is not necessary to *assume* that the α-spin electrons occupy the same orbitals as the β-spin electrons. It can be proved that in a closed-shell system they *must* occupy the same orbitals in order to minimize the total energy; see J. C. Slater, 1960b, p. 2.

(10.9), we may write

$$E = 2 \sum_{i=1}^{n} \langle \psi_i | \mathbf{H} | \psi_i \rangle + \sum_{i=1}^{n} \sum_{j=1}^{n} \langle \psi_i | \mathbf{G}_j | \psi_i \rangle \tag{10.13}$$

where

$$\mathbf{G}_j = 2\mathbf{J}_j - \mathbf{K}_j \tag{10.14}$$

Note that

$$\langle \psi_i | \mathbf{G}_j | \psi_i \rangle = G_{ij} = G_{ji} = \langle \psi_j | \mathbf{G}_i | \psi_j \rangle \tag{10.15}$$

Let us now consider a small variation in E. If we regard the basis set as fixed, the variation can be carried out only through changes in the coefficients c_{ip}. On the other hand, if the energy is at a true minimum, $\delta E = 0$ for any such variation. We obtain

$$\delta E = 2 \sum_{i=1}^{n} (\langle \delta\psi_i | \mathbf{H} | \psi_i \rangle + \langle \psi_i | \mathbf{H} | \delta\psi_i \rangle)$$
$$+ \sum_{i=1}^{n} \sum_{j=1}^{n} (\langle \delta\psi_i | \mathbf{G}_j | \psi_i \rangle + \langle \psi_i | \mathbf{G}_j | \delta\psi_i \rangle + \langle \psi_i | \delta\mathbf{G}_j | \psi_i \rangle) \tag{10.16}$$

where the usual rule for differentiation of a product has been followed. We now eliminate the term involving $\delta\mathbf{G}_j$ by noting that

$$\langle \delta\psi_i | \mathbf{G}_j | \psi_i \rangle + \langle \psi_i | \mathbf{G}_j | \delta\psi_i \rangle = \langle \psi_j | \delta\mathbf{G}_i | \psi_j \rangle \tag{10.17}$$

Carrying out the double summation and substituting in (10.16), we find

$$\delta E = 2 \sum_{i=1}^{n} \left(\langle \delta\psi_i | \mathbf{H} | \psi_i \rangle + \langle \psi_i | \mathbf{H} | \delta\psi_i \rangle + \langle \delta\psi_i | \sum_{j=1}^{n} \mathbf{G}_j | \psi_i \rangle \right.$$
$$\left. + \langle \psi_i | \sum_{j=1}^{n} \mathbf{G}_j | \delta\psi_i \rangle \right) \tag{10.18}$$

The second and fourth terms are simply the complex conjugates of the first and third; that is,

$$\langle \psi_i | \mathbf{H} | \delta\psi_i \rangle = \langle \delta\psi_i | \mathbf{H} | \psi_i \rangle^* \tag{10.19}$$

since all operators are Hermitian. Therefore,

$$\delta E = 2 \sum_{i=1}^{n} \left(\langle \delta\psi_i | \mathbf{H} + \sum_{j=1}^{n} \mathbf{G}_j | \psi_i \rangle + \langle \delta\psi_i | \mathbf{H} + \sum_{j=1}^{n} \mathbf{G}_j | \psi_i \rangle^* \right) \tag{10.20}$$

The above relation is subject to the constraint (10.12), hence

$$\langle \delta\psi_i | \psi_j \rangle + \langle \psi_i | \delta\psi_j \rangle = 0 \tag{10.21}$$

Note that the first term involves a variation in ψ_i, whereas the second involves a variation in ψ_j. If we now define an arbitrary set of multipliers ϵ_{ij} and then sum over both i and j, it is clear that

$$-2 \sum_{i=1}^{n} \sum_{j=1}^{n} \epsilon_{ji}(\langle \delta\psi_i|\psi_j\rangle + \langle\psi_i|\delta\psi_j\rangle) = 0 \qquad (10.22)$$

since each term in the sum is zero. Making use of the relation

$$\langle\psi_i|\delta\psi_j\rangle = \langle\delta\psi_j|\psi_i\rangle^* \qquad (10.23)$$

we may rewrite (10.22) as

$$-2 \sum_{i=1}^{n} \sum_{j=1}^{n} \epsilon_{ji}\langle\delta\psi_i|\psi_j\rangle - 2 \sum_{i=1}^{n} \sum_{j=1}^{n} \epsilon_{ij}\langle\delta\psi_i|\psi_j\rangle^* = 0 \qquad (10.24)$$

where we have interchanged the dummy indices i and j in the second double sum. According to Lagrange's method of undetermined multipliers, we now add (10.20) and (10.24) to obtain

$$\delta E = 0 = 2 \sum_{i=1}^{n} \left(\langle\delta\psi_i|\mathbf{H} + \sum_{j=1}^{n} \mathbf{G}_j|\psi_i\rangle - \sum_{j=1}^{n} \epsilon_{ji}\langle\delta\psi_i|\psi_j\rangle \right)$$
$$+ 2 \sum_{i=1}^{n} \left(\langle\delta\psi_i|\mathbf{H} + \sum_{j=1}^{n} \mathbf{G}_j|\psi_i\rangle^* - \sum_{j=1}^{n} \epsilon_{ij}\langle\delta\psi_i|\psi_j\rangle^* \right) \qquad (10.25)$$

Lagrange's method now states that in order for δE to equal zero, each term in the sum must vanish. Furthermore, since the variations are arbitrary, their "coefficient" must in turn vanish, and we obtain

$$(\mathbf{H} + \sum_{j=1}^{n} \mathbf{G}_j)|\psi_i\rangle = \sum_{j=1}^{n} \epsilon_{ji}|\psi_j\rangle$$
$$(\mathbf{H} + \sum_{j=1}^{n} \mathbf{G}_j)^*|\psi_i\rangle^* = \sum_{j=1}^{n} \epsilon_{ij}|\psi_j\rangle^* \qquad \blacktriangleright(10.26)$$

Taking the complex conjugate of the second equation and subtracting it from the first, we find that

$$\sum_{j=1}^{n} (\epsilon_{ji} - \epsilon_{ij}^*)|\psi_j\rangle = 0 \qquad (10.27)$$

Thus $\epsilon_{ij} = \epsilon_{ji}^*$ for all i,j, and the matrix $\boldsymbol{\epsilon}$ whose elements are ϵ_{ij} is Hermitian.

Equation (10.26) is not quite an eigenvalue relation since ψ_i is transformed into a sum over ψ_j. However, the sum can be diagonalized. Consider the following row vector $\boldsymbol{\psi}$ whose elements are the ψ_i:

$$\boldsymbol{\psi} = (\psi_1\psi_2\psi_3 \cdots \psi_n) \qquad \blacktriangleright(10.28)$$

Then (10.26) may be written in the form

$$\mathbf{F}\boldsymbol{\psi} = \boldsymbol{\psi}\boldsymbol{\varepsilon} \qquad\qquad \blacktriangleright (10.29)$$

or

$$\mathbf{F}'\boldsymbol{\psi}' = \boldsymbol{\psi}'\boldsymbol{\varepsilon}' \qquad\qquad \blacktriangleright (10.30)$$

where we have arbitrarily added primes to all quantities for reasons to be seen. Since $\boldsymbol{\varepsilon}'$ is Hermitian, there exists a unitary matrix $\mathbf{U}$ such that

$$\boldsymbol{\varepsilon} = \mathbf{U}\dagger\boldsymbol{\varepsilon}'\mathbf{U} \qquad\qquad \blacktriangleright (10.31)$$

is a diagonal matrix. Defining $\boldsymbol{\psi} = \boldsymbol{\psi}'\mathbf{U}$, we can easily prove (Prob. 10.1) that

$$\mathbf{F}'\boldsymbol{\psi} = \boldsymbol{\psi}\boldsymbol{\varepsilon} \qquad\qquad \blacktriangleright (10.32)$$

This is now a true (pseudo-) eigenvalue equation, with components of the form

$$\mathbf{F}'\psi_i = \epsilon_i\psi_i \qquad\qquad \blacktriangleright (10.33)$$

where $\epsilon_i \equiv \epsilon_{ii}$. However, it is not immediately obvious that this diagonal form still minimizes the energy, nor is the meaning of $\mathbf{F}'$ entirely well defined. However, we note that the unitary transformation $\boldsymbol{\psi} = \boldsymbol{\psi}'\mathbf{U}$ merely adds certain linear combinations of rows of the Slater determinant Ψ without changing its value; that is, $\Psi = \Psi'$. Furthermore, $\mathbf{F} = \mathbf{F}'$. The proof is not difficult; $\mathbf{H} = \mathbf{H}'$ since this operator does not involve the ψ_i. As for the two-electron operator $\mathbf{G}' = \sum\limits_{j=1}^{n} \mathbf{G}'_j$, we note that

$$\sum_{j=1}^{n} \mathbf{G}'_j = \sum_{j=1}^{n} (2\mathbf{J}'_j - \mathbf{K}'_j) = \sum_{j=1}^{n}\sum_{p=1}^{m}\sum_{q=1}^{m} c'^{*}_{jp}c'_{jq}(2\mathbf{J}_{pq} - \mathbf{K}_{pq}) \qquad (10.34)$$

where we have expanded the orbital operators in terms of constituent "basis operators" whose nature need not be further defined at this point. The coefficients c'_{jp} are related to the unprimed coefficients c_{jp} by the equation

$$c_{ip} = \sum_{j=1}^{n} c'_{jp}U_{ji} \qquad\qquad (10.35)$$

or

$$c'_{ip} = \sum_{j=1}^{n} c_{jp}U^{*}_{ij} \qquad\qquad (10.36)$$

Substituting in (10.34), we obtain

$$\sum_{j=1}^{n} \mathbf{G}'_j = \sum_{j=1}^{n}\sum_{p=1}^{m}\sum_{q=1}^{m}\sum_{k=1}^{n}\sum_{l=1}^{n} c^{*}_{kp}c_{lq}U_{jk}U^{*}_{jl}(2\mathbf{J}_{pq} - \mathbf{K}_{pq}) \qquad (10.37)$$

The sum over j introduces a factor of δ_{kl}, hence

$$\sum_{j=1}^{n} \mathbf{G}'_j = \sum_{k=1}^{n} \sum_{p=1}^{m} \sum_{q=1}^{m} c^*_{kp} c_{kq}(2\mathbf{J}_{pq} - \mathbf{K}_{pq}) \tag{10.38}$$

The right-hand side is by definition equal to the total two-electron operator in the unprimed scheme, so

$$\sum_{j=1}^{n} \mathbf{G}'_j = \sum_{j=1}^{n} \mathbf{G}_j \tag{10.39}$$

This completes the proof; we have shown that

$$\mathbf{F}\boldsymbol{\psi} = \boldsymbol{\psi}\boldsymbol{\varepsilon} \qquad\qquad \blacktriangleright (10.40)$$

or, in component form,

$$\mathbf{F}\psi_i = \epsilon_i \psi_i \tag{10.41}$$

The SCF equations therefore follow directly from the variation principle. As a by-product, we have proved that an arbitrary unitary transformation on a Slater determinant alters neither the determinant itself nor the resultant Fock operator. This result is often used to demonstrate that the molecular orbital picture can be transformed into an equivalent "localized orbital" scheme in which the standard orbitals are chosen to be those most nearly resembling atomic functions or certain "bond" functions.

From the definition of $\mathbf{F}$, Eqs. (10.10) and (10.11) immediately follow. However, it is not really a straightforward matter to solve for the orbital energies and coefficients since $\mathbf{F}$ involves the coefficients via (10.38). The process involves an iterative determination of the exact solution. An arbitrary set of coefficients is chosen, and $\mathbf{F}$ is determined and used to calculate a set of orbitals and orbital energies. The *new* orbitals are used to determine a *new* $\mathbf{F}$, and so on until convergence. There is actually no guarantee that this process does converge, and, divergent cases, although rare, have been found. However, a number of procedures‡ have been described which speed up the convergence of the process, and so there is not any real difficulty.

It is possible to write Eq. (10.1) in matrix form as

$$\boldsymbol{\psi} = \hat{\boldsymbol{\varphi}}\mathbf{C} \qquad\qquad \blacktriangleright (10.42)$$

where $\mathbf{C}$ is a *rectangular* matrix of dimension $m \times n$, n being the number of occupied orbitals ψ_i and m being the number of basis functions φ_p. Obviously $m > n$. Note that $\hat{\boldsymbol{\varphi}}$ is a row vector

$$\hat{\boldsymbol{\varphi}} = (\varphi_1 \varphi_2 \varphi_3 \cdots \varphi_m) \qquad\qquad \blacktriangleright (10.43)$$

‡ R. McWeeny, 1956; C. C. J. Roothaan and P. S. Bagus, 1963.

as is ψ. Hence

$$\psi_i = \sum_{p=1}^{m} \varphi_p C_{pi} \qquad \blacktriangleright (10.44)$$

where the conventional order of the subscripts has been changed to agree with matrix notation requirements.‡ We may also define a matrix $\mathbf{S}$ and a matrix F_{pq} such that

$$S_{pq} = \langle \varphi_p | \varphi_q \rangle$$
$$F_{pq} = \langle \varphi_p | \mathbf{F} | \varphi_q \rangle \qquad (10.45)$$

Substituting in (10.40), we then obtain

$$\mathbf{F}\hat{\varphi}\mathbf{C} = \hat{\varphi}\mathbf{C}\boldsymbol{\varepsilon} \qquad (10.46)$$

or

$$\hat{\varphi}\dagger\mathbf{F}\hat{\varphi}\mathbf{C} = \hat{\varphi}\dagger\hat{\varphi}\mathbf{C}\boldsymbol{\varepsilon} \qquad (10.47)$$

If we now integrate over all coordinates, we obtain

$$\mathbf{F}\mathbf{C} = \mathbf{S}\mathbf{C}\boldsymbol{\varepsilon} \qquad \blacktriangleright (10.48)$$

or, in component form,

$$\sum_{q=1}^{m} (F_{pq} - S_{pq}\epsilon_i)C_{qi} = 0 \qquad i = 1, 2, \ldots, n \qquad \blacktriangleright (10.49)$$

The orbital energies therefore can be found by solving the secular determinant

$$|F_{pq} - S_{pq}\epsilon_i| = 0 \qquad \blacktriangleright (10.50)$$

for the eigenvalues ϵ_i. A method for determining the corresponding coefficients is outlined in Appendix 2.

The solution to the secular determinant (10.50) actually produces m eigenvalues ϵ_i and m orbitals ψ_i. The extra $m - n$ are *virtual* orbitals, in the sense that they are empty and are not part of the Slater determinant. Nevertheless, they are of great importance in discussing excited states, as we will show in Sec. 10.3. Note that this means we can define $\mathbf{C}$ to be square ($m \times m$) and let ψ have m components. It is clear that only the n *lowest* eigenvalues ϵ_i correspond to occupied orbitals.

PROBLEMS

10.1 Verify (10.32).
Hint: Begin by showing that $\boldsymbol{\varepsilon}' = \mathbf{U}\boldsymbol{\varepsilon}\mathbf{U}\dagger$.

10.2 Show that $\mathbf{C}\dagger\mathbf{S}\mathbf{C} = \mathbf{I}$, the $m \times m$ unit matrix.

10.3 Use the result proved in the previous problem to show that $\boldsymbol{\varepsilon} = \mathbf{C}\dagger\mathbf{F}\mathbf{C}$.

‡ Thus, $C_{pi} = c_{ip}$.

10.2 OPEN-SHELL EQUATIONS

The Roothaan-Hartree-Fock equations for an open shell are a great deal more complicated than those for a closed shell, and, considering the relative complexity of the derivation given in the previous section, we shall omit a derivation. Physically, as noted in Chap. 6, the difficulty arises from the different exchange potentials seen by up-spin and down-spin electrons; mathematically this problem appears as off-diagonal Lagrangian multipliers ϵ_{ij} which are hard to eliminate.

At first sight it might appear obvious that we should define two sets of orbitals, one for up-spin electrons and one for down-spin electrons. Thus, for example, lithium ($1s^2 2s$) could have two different $1s$ orbitals. The trouble with this approach (known as the unrestricted Hartree-Fock or UHF method) is that the resulting Slater determinant is not an eigenfunction of $\mathbf{S}^2$. This is not an overriding objection—one can use "projection operators" to get out a function‡ which is *very nearly* an eigenfunction of $\mathbf{S}^2$—but the equations for this method are hard to generalize. In the restricted Hartree-Fock (RHF) approach, one writes the basic Slater determinant of highest possible spin as

$$\Psi = |\psi_1^+(1)\psi_1^-(2)\psi_2^+(3)\psi_2^-(4) \cdots \psi_{n_c}^+(2n_c - 1)\psi_{n_c}^-(2n_c)\psi_{n_c+1}^+(2n_c + 1)$$
$$\cdots \psi_{n_c+n_o}^+(2n_c + n_o)| \tag{10.51}$$

The first n_c orbitals are doubly occupied and form the closed-shell part of the system. The next n_o orbitals all contain electrons with α spin. This determinant *is* an eigenfunction of $\mathbf{S}^2$, with $S = M_S = \frac{1}{2}n_o$. The Roothaan approach follows the RHF formalism, but it is *not* restricted to systems in which all the open-shell electrons contain α-spin electrons.

The energy of an *arbitrary* open-shell determinant can be written down at once, using the methods developed in Chap. 6. We obtain

$$E = \left[2 \sum_{k=1}^{n_c} H_k + \sum_{k=1}^{n_c} \sum_{l=1}^{n_c} (2J_{kl} - K_{kl}) \right]$$
$$+ \left[\sum_{k=1}^{n_c} \sum_{m=n_c+1}^{n_o} (2J_{km} - K_{km}) \right]$$
$$+ \left[\sum_{m=n_c+1}^{n_o} H_m + \sum_{m=n_c+1}^{n_o} \sum_{n=n_c+1}^{n_o} (J_{mn} - K'_{mn}) \right] \qquad \blacktriangleright (10.52)$$

Following Roothaan, we see that subscripts k and l refer to the closed shells, m and n to the open shells. Note that the first part of (10.52) is the energy of the closed-shell in (10.51), the third part is the energy of the open shell, and the second part is the interaction between the two sets of shells. The determinant need *not* have α spin for all electrons in

‡ G. Berthier, 1964.

the open shell; we take K'_{mn} as zero when the two orbitals m and n contain electrons with opposite spin.

Ordinarily, we are not interested in the energy of a *single* determinant *unless* all electrons have α spin, since otherwise the determinant will not necessarily be an eigenfunction of $\mathbf{S}^2$. However, we *are* interested in the energies of linear combinations of Slater determinants which are suitable eigenfunctions of the spin. The energy of such a linear combination is, in general, not so easily written down. However, as noted by Roothaan, for a large number of systems it is possible to write this energy as

$$E = 2 \sum_{k=1}^{n_c} H_k + \sum_{k=1}^{n_c} \sum_{l=1}^{n_c} (2J_{kl} - K_{kl}) + 2 \sum_{k=1}^{n_c} \sum_{m=1}^{n_s} (2J_{km} - K_{km})$$

$$+ f\left[2 \sum_{m=1}^{n_s} H_m + f \sum_{n=1}^{n_s} \sum_{n=1}^{n_s} (2aJ_{mn} - bK_{mn}) \right] \quad \blacktriangleright (10.53)$$

Here f is the fractional occupancy $(0 < f < 1)$ of the open shell, and a and b are constants to be determined. Note that the sums over m and n now extend not over the number of electrons in the open shell, but rather over the number of degenerate orbitals belonging to an irreducible representation. Thus, $n_s = 3$ for the t_{2g} shell in octahedral symmetry, $n_s = 2$ for the e_g shell, and $n_s = 1$ for any spatially nondegenerate shell. For the t_{2g}^4 configuration, $f = 2/3$; for e_g^3, $f = 3/4$. The constants a and b depend on the specific state of the given configuration, and K_{mn} is nonzero.

We now define coulomb operators $\mathbf{J}$, exchange operators $\mathbf{K}$, coulomb coupling operators $\mathbf{L}$, and exchange coupling operators $\mathbf{M}$ in the following way:

$$\mathbf{J}_c = \sum_{k=1}^{n_c} \mathbf{J}_k \qquad \mathbf{K}_c = \sum_{k=1}^{n_c} \mathbf{K}_k$$

$$\mathbf{J}_o = f \sum_{m=1}^{n_s} \mathbf{J}_m \qquad \mathbf{K}_o = f \sum_{m=1}^{n_s} \mathbf{K}_m$$

$$\mathbf{L}_i|\psi\rangle = \langle\psi_i|\mathbf{J}_o|\psi\rangle|\psi_i\rangle + \langle\psi_i|\psi\rangle\mathbf{J}_o|\psi_i\rangle$$

$$\mathbf{M}_i|\psi\rangle = \langle\psi_i|\mathbf{K}_o|\psi\rangle|\psi_i\rangle + \langle\psi_i|\psi\rangle\mathbf{K}_o|\psi_i\rangle \qquad \blacktriangleright (10.54)$$

$$\mathbf{L}_c = \sum_{k=1}^{n_c} \mathbf{L}_k \qquad \mathbf{M}_c = \sum_{k=1}^{n_c} \mathbf{M}_k$$

$$\mathbf{L}_o = f \sum_{m=1}^{n_s} \mathbf{L}_m \qquad \mathbf{M}_o = f \sum_{m=1}^{n_s} \mathbf{M}_m$$

Using these definitions, we define two different Fock operators

$$\mathbf{F}_c = \mathbf{H} + 2\mathbf{J}_c - \mathbf{K}_c + 2\mathbf{J}_o - \mathbf{K}_o + 2\alpha\mathbf{L}_o - \beta\mathbf{M}_o$$
$$\mathbf{F}_o = \mathbf{H} + 2\mathbf{J}_c - \mathbf{K}_c + 2a\mathbf{J}_o - b\mathbf{K}_o + 2\alpha\mathbf{L}_c - \beta\mathbf{M}_c$$

▶(10.55)

where

$$\alpha = \frac{1-a}{1-f} \qquad \beta = \frac{1-b}{1-f}$$

▶(10.56)

Roothaan has shown that these two operators satisfy pseudo-eigenvalue equations of the same type as for a closed-shell system; thus

$$\mathbf{F}_c\psi_k = \epsilon_k\psi_k$$
$$\mathbf{F}_o\psi_m = \epsilon_m\psi_m$$

▶(10.57)

with all off-diagonal Lagrangian multipliers eliminated. There are some differences in the way the orbitals are obtained, however. The operator $\mathbf{F}_c$ generates a set of functions ψ_k including a number of virtual (empty) orbitals. Only the subset corresponding to closed shells is considered occupied; that is, in the iterative SCF procedure, the expansion coefficients produced by solving the first of Eqs. (10.57) can be used in constructing *new* Fock operators only when these coefficients correspond to *occupied closed* shells. Similarly, $\mathbf{F}_o$ generates a set of functions ψ_m including orbitals nominally doubly occupied, but only the subset which cooresponds to the open shells is used in the SCF procedure. The Roothaan method guarantees that all occupied closed-shell orbitals are orthogonal to all occupied open-shells orbitals; furthermore,

$$\langle\psi_k|\psi_l\rangle = \delta_{kl}$$
$$\langle\psi_m|\psi_n\rangle = \delta_{mn}$$

(10.58)

However, the relation

$$\langle\psi_m|\psi_k\rangle = \delta_{mk}$$

(10.59)

is *not* necessarily true unless ψ_m and ψ_k both represent occupied orbitals.

The coupling operators transform open-shell orbitals into closed-shell orbitals, and vice versa. In particular,

$$\mathbf{L}_c|\psi_m\rangle = \sum_{k=1}^{n_c} |\psi_k\rangle\langle\psi_k|\mathbf{J}_o|\psi_m\rangle$$

$$\mathbf{M}_c|\psi_m\rangle = \sum_{k=1}^{n_c} |\psi_k\rangle\langle\psi_k|\mathbf{K}_o|\psi_m\rangle$$

$$\mathbf{L}_o|\psi_k\rangle = f\sum_{m=1}^{n_s} |\psi_m\rangle\langle\psi_m|\mathbf{J}_o|\psi_k\rangle$$

$$\mathbf{M}_o|\psi_k\rangle = f\sum_{m=1}^{n_s} |\psi_m\rangle\langle\psi_m|\mathbf{K}_o|\psi_k\rangle$$

(10.60)

These operators take a particularly simple form if the irreducible representation to which the orbitals of the open shell belong is different from all the irreducible representations to which the closed-shell orbitals belong. Typical examples are atomic ions with open $2p$ or open $3d$ shells, but *not* ions with open $2s$, $3p$, or $4d$ shells. In such cases the orthogonality of the closed and open shells is automatic, and all integrals of the above coupling operators are zero. Explicitly,

$$\langle\psi_m|\mathbf{L}_c|\psi_m\rangle = \sum_{k=1}^{n_c} \langle\psi_m|\psi_k\rangle\langle\psi_k|\mathbf{J}_o|\psi_m\rangle \tag{10.61}$$

which is zero because ψ_m and ψ_k belong to different irreducible representations. (In the course of an SCF iteration in which ψ_m and ψ_k belong to the same irreducible representation, $\langle\psi_m|\psi_k\rangle$ is not zero. But when self-consistency has been reached, the integral must vanish.) Therefore, in this special case,

$$\begin{aligned} \mathbf{F}_c &= \mathbf{H} + 2\mathbf{J}_c - \mathbf{K}_c + 2\mathbf{J}_o - \mathbf{K}_o \\ \mathbf{F}_o &= \mathbf{H} + 2\mathbf{J}_c - \mathbf{K}_c + 2a\mathbf{J}_o - b\mathbf{K}_o \end{aligned} \qquad \blacktriangleright (10.62)$$

These equations make obvious good physical sense; the closed shell "sees" the coulombic potential of all electrons plus the average exchange potential, and the open shell "sees" a somewhat different potential, depending on the spins of the electrons in the open shell, as reflected in the values of a and b.

Not all open-shell situations‡ can be fit to (10.55). Roothaan has discussed a number of special cases. In particular, for a configuration with $f = \frac{1}{2}$ and maximum spin (the Hund's Rule ground state), $a = 1$, $b = 2$, $\alpha = 0$, and $\beta = -2$ always. This rule can be extended to systems with several open shells *each* having $f = \frac{1}{2}$ and maximum spin, therefore including the lowest triplet excited states of molecules with closed-shell ground states and nondegenerate orbitals.

In the Roothaan formalism, the energy preserves the simple form of (10.11) even for open shells; thus

$$E = \sum_{k=1}^{n_c} (H_k + \epsilon_k) + f \sum_{m=1}^{n_s} (H_m + \epsilon_m) \qquad \blacktriangleright (10.63)$$

Of course, the sums only extend over occupied orbitals. The above represents a considerable simplification of the energy formula. It may be of some use in semiempirical work, where it is often impossible to determine the individual J's and K's although the ϵ's are known.

We illustrate the above discussion with a calculation on the ^3P ground term of the $2p^4$ configuration. There is but one open shell of p

‡ Open d shells cannot, in general, be treated by this method. A far-reaching generalization has been described by Roothaan and Bagus, 1963.

symmetry, and so we use (10.62). We state without proof that $a = \frac{15}{16}$, and $b = \frac{9}{8}$. Clearly, $f = \frac{2}{3}$. If we lump together the operators $\mathbf{J}_c$ and $\mathbf{K}_c$ with $\mathbf{H}$ and call the resulting operator $\mathbf{U}$, we have

$$\epsilon_p = \langle \varphi_p | \mathbf{U} | \varphi_p \rangle + \frac{1}{8} \langle \varphi_p | (15\mathbf{J}_o - 9\mathbf{K}_o) | \varphi_p \rangle \tag{10.64}$$

where

$$\mathbf{J}_o = \frac{2}{3}(\mathbf{J}_1 + \mathbf{J}_0 + \mathbf{J}_{-1})$$
$$\mathbf{K}_o = \frac{2}{3}(\mathbf{K}_1 + \mathbf{K}_0 + \mathbf{K}_{-1}) \tag{10.65}$$

Using Table 6.5, we obtain

$$\epsilon_p = U + \frac{5}{4}(3F^0) - \frac{3}{4}(F^0 + \frac{10}{25}F^2)$$
$$= U + 3F^0 - \frac{3}{10}F^2 \tag{10.66}$$

Let us now consider the same calculation in an "unrestricted" Hartree-Fock scheme. The $M_S = 1$, $M_L = -1$ component of the ^3P term is the Slater determinant

$$|^3P\ 1\ -1\rangle = |1^+\ 0^+\ -1^2\rangle \tag{10.67}$$

We can define a *separate* orbital energy for each of the six spin orbitals, according to the relation

$$\epsilon_i = U + \sum_j (J_{ij} - K'_{ij}) \tag{10.68}$$

where the sum over j extends over all occupied spin orbitals and where K'_{ij} is zero if the spin orbital j has spin opposite to spin orbital i. For example,

$$\epsilon_{1^+} = U + J_{10} + 2J_{1-1} - K_{10} - K_{1-1}$$
$$= U + 3F^0 - \frac{9}{25}F^2 \tag{10.69}$$

Similarly,

$$\epsilon_{0^+} = U + 3F^0 - \frac{12}{25}F^2$$
$$\epsilon_{-1^+} = U + 3F^0 - \frac{9}{25}F^2 \tag{10.70}$$
$$\epsilon_{-1^-} = U + 3F^0$$

Averaging the orbital energies of the occupied orbitals, we find

$$\bar{\epsilon}_p = U + 3F^0 - \frac{3}{10}F^2 \tag{10.71}$$

which is identical to the Roothaan ϵ_p. This illustrates the following important principle: the Roothaan procedure "spherically averages" the Fock operator so it belongs to the A_1 representation of the appropriate atomic or molecular symmetry group. This preserves the degeneracy of the open-shell orbitals, a feature not present in the unrestricted Hartree-Fock method. In general, the Roothaan procedure produces energies which are identical to the "average orbital energy" defined as above.

Note, however, that in a true UHF calculation the six p spin orbitals would not necessarily have the same shape, hence the average of the UHF orbital energies would no longer be the same as the Roothaan ϵ.

Very few actual molecular calculations have been done using the Roothaan open-shell equations. This is not so much due to difficulty in applying them as to the intrinsic difficulty inherent in any molecular Hartree-Fock problem (see Chap. 7.) Roothaan's methods have been widely used to treat atomic systems, and in particular the comprehensive calculations of Clementi mentioned in Chap. 6 follow the Roothaan formalism. Explicit methods of calculations, including a description of the computer programs together with the details of the extension of the Roothaan theory to several simultaneously open shells, have been given by Roothaan and Bagus.

PROBLEMS

10.4 Verify the entire sequence of steps leading to (10.71) and starting with (10.67).

10.5 Show that the RHF and UHF methods give identical results when applied to the ^4S ground state of the $2p^3$ configuration.

10.6 Derive Eq. (10.63).

10.3 EXCITED STATES OF CLOSED-SHELL SYSTEMS

Generally speaking, if a given term is the lowest belonging to a particular irreducible representation and spin, there is no theoretical difficulty in applying Roothaan's equations and minimizing its energy. However, in closed-shell molecules with little symmetry, we are often interested in the energy of an excited $^1A_1$ state. Direct minimization of the energy of this state is impossible unless we guarantee that the excited state is orthogonal to the ground state; this is not usually easy.

Nevertheless, it might seem best to proceed to a separate minimization of the energy of each excited state of interest. Then, hopefully, the energy difference between ground and excited states will be accurately predicted as the difference between the corresponding Hartree-Fock energies. This presumes that the correlation energy will remain roughly constant for all states. However, this hope turns out not to be well founded, and separately minimizing energies of excited states apparently gives no better agreement with experiment than the procedure we describe next. Actually, there is not much firm evidence in favor of either method; the following method at least has the virtue of simplicity and ease.

Consider the Slater determinant given by (10.2). When we solve the Roothaan-Hartree-Fock equations for a closed shell, we obtain $m - n$ *extra* solutions, where m is the number of basis functions and n is the number of doubly occupied orbitals. These extra virtual orbitals can be used to construct excited states. Let us assume for simplicity that all orbitals are spatially nondegenerate. Then the four Slater determinants

$$
\begin{aligned}
\Psi_1 &= |\psi_1^+(1)\psi_1^-(2) \cdot \cdot \cdot \\
&\quad \psi_{n-1}^+(2n-3)\psi_{n-1}^-(2n-2)\psi_n^+(2n-1)\psi_{n+1}^+(2n)| \\
\Psi_2 &= |\psi_1^+(1)\psi_1^-(2) \cdot \cdot \cdot \\
&\quad \psi_{n-1}^+(2n-3)\psi_{n-1}^-(2n-2)\psi_n^-(2n-1)\psi_{n+1}^-(2n)| \\
\Psi_3 &= |\psi_1^+(1)\psi_1^-(2) \cdot \cdot \cdot \\
&\quad \psi_{n-1}^+(2n-3)\psi_{n-1}^-(2n-2)\psi_n^+(2n-1)\psi_{n+1}^-(2n)| \\
\Psi_4 &= |\psi_1^+(1)\psi_1^-(2) \cdot \cdot \cdot \\
&\quad \psi_{n-1}^+(2n-3)\psi_{n-1}^-(2n-2)\psi_n^-(2n-1)\psi_{n+1}^+(2n)|
\end{aligned}
$$

$\blacktriangleright$ (10.72)

can be combined to give both singlet and triplet excited states. Note that a single electron in ψ_n has been excited to ψ_{n+1}. Suppose ψ_n transforms as Γ_p and ψ_{n+1} transforms as Γ_q for an arbitrary symmetry group and that $\Gamma_p \times \Gamma_q = \Gamma_r$. Then the states can be labeled $^1\Gamma_r$ and $^3\Gamma_r$. The $M_S = 1$ component of the triplet can be written down by inspection as

$$
|{}^3\Gamma_r, 1\rangle = \Psi_1 \qquad\qquad \blacktriangleright (10.73)
$$

Using the shift operator $\mathbf{S}^- = \mathbf{s}_{2n-1}^- + \mathbf{s}_{2n}^-$, we obtain

$$
\begin{aligned}
\mathbf{S}^-|{}^3\Gamma_r, 1\rangle &= \sqrt{2}\,\hbar|{}^3\Gamma_r, 0\rangle \\
&= \hbar(\Psi_3 + \Psi_4)
\end{aligned}
\qquad (10.74)
$$

Therefore,

$$
|{}^3\Gamma_r, 0\rangle = \frac{1}{\sqrt{2}}\,(\Psi_3 + \Psi_4) \qquad\qquad \blacktriangleright (10.75)
$$

Furthermore, applying the shift operators again, or simply by inspection, we have

$$
|{}^3\Gamma_r, -1\rangle = \Psi_2 \qquad\qquad (10.76)
$$

By the orthogonality of the singlet with the triplet,

$$
|{}^1\Gamma_r\rangle = \frac{1}{\sqrt{2}}\,(\Psi_3 - \Psi_4) \qquad\qquad \blacktriangleright (10.77)
$$

There is an arbitrary phase choice involved for the singlet, which we set equal to $(-1)^{1/2 - m_{s(2n-1)}}$.

Using the rules developed in Chap. 6, it is easily verified that the singlet (and triplet) wavefunctions are orthogonal to the ground state. This follows whether or not they belong to the same irreducible representation. We cannot easily minimize the energies of these two excited states without destroying this orthogonality, however, so we leave them as they are, with orbitals determined by the *ground-state Fock operator*.

We next evaluate the energy of these states. Taking the $M_S = 1$ component of the triplet, we obtain

$$\mathcal{3C}_{11} \equiv \langle \Psi_1 | \mathcal{3C} | \Psi_1 \rangle = 2 \sum_{i=1}^{n-1} H_i + \sum_{i=1}^{n-1} \sum_{j=1}^{n-1} (2J_{ij} - K_{ij}) + H_n$$
$$+ H_{n+1} \sum_{i=1}^{n-1} \sum_{j=n}^{n+1} (2J_{ij} - K_{ij}) + J_{n,n+1} - K_{n,n+1} \quad (10.78)$$

As usual, H_i represents the one-electron energy of the ith orbital. Comparing $\mathcal{3C}_{11}$ to the energy of the ground state

$$E_{GS} = 2 \sum_{i=1}^{n} H_i + \sum_{i=1}^{n} \sum_{j=1}^{n} (2J_{ij} - K_{ij}) \quad (10.79)$$

we obtain

$$\mathcal{3C}_{11} = E_{GS} + (H_{n+1} - H_n) + \sum_{i=1}^{n-1} [2(J_{i,n+1} - J_{in})$$
$$- (K_{i,n+1} - K_{in})] - J_{nn} + J_{n,n+1} - K_{n,n+1} \quad (10.80)$$

This may be further simplified by noting that

$$\epsilon_n = H_n + \sum_{i=1}^{n-1} (2J_{in} - K_{in}) + J_{nn}$$
$$\epsilon_{n+1} = H_{n+1} + \sum_{i=1}^{n-1} (2J_{i,n+1} - K_{i,n+1}) + 2J_{n,n+1} - \mathrm{K}_{n,n+1} \quad (10.81)$$

Combining, we find

$$\mathcal{3C}_{11} = E_{GS} + (\epsilon_{n+1} - \epsilon_n) - J_{n,n+1} = E(^3\Gamma_r) \quad (10.82)$$

The energy of the singlet may be obtained in a similar way. We state without proof (Prob. 10.7) that

$$\mathcal{3C}_{33} = \mathcal{3C}_{11} + K_{n,n+1} = \mathcal{3C}_{44} \quad (10.83)$$

The off-diagonal element $\mathcal{3C}_{34}$ may be calculated using the formulas of Table 6.4. The result is

$$\mathcal{3C}_{34} = -K_{n,n+1} \quad (10.84)$$

Substituting in (10.77), we find

$$E(^1\Gamma_r) = \frac{1}{2}(\mathcal{3C}_{33} + \mathcal{3C}_{44} - 2\mathcal{3C}_{34}) = E_{GS} + (\epsilon_{n+1} - \epsilon_n)$$
$$- J_{n,n+1} + 2K_{n,n+1} \quad (10.85)$$

The formulas we have developed represent a rather special case, in which a *single* electron is transferred from the highest occupied orbital

(ψ_n) to the lowest virtual orbital (ψ_{n+1}). Such states are called *singly* excited with respect to the ground state. Let us abbreviate the resulting states according to their orbital origin rather than according to the irreducible representation to which they belong, namely, $^3\Psi_{n\to n+1}$ and $^1\Psi_{n\to n+1}$. Then we can write the following as general relations:

$$\langle {}^3\Psi_{i\to k}|\mathcal{3C}|{}^3\Psi_{i\to k}\rangle - E_{GS} = \epsilon_k - \epsilon_i - J_{ik}$$
$$\langle {}^1\Psi_{i\to k}|\mathcal{3C}|{}^1\Psi_{i\to k}\rangle - E_{GS} = \epsilon_k - \epsilon_i - J_{ik} + 2K_{ik}$$

▶(10.86)

Here ψ_i is any orbital occupied in the ground (closed-shell) state, and ψ_k any virtual orbital. We stress again that all quantities are to be calculated with the ground-state Fock operator; an open-shell treatment is in no way involved.

It is also possible to define states which are doubly, triply, . . . excited with respect to the ground state and to determine their energies in this scheme. There is nothing especially difficult about this, but, generally speaking, multiply excited states are of less *interest* since they are spectroscopically forbidden as transitions from the ground state. However, as noted in the next section, such states can interact and mix with singly excited states, thereby affecting their energies and properties.

Let us now consider the state produced by *removing* an electron from ψ_n, the highest occupied orbital. The resulting state is a doublet, with symmetry Γ_p, since we assumed that ψ_n transforms as Γ_p. The $M_S = \frac{1}{2}$ component may be written

$$|{}^2\Gamma_p\, \tfrac{1}{2}\rangle = |\psi_1^+(1)\psi_{\bar{1}}^-(2)\psi_2^+(3)\psi_{\bar{2}}^-(4)\,\cdots\,\psi_n^+(2n-1)|$$

(10.87)

The energy is easily evaluated, and we get

$$E(^2\Gamma_p) = 2\sum_{i=1}^{n-1} H_i + \sum_{i=1}^{n-1}\sum_{j=1}^{n-1}(2J_{ij} - K_{ij})$$
$$+ \sum_{i=1}^{n-1}(2J_{in} - K_{in}) + H_n$$

(10.88)

In terms of the energy of the closed-shell ground state,

$$E(^2\Gamma_p) = E_{GS} - \sum_{i=1}^{n-1}(2J_{in} - K_{in}) - J_{nn} - H_n$$

▶(10.89)

$$= E_{GS} - \epsilon_n$$

In this sense, $-\epsilon_n$ is equal to the ionization potential of the neutral atom or molecule, a result known as Koopmans' theorem. However, the proof we have given contains grave defects. We have assumed that the orbitals of the ion are the same as the orbitals of the closed-shell system, an assumption not only unlikely, but proved seriously incorrect in actual

Hartree-Fock calculations on ionized systems. The same objection applies to our treatment of excited states; they do not *in fact* have the same orbital forms as the ground state. Nevertheless, Koopmans' theorem holds with astonishing accuracy, and calculations of excited-state energies by the method outlined above seem to give good results. The reason for this is not well understood, beyond the obvious statement that the energy change produced by the distortion of the orbitals of the ion is compensated by a change in correlation energy. Thus, although any assumption of constant orbital behavior in ground and excited states is clearly wrong, the assumption works quite well.

Koopmans' theorem is often turned around and applied to the electron affinity of an atom or molecule. By the argument given, $-\epsilon_{n+1}$ should be equal to the electron affinity. However, the theorem fails badly in practice, presumably because here the orbital distortion energy and change in correlation energy add rather than cancel. Thus, the assumption of constant orbital behavior must be used with great care.

PROBLEMS

10.7 Prove both (10.83) and (10.84).

◆ **10.8** Work out formulas corresponding to (10.86) for doubly excited configuraitons. (For the results, see Murrell and McEwen, 1956.)

10.4 CONFIGURATION INTERACTION

No single Slater determinant can ever be an exact solution to the Schrödinger equation for an atom or molecule since the orbital picture invariably neglects the instantaneous correlation between the motions of the electrons. In theory, this difficulty can be circumvented by taking a linear combination of a large number of determinants, but it is not clear how to best choose the extra configurations so as to speed the convergence of the energy minimization. One straightforward way of constructing excited configurations was outlined in the previous section; the method has the advantage that all excited configurations are orthogonal. The disadvantage, which is well known, is that the convergence of this process is too slow to be useful in calculating correlation energies or "nearly exact" ground-state wavefunctions. Nevertheless, *limited* configuration interaction of this sort has its uses, and we outline the procedure here.

We have already seen how to calculate the energies of the singly excited configurations. We next inquire into the magnitude of the off-diagonal matrix element *between* two such configurations. First consider the matrix element between the ground-state function Ψ of (10.2) and an arbitrary singly excited singlet such as $^1\Gamma_r$ of (10.77). Using the rules of Table 6.4, we find

$$\langle {}^1\Gamma_r|\mathfrak{IC}|\Psi\rangle = \sqrt{2}\left\{ \sum_{i=1}^{n-1}\left[2\langle\psi_i(1)\psi_n(2)|\frac{1}{r_{12}}|\psi_i(1)\psi_{n+1}(2)\rangle \right.\right.$$

$$\left. - \langle\psi_i(1)\psi_n(2)|\frac{1}{r_{12}}|\psi_{n+1}(1)\psi_i(2)\rangle\right] + \langle\psi_n(1)\psi_n(2)|\frac{1}{r_{12}}|\psi_n(1)\psi_{n+1}(2)\rangle$$

$$\left. + \langle\psi_n|\mathbf{H}|\psi_{n+1}\rangle\right\} \quad (10.90)$$

We now argue that this matrix element is zero. Consider the element

$$\langle\psi_n|\mathbf{F}|\psi_{n+1}\rangle \quad (10.91)$$

This element is zero since the ψ_i are all eigenfunctions of $\mathbf{F}$. Expanding it in terms of the constituent operators of $\mathbf{F}$ we obtain

$$\langle\psi_n|\mathbf{F}|\psi_{n+1}\rangle = \langle\psi_n|\mathbf{H}|\psi_{n+1}\rangle + \sum_{i=1}^{n-1}\langle\psi_n|(2\mathbf{J}_i - \mathbf{K}_i)|\psi_{n+1}\rangle$$

$$+ \langle\psi_n|\mathbf{J}_n|\psi_{n+1}\rangle$$

$$= \langle\psi_n|\mathbf{H}|\psi_{n+1}\rangle + \sum_{i=1}^{n-1}(2\langle\psi_i\psi_n\|\psi_i\psi_{n+1}\rangle$$

$$- \langle\psi_i\psi_n\|\psi_{n+1}\psi_i\rangle) + \langle\psi_n\psi_n\|\psi_n\psi_{n+1}\rangle \quad (10.92)$$

$$= \frac{1}{\sqrt{2}}\langle{}^1\Gamma_r|\mathbf{H}|\Psi\rangle$$

Thus all singly excited singlet states have zero matrix elements with the ground state. This result is known as Brillouin's theorem and was discussed in Sec. 6.9. As noted therein, only configurations doubly excited with respect to the ground configuration have nonzero matrix elements with it.

Next we return to the problem of evaluating the off-diagonal matrix element between two configurations singly excited with respect to the ground state. Following the notation introduced in the previous section, we have

$$\langle{}^1\Psi_{i\to k}|\mathfrak{IC}|{}^1\Psi_{j\to l}\rangle = 2\langle\psi_j\psi_k\|\psi_l\psi_i\rangle - \langle\psi_j\psi_k\|\psi_i\psi_l\rangle$$
$$\langle{}^3\Psi_{i\to k}|\mathfrak{IC}|{}^3\Psi_{j\to l}\rangle = -\langle\psi_j\psi_k\|\psi_i\psi_l\rangle \qquad \blacktriangleright(10.93)$$

We leave the proofs as an exercise (Prob. 10.9). The formulas do not apply when $i = j$ and $k = l$ since this represents a *diagonal* energy term. It should be noted that these formulas imply a definite choice of phase in the way the determinants are set up. Specifically, with the ground-state determinant defined as in (10.2), we take the determinants $k \rightarrow l$ as

$$|\psi_1(1)\alpha_1\psi_1(2)\beta_2 \cdots \psi_k(2k - 1)\xi_{2k-1}\psi_l(2k)\xi_{2k}$$
$$\cdots \psi_n(2n - 1)\alpha_{2n-1}\psi_n(2n)\beta_{2n}| \quad (10.94)$$

Depending on the values of the spin functions ξ_{2k-1} and ξ_{2k}, the above function gives rise to four different Slater determinants. To completely settle all phase choices, we always take the triplet function with $M_S = 0$ as the combination $\alpha_{2k-1}\beta_{2k} + \beta_{2k-1}\alpha_{2k}$ and the singlet as $\alpha_{2k-1}\beta_{2k} - \beta_{2k-1}\alpha_{2k}$.

Thus far we have neglected multiply-excited configurations. These can mix with the singly excited configurations to alter their energies still further. Ordinarily, for low-lying excited states, one hopes that the multiply-excited states lie at energies too high to have a significant interaction with the state of interest, but this assumption may not always be justified. Formally, it is an easy matter to take these extra states into account, but naturally there is a good deal of additional labor involved.

Limited configuration interaction is commonly used to treat the mixing of degenerate or nearly degenerate excited states of the same symmetry. This use is of especial importance in alternant hydrocarbons (Sec. 11.4), where the simple theories demand certain degeneracies in the excited states. Without limited configuration interaction, the results of the simple treatments are therefore meaningless since the actual interaction of the "degenerate" states is quite strong.

PROBLEMS

10.9 Prove the general relations of (10.93).

10.10 Show that two configurations belonging to different irreducible representations cannot interact.

10.11 Use the Roothaan expansion (10.1) to determine the matrix elements of (10.93) in terms of expansion coefficients and matrix elements over basis functions.

10.5 BOND ORDERS, CHARGE DENSITIES, AND SPIN DENSITIES

LCAO–MO calculations are ordinarily carried out with minimal basis sets, that is, with one Slater-type function per atomic orbital. As noted in Chap. 7, this is primarily because the work involved in a calculation depends crucially on the number of basis functions, so one naturally tries to keep the basis size small. (A second factor is simply that no workable approximate methods have been developed for systems with larger basis sets.) In such calculations, it is useful to define a quantity known as the bond order, according to the equation

$$B_{pq} = (1 + S_{pq}) \sum_i N(i)c_{ip}^* c_{iq} \qquad \blacktriangleright (10.95)$$

As written, the formula applies to both open- and closed-shell systems; $N(i)$ is the number of electrons in molecular orbital i. It can be used in the unrestricted Hartree-Fock scheme as well as the restricted since, in the former case, each spin orbital has its own set of coefficients.

The above definition of bond order was proposed by Mulliken.[‡] Other definitions may be found in the literature, but they are basically similar. For example, the factor S_{pq} is often omitted.[§] However, all have in common a strong tendancy to correlate with experimental bond lengths, and it is theoretical interest to examine why this is so. First, we note that B_{pq} is a sum of orbital bond orders $b_{pq}^{(i)}$,

$$B_{pq} = (1 + S_{pq}) \sum_i N(i) b_{pq}^{(i)} \tag{10.96}$$

where

$$b_{pq}^{(i)} = c_{ip}^* c_{iq}$$

The orbital bond order can be positive, negative, or zero, depending on the relative signs of c_{ip} and c_{iq}. If molecular orbital ψ_i has a nodal plane between atoms p and q, then $b_{pq}^{(i)}$ will be negative. Thus, $b_{pq}^{(i)}$ corresponds directly to the conventional notion of bond strength, being positive for an orbital bonding the two atoms and negative for an antibonding region. Its magnitude depends on the magnitude of the molecular orbital in the region of the two atoms; $|b_{pq}^{(i)}|$ will be small if the probability that the electron is near *other* atoms is large. The bonding or antibonding effect depends on the number of electrons available to produce it. Thus, the sum of $b_{pq}^{(i)}$ over the occupied molecular orbitals is a measure of the magnitude of the total wavefunction *between* the nuclei p and q.

B_{pq} is only well defined for atoms which are neighbors. For non-bonded atoms, calculated B_{pq} are erratically positive and negative, in magnitude not much smaller than for neighbors. This probably indicates that the effect is also proportional to S_{pq}, and perhaps a definition replacing $1 + S_{pq}$ by S_{pq} in (10.95) would be more suitable. However, it has the disadvantage that the bond order would be zero for systems in which the calculation has been simplified by the approximation $S_{pq} = 0$.

The analysis just given is applicable as such only for systems which have but one bond between each pair of atoms, and thus a single orbital per atom in the framework of a minimum basis set. For systems with multiple-bonded atoms, it is only necessary to sum over the different possible bonds to get the total bond order. Such an analysis for N_2 or O_2 indeed gives a bond order of about 3 for the former and 2 for the latter, in

[‡] R. S. Mulliken, 1955.
[§] Most calculations on large systems are carried out under the assumption $S_{pq} = 0$, so the two definitions are identical in such cases (see Chap. 11).

agreement with the more elementary analysis of Chap. 7. The reader is referred back to that chapter for a comparison of definitions.

Mulliken has also defined a quantity known as the orbital overlap population,

$$p_{pq}^{(i)} = c_{ip}^* c_{iq} S_{pq} \tag{10.97}$$

Summed over the occupied orbitals, we obtain

$$P_{pq} = S_{pq}(2 - \delta_{pq}) \sum_i N(i) c_{ip}^* c_{iq} \qquad \blacktriangleright (10.98)$$

This quantity is known as the overlap population between orbitals p and q. The definition and introduction of the factor $2 - \delta_{pq}$ may be understood as follows. Taking a given molecular orbital ψ_i and normalizing it to unity, we have the equation

$$1 = \sum_{p=1}^{m} |c_{ip}|^2 + \sum_{p=1}^{m} \sum_{q \neq p} c_{ip}^* c_{iq} S_{pq} \tag{10.99}$$

Since each off-diagonal element occurs twice (assuming the coefficients are all real, or simply taking only the real parts), this can be rewritten as

$$1 = \sum_{p=1}^{m} c_{ip}^2 + 2 \sum_{p=1}^{m-1} \sum_{q>p} c_{ip} c_{iq} S_{pq} \tag{10.100}$$

Each term thus represents a probability density, although, since orbitals p and q overlap, there are terms representing the probability that the electron is "simultaneously" on both centers. Summing over all occupied molecular orbitals, we obtain the total number of electrons in the system as

$$N = \sum_{p=1}^{m} P_{pp} + \sum_{p=1}^{m-1} \sum_{q>p} P_{pq} \tag{10.101}$$

Another quantity defined by Mulliken is the *gross* atomic population,

$$Q_p = P_{pp} + \frac{1}{2} \sum_{q \neq p} P_{pq} \qquad \blacktriangleright (10.102)$$

Note that

$$N = \sum_{p=1}^{m} Q_p \tag{10.103}$$

Essentially, the overlap population density has been divided equally between the two atoms, with half assigned to each center. The justification for assigning one-half, and not some other fraction, has been given by Mulliken; basically, this assignment minimizes the chance that any one orbital will be assigned much more than two electrons.

Overlap populations and gross atomic populations have been widely used in valence theory. The reactivity of a given atom to electrophilic or nucleophilic attack has often been correlated with predicted gross populations, and it is clear that it is possible to draw a connection between bond length and overlap population via the concept of the bond order. Additional correlations between bond order and reactivity have also been attempted. However, it must be remembered that overlap populations cannot be observed directly, so any correlation between populations and an observed quantity involves additional theoretical constructs. Nevertheless, since populations are one-electron properties, one expects them to be predicted with relatively high accuracy from SCF theory.

A quantity closely connected to the Mulliken population is the spin density. This may be defined as

$$S_{pq} = S_{pq}(2 - \delta_{pq}) \left[\sum_{i=1}^{n_\alpha} N(i) c_{ip}^* c_{iq} - \sum_{i=1}^{n_\beta} N'(i) c_{ip}'^* c_{ip}' \right] \qquad \blacktriangleright (10.104)$$

The first sum extends only over those molecular orbitals containing α-spin electrons, and the second sum is over β-spin orbitals. In the restricted Hartree-Fock scheme, $c_{ip} = c_{ip}'$ for the closed shells, and we may replace the above expression by one in which the sum extends only over open shells. Note that all spin densities are zero in a closed-shell system.

Spin densities are of great interest theoretically since they are connected very closely with observable quantities in electron spin and nuclear magnetic resonance. What one actually measures in these experiments is the spin density at or very close to the nucleus of the atom in question. This may be theoretically correlated with S_{pp} (or perhaps $S_{pp} + \frac{1}{2} \sum_{q \neq p} S_{pq}$), but the connection is not perfect. However, there has been little difficulty in finding semiempirical relationships between calculated spin densities and experimental measurements. One experimental result of great theoretical interest is the existence of *negative* spin densities, corresponding to a *local* spin density of opposite sign to the total spin density of the system. This corresponds to a local β density even while the system has a net α electron. In the RHF scheme, negative spin densities cannot be predicted unless configuration interaction is invoked; this is the main reason why the UHF theory is preferred by many workers.

Concepts such as charge densities, bond orders, and spin densities are not restricted to molecular orbital theory. They may in fact be used in conjunction with any theoretical method employing a finite basis set. In particular, we can always define a density matrix **D** such that each element represents the coefficient of the square of the total wavefunction for a particular product of basis functions.‡ For the molecular orbital scheme,

$$D_{pq} = \sum_i N(i) c_{ip}^* c_{iq} \qquad \blacktriangleright (10.105)$$

or, for a closed shell,

$$D_{pq} = 2 \sum_{i=1}^n c_{ip}^* c_{iq} \qquad (10.106)$$

As an example, let us compare the density matrix (also commonly referred to as the first-order density matrix) for the H_2 molecule in both the MO and VB schemes. With basis functions $\varphi_{1s} = N e^{-r}$ (see Prob. 7.3), the overlap integral at 1.4 au is 0.75294, so, from Eq. (7.10),

$$\psi = \left(\frac{1}{3.50588} \right)^{1/2} (\varphi_{1s}^A + \varphi_{1s}^B) \qquad (10.107)$$

The elements of the molecular orbital density matrix are therefore

$$D_{12} = D_{21} = D_{11} = D_{22} = \frac{1}{3.50588} = 0.5705 \qquad (10.108)$$

Note that from (10.96) the elements of the Mulliken bond-order and charge-density matrices are

$$\begin{aligned}
B_{12} &= 1.00 \\
P_{11} &= P_{22} = 0.5705 \\
P_{12} &= 0.8591 \\
Q_1 &= Q_2 = 1
\end{aligned} \qquad (10.109)$$

In the valence-bond scheme (see Prob. 10.12),

$$\begin{aligned}
D_{11} &= D_{22} = 0.6382 \\
D_{12} &= 0.4805 \\
B_{12} &\equiv (1 + S_{12})D_{12} = 0.8423 \\
P_{12} &\equiv S_{12}D_{12} = 0.7236 \\
P_{11} &= P_{22} \equiv D_{11} = 0.6382 \\
Q_1 &= Q_2 = 1
\end{aligned} \qquad (10.110)$$

The reader will note that although the valence-bond function has a lower energy than the corresponding MO function, it has smaller bond order and overlap population. The extra energy is gained through correlation; it must be remembered that the electrons are partially correlated in the VB function, but not at all in the MO function. The example therefore illustrates both the danger inherent in any comparison of bond orders in

‡ For a more precise definition see R. McWeeny, 1959, 1960.

different theoretical schemes as well as the great importance of correlation energy.

It is possible to define a quantity known as the second-order density matrix‡ which is a direct measure of interelectronic distances. Higher-order density matrices may be defined as well, but it can be shown that the energy of any system within any theoretical scheme depends only on the second-order density matrix. (The value of any one-electron operator can be determined from the first-order density matrix alone.) Theoretically speaking, it is an easier task to solve for this second-order density than for the exact wavefunction, and so some attention has been paid to the problem of finding the former quantity directly (the n-representability problem). Not much headway has been made, however, and so we terminate our discussion without precisely defining just what is meant by an nth-order density matrix.

The first-order density matrix is especially important in the Hartree-Fock scheme (no electronic correlation) since $\mathbf{D}$ entirely determines the Fock Hamiltonian and the energy. This may be seen by combining (6.100) with (6.101). We obtain

$$
\begin{aligned}
E =\ & 2 \sum_{i=1}^{n} \sum_{p=1}^{m} \sum_{q=1}^{m} c_{ip}^{*} c_{iq} h_{pq} \\
& + \sum_{i=1}^{n} \sum_{j=1}^{n} \sum_{p=1}^{m} \sum_{q=1}^{m} \sum_{r=1}^{m} \sum_{s=1}^{m} c_{ip}^{*} c_{jq}^{*} c_{ir} c_{js} \left(2\langle pq\|rs\rangle - \langle pq\|sr\rangle\right) \\
=\ & \sum_{p=1}^{m} \sum_{q=1}^{m} D_{pq} h_{pq} + \sum_{p=1}^{m} \sum_{q=1}^{m} \sum_{r=1}^{m} \sum_{s=1}^{m} D_{pr} D_{qs} \\
& \qquad\qquad \left(2\langle pq\|rs\rangle - \langle pq\|sr\rangle\right) \quad (10.111)
\end{aligned}
$$

This result does not conflict with the statement made in the previous paragraph concerning the importance of the *second-order* density matrix to the energy—the Hartree-Fock scheme is unique in that the second-order density matrix (and all higher-density matrices) are completely determined by $\mathbf{D}$.

In open-shell systems a spin-density or open-shell (first-order) density matrix may be defined by analogy with $\mathbf{D}$. The energy for such systems can be set up in a manner analogous to (10.111), and this is in fact, the most general way of handling open-shell systems in the RHF formalism.

Still another application of the first-order density matrix is the construction of *natural orbitals*. These are orbitals which although not providing as good a ground-state wavefunction as the Roothaan-Hartree-Fock molecular orbitals, provide excited configurations that mix well

‡ R. McWeeny, 1959, 1960.

with the ground configuration to give faster convergence in the configuration interaction. Natural orbitals are constructed by diagonalizing the Hartree-Fock density matrix. Since D is Hermitian, there exists a unitary matrix U such that

$$\mathbf{n} = \mathbf{U}\dagger\mathbf{D}\mathbf{U} \qquad \blacktriangleright (10.112)$$

is a diagonal matrix. The diagonal elements or eigenvalues of this matrix $n_i = n_{ii}$ are called the occupation numbers; Löwdin (1955) has shown that $0 \leq n_i \leq 1$.

Using the transformation matrix U, we can define a new set of orbitals χ_i according to the equation

$$\chi = \psi\mathbf{U} \qquad (10.113)$$

or, in component form,

$$\chi_i = \sum_j \psi_j U_{ji} \qquad \blacktriangleright (10.114)$$

From these *natural orbitals* one can construct a Slater determinant for the ground state, and, using the virtual natural orbitals, Slater determinants for excited configurations. These may then be mixed together through configuration interaction, with somewhat faster convergence than with the Roothaan-Hartree-Fock orbitals.

PROBLEMS

10.12 Verify (10.110) for the density matrix in the valence-bond scheme.
Hint: Begin with Eq. (7.13), consider the square of the total valence-bond wavefunction, and integrate over the coordinates of electron 2. This gives the probability that *an* electron is in a certain volume element. To convince yourself of this, follow the same procedure for the molecular orbital state function, and show that it leads to the same results as Eq. (10.106).

10.13 Consider the state function given by Eq. (7.22), which represents the molecular orbital function plus configuration interaction or, equivalently, the valence-bond function including ionic terms. With the basis set given in the present chapter ($S_{12} = 0.75294$) and assuming $C = 0.25$, find the density matrix, bond orders, etc.

10.14 What would be the result of diagonalizing the SCF density matrix for H_2 when the basis set is restricted to the hydrogen atomic $1s$ orbitals as in this section?

◆ **10.15** Consider the basis functions for the helium atom given in Eq. (6.111): $\chi_1 = N_1 e^{-1.44608r}$ and $\chi_2 = N_2 e^{-2.86222r}$. Using the methods developed in Sec. 5.4, it is possible to derive the integrals

$$\langle \chi_1 | \chi_2 \rangle = S_{12} = \frac{8(\zeta_1\zeta_2)^{3/2}}{(\zeta_1 + \zeta_2)^3}$$

$$\langle \chi_1 | -\tfrac{1}{2}\nabla^2 - \frac{\zeta_2}{R} | \chi_2 \rangle = -\tfrac{1}{2}\zeta_2{}^2 S_{12}$$

$$\langle \chi_1 | \frac{1}{R} | \chi_2 \rangle = \frac{4(\zeta_1\zeta_2)^{3/2}}{(\zeta_1 + \zeta_2)^2}$$

$$\langle \chi_1\chi_2 | 1/r_{12} | \chi_3\chi_4 \rangle \equiv \langle 12 \| 34 \rangle = 32(\zeta_1\zeta_2\zeta_3\zeta_4)^{3/2}$$

$$\left\{ \frac{1}{(\zeta_1 + \zeta_2 + \zeta_3 + \zeta_4)^3} \left[\frac{1}{(\zeta_2 + \zeta_4)} \cdot \frac{1}{(\zeta_1 + \zeta_3)} \right] \right.$$

$$\left. + \frac{1}{(\zeta_1 + \zeta_2 + \zeta_3 + \zeta_4)} \left[\frac{1}{(\zeta_2 + \zeta_4)^2} \cdot \frac{1}{(\zeta_1 + \zeta_3)^2} \right] \right\}$$

Evaluating these integrals, one obtains $S_{12} = 0.84239$, $H_{11} = -1.84657$, $H_{22} = -1.62828$, $H_{12} = -1.88595$, $\langle 11 \| 11 \rangle = 0.90380$, $\langle 11 \| 12 \rangle = 0.90405$, $\langle 12 \| 12 \rangle = 1.17493$, $\langle 11 \| 22 \rangle = 0.95540$, $\langle 12 \| 22 \rangle = 1.28908$, $\langle 22 \| 22 \rangle = 1.78889$. Use these integrals to evaluate the energy of the function given in (6.111), and confirm that this function leads to a diagonal Fock matrix as developed in Sec. 10.1. Confirm that the total energy of helium is -2.86167 au in the Hartree-Fock limit.

11

APPROXIMATE TREATMENTS
OF LARGE MOLECULES

But be ye doers of the word, and
not hearers only, deceiving your
own selves.
James, i, 22

11.1 THE π-ELECTRON APPROXIMATION

Ethylene has 16 electrons, butadiene 30, and benzene 42. An SCF calculation of Hartree-Fock accuracy even on ethylene is difficult in the
extreme, and for benzene this calculation is, with present technology,
completely out of the question. If orbital theory is to have any relevance
to practical chemistry and, in particular, to the chemistry of aromatic
systems, approximations are essential. Fortunately, most aromatic systems have their carbon atoms lying in a plane, which generally means that
the π electrons are in molecular orbitals belonging to different irreducible
representations than the σ electrons. In such cases the Fock operator
has no off-diagonal matrix elements coupling the σ and π orbitals, so they
may be treated semi-independently. Of course, in order to determine the
Fock operator exactly, it is necessary to know the molecular orbitals for

both the σ and π shells. However, in the framework of constant orbital behavior discussed in Sec. 10.3, excitation energies to virtual π orbitals (usually designated π^*) do not depend on the σ except through the ground-state Fock operator and the associated π-electron orbital energies.

In the π-electron approximation we go one step farther and parametrize the behavior of the σ-orbital part of the molecule. We therefore write the Fock operator as

$$\mathbf{F} = -\frac{\nabla^2}{2} + \sum_{k=1}^{K} \mathbf{V}_k + \sum_{i=1}^{n} (2\mathbf{J}_i - \mathbf{K}_i) \qquad \blacktriangleright (11.1)$$

The sum over i extends only over the π orbitals. The $\mathbf{V}_k$ are the "core" attraction operators; that is, they contain the nuclear attraction plus the coulombic and exchange interactions with the σ electrons. In benzene, for example, $\mathbf{V}_k$ will include both a carbon and a hydrogen nucleus and their associated σ electrons. As written, (11.1) is not of necessity approximate since the sum over operators $\mathbf{V}_k$ could be expanded into a theoretically correct sum. However, in actual applications, one does not attempt an accurate calculation of the matrix elements of $\mathbf{V}_k$ between basis orbitals.

Let us now examine the three molecules mentioned above in somewhat greater detail. We will not for the present attempt a calculation using the Fock operator of (11.1), but we will try to discover how much information can be extracted from simple molecular orbital theory if the σ core is ignored. We restrict our attention to minimum-basis-set LCAO calculations and use a single $2p$ Slater-type orbital on carbon. All three molecules are planar, and for ethylene and benzene, we take the z axis perpendicular to the molecular plane. The π orbitals are thus to be constructed solely from a carbon $2p_z$ basis. In ethylene (symmetry group D_{2h}), there are only two possible such orbitals,

$$\psi_u = \frac{N}{\sqrt{2}} (\varphi_{2p_z}^A + \varphi_{2p_z}^B)$$

$$\psi_g = \frac{N'}{\sqrt{2}} (\varphi_{2p_z}^A - \varphi_{2p_z}^B) \qquad (11.2)$$

The two π electrons (one from each carbon atom) both go into ψ_u, which has no nodal plane between the two atoms. The situation is entirely analogous to that of the hydrogen molecule; in fact, the density matrix analysis of the previous section holds in detail, so $B_{12} = 1$. The ethylene molecule is therefore the prototype for a π bond. Note that in the π-electron approximation we ignore single-bond contribution from the σ orbitals; ethylene has, of course, a true double bond.

In butadiene, the situation is more complex. There are, first of all, two distinct forms of the molecule (Fig. 11.1)—*trans*-butadiene, having

Fig. 11.1 Configuration of the nuclei in *trans-* and *cis-*butadiene.

C_{2h} symmetry, and *cis*-butadiene, belonging to C_{2v}. According to convention, the z axis is the axis of high symmetry, and thus it is chosen differently for the two forms of the molecule. The group $C_{2h} = C_2 \times S_2$ consists of four operations, E, C_2, σ_h, and the inversion i. Using Table 8.1, we can show that the π orbitals $(2p_z)$ transform as $2A_u + 2B_g$. Appropriate linear combinations may be written down by inspection as follows:

$$\chi_{1a_u} = \frac{N_{1a}}{\sqrt{2}} (\varphi_2 + \varphi_3)$$

$$\chi_{2a_u} = \frac{N_{2a}}{\sqrt{2}} (\varphi_1 + \varphi_4)$$

$$\chi_{1b_g} = \frac{N_{1b}}{\sqrt{2}} (\varphi_1 - \varphi_4)$$

$$\chi_{2b_g} = \frac{N_{2b}}{\sqrt{2}} (\varphi_2 - \varphi_3)$$

(11.3)

In *cis*-butadiene, axes must be chosen with greater care, since the π orbitals do not lie along the high-symmetry axis. Taking the π orbitals as $2p_y$ (Fig. 11.1), it is easily shown that they transform as $2A_2 + 2B_2$. (If we had reversed the x and y axes, the π orbitals would have transformed as $2A_2 + 2B_1$.) The linear combinations given above are still appropriate but now are to be labeled χ_{1b_2}, χ_{2b_2}, χ_{1a_2}, and χ_{2a_2}, respectively. In both molecules there are two orbitals belonging to each of the irreducible representations, so the molecular orbitals are not determined entirely by symmetry. In fact, from the symmetry orbitals we can construct general molecular orbitals according to the scheme

$$\psi_{1b_2} = c_1\chi_{1b_2} + c_2\chi_{2b_2}$$
$$\psi_{2b_2} = c_3\chi_{1b_2} - c_4\chi_{2b_2}$$
$$\psi_{1a_2} = c_5\chi_{1a_2} + c_6\chi_{2a_2}$$
$$\psi_{2a_2} = c_7\chi_{1a_2} - c_8\chi_{2a_2}$$

(11.4)

where we have chosen the C_{2v} notation for convenience. Taking all coefficients as positive, the first molecular orbital has zero nodes, the second has two nodes, the third has one node, and the last has three nodes. We therefore expect the order ψ_{1b_2}, ψ_{1a_2}, ψ_{2b_2}, ψ_{2a_2}, with ground-state Slater determinant

$$\Psi = |\psi_{1b_2}^+\psi_{1b_2}^-\psi_{1a_2}^+\psi_{1a_2}^-| \tag{11.5}$$

The lowest possible excited state is produced by removing an electron from ψ_{1a_2} and placing it in ψ_{2b_2}; this state has B_1 symmetry ($A_2 \times B_2 = B_1$) and is allowed as an x-polarized transition since the dipole moment operator x transforms as B_1 ($A_1 \times B_1 \times B_1$ contains A_1). Thus, although group theory cannot give us the values of the coefficients in (11.4), it can be used to order the molecular orbitals and determine the polarizations of transitions to the lowest excited states.

Benzene belongs to D_{6h}, and the π orbitals transform as $A_{2u} + B_{2g} + E_{1g} + E_{2u}$. Since no irreducible representation occurs more than once in the decomposition, the molecular orbitals are completely determined by symmetry when we use a minimum basis set. The a_{2u} and b_{2g} combinations are easily shown to be

$$\psi_{a_{2u}} = \frac{N_a}{\sqrt{6}} (\varphi_1 + \varphi_2 + \varphi_3 + \varphi_4 + \varphi_5 + \varphi_6)$$

$$\psi_{b_{2g}} = \frac{N_b}{\sqrt{6}} (\varphi_1 - \varphi_2 + \varphi_3 - \varphi_4 + \varphi_5 - \varphi_6) \qquad \blacktriangleright (11.6)$$

These are, respectively, the lowest and highest molecular orbitals, having zero and six nodes. For the two degenerate representations, there is some arbitrariness in how the orbitals are chosen; the choice essentially depends on where the x and y axes are located on the molecule. Following convention (see Fig. 11.2), we take

$$\psi_{e_{1gx}} = \frac{N_{e1}}{\sqrt{12}} (2\varphi_1 + \varphi_2 - \varphi_3 - 2\varphi_4 - \varphi_5 + \varphi_6)$$

$$\psi_{e_{1gy}} = \frac{N_{e1}}{2} (\varphi_2 + \varphi_3 - \varphi_5 - \varphi_6)$$

$$\psi_{e_{2uxy}} = \frac{N_{e2}}{\sqrt{12}} (2\varphi_1 - \varphi_2 - \varphi_3 + 2\varphi_4 - \varphi_5 - \varphi_6) \qquad \blacktriangleright (11.7)$$

$$\psi_{e_{2u(x^2-y^2)}} = \frac{N_{e2}}{2} (-\varphi_2 + \varphi_3 - \varphi_5 + \varphi_6)$$

We have labeled the components of the degenerate orbitals without regard to parity; for example, (x,y) actually forms a basis for the E_{1u} representation, not the E_{1g}. Note that the e_{1g} orbitals have two nodes

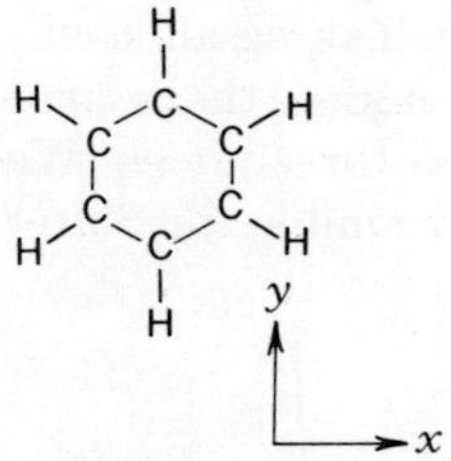

Fig. 11.2 Nuclear configuration for the benzene molecule.

each, whereas the e_{2u} orbitals have four. The resulting orbital pattern is sketched in Fig. 11.3.

A discussion of the excited states of benzene is somewhat complicated. The lowest states, produced by the excitation $\psi_{e_{1g}} \rightarrow \psi_{e_{2u}}$, are‡ $^3B_{1u}$, $^3B_{2u}$, $^3E_{1u}$, $^1B_{2u}$, $^1B_{2u}$, and $^1E_{1u}$. (The only other states of the same symmetry produced by $\pi \rightarrow \pi^*$ excitations are the $^3B_{1u}$ and $^1B_{1u}$ arising from the transition $\psi_{a_{2u}} \rightarrow \psi_{b_{2g}}$. These states occur at relatively high energies, and so configuration interaction with them may be ignored. However, other excited states, arising from $\sigma \rightarrow \pi^*$, $\pi \rightarrow \sigma^*$, and $\sigma \rightarrow \sigma^*$ transitions, may interact with our lowest states. Of course, these interactions cannot be taken into account within the framework of the π-electron approximation.) The easiest method of working out the relative energies of these $\pi \rightarrow \pi^*$ states is via a table of group coupling coefficients. Ignoring parity, the coordinates x and y transform as E_1, whereas $x^2 - y^2$ and $xy + yx$ transform as E_2. Let us define $\epsilon = x^2 - y^2$ and $\xi = xy + yx$. We need coefficients in the expansion

$$|E_1 E_2 \Gamma \gamma\rangle = \sum_{\alpha,\beta} \langle E_1 E_2 \alpha \beta | E_1 E_2 \Gamma \gamma \rangle | E_1 E_2 \alpha \beta \rangle \tag{11.8}$$

Note that in the group D_{6h}, $e_{1g} \times e_{2u} = b_{1u} + b_{2u} + e_{1u}$. See also Prob. 11.2.

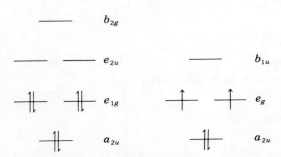

Fig. 11.3 Orbital energy patterns in benzene and *cyclo*-butadiene.

for $\Gamma = E_1$, B_1, and B_2. Explicitly (Prob. 11.3),

$$|E_1 E_2 B_1\rangle = \frac{1}{\sqrt{2}} (x\epsilon - y\xi)$$

$$|E_1 E_2 B_2\rangle = \frac{1}{\sqrt{2}} (x\xi + y\epsilon)$$

$$|E_1 E_2 E_1 x\rangle = \frac{1}{\sqrt{2}} (x\epsilon + y\xi)$$

$$|E_1 E_2 E_1 y\rangle = \frac{1}{\sqrt{2}} (x\xi - y\epsilon)$$

▶(11.9)

The triplet states are the most easily written out. For example, consider the four Slater determinants with $M_S = 1$,

$$\Psi_1 = |\psi_{a_{2u}}^+ \psi_{a_{2u}}^- \psi_x^+ \psi_\xi^+ \psi_y^+ \psi_y^-|$$
$$\Psi_2 = |\psi_{a_{2u}}^+ \psi_{a_{2u}}^- \psi_x^+ \psi_\epsilon^+ \psi_y^+ \psi_y^-|$$
$$\Psi_3 = |\psi_{a_{2u}}^+ \psi_{a_{2u}}^- \psi_x^+ \psi_x^- \psi_y^+ \psi_\xi^+|$$
$$\Psi_4 = |\psi_{a_{2u}}^+ \psi_{a_{2u}}^- \psi_x^+ \psi_x^- \psi_y^+ \psi_\epsilon^+|$$

▶(11.10)

(Note that the occupancy of the closed σ core is not indicated in the determinants.) Then, using the coupling coefficients, we have

$$|{}^3 E_{1u}\, 1\, x\rangle = \frac{1}{\sqrt{2}} (\Psi_2 + \Psi_3)$$

$$|{}^3 E_{1u}\, 1\, y\rangle = \frac{1}{\sqrt{2}} (\Psi_1 - \Psi_4)$$

$$|{}^3 B_{1u}\, 1\rangle = \frac{1}{\sqrt{2}} (\Psi_2 - \Psi_3)$$

$$|{}^3 B_{2u}\, 1\rangle = \frac{1}{\sqrt{2}} (\Psi_1 + \Psi_4)$$

▶(11.11)

Other triplet components with $M_S < 1$, and the corresponding singlet states, may be determined using the shift operator $\mathbf{S}^-$.

The energies of these states may be found using the methods of Sec. 10.3. Unfortunately, the results quoted there were not derived for situations in which there is orbital degeneracy. Nevertheless, they can be used without modification if we consider the "configuration interaction" between states which are degenerate according to the ordinary theory. Evaluating first the diagonal energy of the four $M_S = 1$ determinants above, we obtain

$$E(\Psi_1) = \Delta\epsilon - J_{x\xi}$$
$$E(\Psi_2) = \Delta\epsilon - J_{x\epsilon}$$
$$E(\Psi_3) = \Delta\epsilon - J_{y\xi}$$
$$E(\Psi_4) = \Delta\epsilon - J_{y\epsilon}$$
$$\Delta\epsilon = \epsilon_{e_{2u}} - \epsilon_{e_{1g}}$$

▶(11.12)

By symmetry, $J_{x\xi} = J_{y\epsilon}$ and $J_{x\epsilon} = J_{y\xi}$ (Prob. 11.3). Therefore,

$$E(\Psi_1) = E(\Psi_4)$$
$$E(\Psi_2) = E(\Psi_3)$$

(11.13)

According to Sec. 10.4, the off-diagonal element between Ψ_1 and Ψ_4 is

$$-\langle\psi_x(1)\psi_\epsilon(2)|\ \frac{1}{r_{12}}\ |\psi_y(1)\psi_\xi(2)\rangle \equiv -\langle x\epsilon\|y\xi\rangle$$

Similarly, the corresponding element between Ψ_2 and Ψ_3 is

$$-\langle\psi_x(1)\psi_\xi(2)|\ \frac{1}{r_{12}}\ |\psi_y(1)\psi_\epsilon(2)\rangle \equiv -\langle x\xi\|y\epsilon\rangle$$

The energies of the three triplet states are thus

$$E(^3E_{1u}) = \Delta\epsilon - J_{x\epsilon} - \langle x\epsilon\|y\xi\rangle$$
$$= \Delta\epsilon - J_{x\xi} + \langle x\epsilon\|y\xi\rangle$$
$$E(^3B_{1u}) = \Delta\epsilon - J_{x\epsilon} + \langle x\epsilon\|y\xi\rangle$$
$$E(^3B_{2u}) = \Delta\epsilon - J_{x\xi} - \langle x\epsilon\|y\xi\rangle$$

▶(11.14)

Note that $\langle x\epsilon\|y\xi\rangle = \langle x\xi\|y\epsilon\rangle$, which implies $J_{x\epsilon} = J_{x\xi} - 2\langle x\epsilon\|y\xi\rangle$. Thus, assuming that the matrix element $\langle x\epsilon\|y\xi\rangle$ is positive, which calculations indicate it is, the theory predicts the $^3B_{2u}$ state to lie lowest in energy, the $^3E_{1u}$ above it by $2\langle x\epsilon\|y\xi\rangle$, and it predicts the $^3B_{1u}$ to be highest, above the $^3B_{2u}$ by $4\langle x\epsilon\|y\xi\rangle$. This gives rise to an interval rule, but it is an unreliable one since configuration interaction is not expected to be totally unimportant; we have neglected all core excitations and excitations to virtual σ^* orbitals.

The singlets may be handled in a similar way. Corresponding to the four Slater determinants of (11.10), there are four singlet linear combinations such as

$$\Psi_1' = \frac{1}{\sqrt{2}}\ (|\psi_{a_{1g}}^+\psi_{a_{1g}}^-\psi_x^+\psi_\xi^+\psi_y^+\psi_y^-| - |\psi_{a_{1g}}^+\psi_{a_{1g}}^-\psi_x^-\psi_\xi^+\psi_y^+\psi_y^-|)$$

▶(11.15)

From (10.86),

$$E(\Psi_1') = E(\Psi_4') = \Delta\epsilon - J_{x\xi} + 2K_{x\xi}$$
$$E(\Psi_2') = E(\Psi_3') = \Delta\epsilon - J_{x\epsilon} + 2K_{x\epsilon}$$

▶(11.16)

The off-diagonal elements between degenerate determinants are $2\langle x\epsilon\|\xi y\rangle - \langle x\epsilon\|y\xi\rangle$ and $2\langle x\xi\|\epsilon y\rangle - \langle x\epsilon\|y\xi\rangle$, respectively. The energies of the singlets are therefore

$$
\begin{aligned}
E(^1E_{1u}) &= \Delta\epsilon - J_{x\epsilon} + 2K_{x\epsilon} + 2\langle x\xi\|\epsilon y\rangle - \langle x\epsilon\|y\xi\rangle \\
&= \Delta\epsilon - J_{x\xi} + 2K_{x\xi} - 2\langle x\epsilon\|\xi y\rangle + \langle x\epsilon\|y\xi\rangle \\
E(^1B_{1u}) &= \Delta\epsilon - J_{x\epsilon} + 2K_{x\epsilon} - 2\langle x\xi\|\epsilon y\rangle + \langle x\epsilon\|y\xi\rangle \\
E(^1B_{2u}) &= \Delta\epsilon - J_{x\xi} + 2K_{x\xi} + 2\langle x\epsilon\|\xi y\rangle - \langle x\epsilon\|y\xi\rangle
\end{aligned}
$$

$\blacktriangleright$ (11.17)

The ordering of the three states cannot be predicted in the absence of a numerical calculation; no interval rule can be determined with the information at hand. Explicit calculations with a Slater-type $2p_z$ basis show the $^1B_{2u}$ is lowest in energy, followed by the $^1B_{1u}$ and the $^1E_{1u}$. Since the dipole operator transforms as E_{1u}, only transitions to the $^1E_{1u}$ state are allowed. Experimentally, a strong band at 54,500 cm^{-1} is preceded by two weak bands, probably the $^1B_{1u}$ and the $^1B_{2u}$ partially allowed by vibronic mixing. The primary phosphorescent emission occurs near 30,000 cm^{-1} and presumably corresponds to radiation from the $^3B_{1u}$.

Yet another application of the π-electron approximation can be made to cyclobutadiene. Assuming this (hypothetical) molecule is planar, it belongs to the group D_{4h}, and its π orbitals transform as $A_{2u} + B_{1u} + E_g$. The molecular orbitals are therefore entirely determined by symmetry, and the consequent orbital pattern is sketched in Fig. 11.3. (The a_{2u} orbital has no nodal planes between carbon nuclei, whereas the b_{1u} has a node between every pair.) Since butadiene has four π electrons, Hund's rule predicts that the ground state of this molecule should be a triplet. We expect a triplet of this sort to be very highly reactive, which is the probable reason that all attempts to synthesize cyclobutadiene have failed. It is possible to show quite generally that all cyclic molecules of the type $(CH)_m$ will be radicals (m odd) or diradicals (triplet ground state) unless $m = 4n + 2$ with n an integer. [Actually, cyclooctatetraene ($m = 8$) is stable, but only because the molecule distorts into a nonplanar configuration (as is to be expected on the basis of the Jahn-Teller theorem; see Sec. 11.9)]. This result is known as the $4n + 2$ rule and was first proved using the Hückel method (*vide infra*). However, the proof is quite independent of any specific calculational approximations.

Simple molecular orbital theory is also useful in valence problems, that is, in the qualitative prediction of bond angles, bond lengths, stabilities, and pathways of chemical reactions. For example, in Sec. 7.3 we were able to predict the trend in bond lengths and bond strengths in homonuclear diatomic molecules. Using group theory, and using the π-electron approximation, valence-theory predictions can be extended to larger molecules. Walsh (1953) has used molecular orbital theory to predict the shapes of simple molecules of the type AB_2, BAC, AB_3, H_2AB and $HAAH$. More recently, Gimarc (1969) has extended this work to include the more complicated molecules $BAAB$—for example, C_2F_2, N_2O_2, and H_2O_2. Perhaps the most beautiful application of molecular orbital theory to valence problems has been made by Woodward and Hoffmann (1965), who have shown that the steric course of virtually every concerted reaction in organic chemistry can be predicted on the basis of molecular orbital theory. As shown by Longuet-Higgins and Abrahamson (1965), these predictions can be made without making detailed calculations, and depend for their success on the simplest molecular orbital concepts, extended to increasingly larger systems through the use of symmetry. Short reviews of this work have been given by Hoffmann and Woodward (1968) and by Vollmer and Servis (1968). Every student should read *at least* one of the reviews, for the "Woodward-Hoffmann rules" provide an extraordinary example of the predictive power of elementary orbital concepts.

In this section we have devoted a considerable amount of detail to a discussion of the π-electron approximation and its applications. This we have done in order to stress the full power of simple orbital theory when married to group theory. Obviously, there are few molecules from which we can abstract as much information as we did for benzene. Nevertheless, many of the successes of the π-electron method are rooted in the same kind of approach that we earlier applied to homonuclear diatomic molecules. Group theory is the bridge by which we can take the most elementary ideas and apply them to relatively more complex systems. In the following sections we discuss specific methods of numerical calculation, in an effort to treat molecules of reduced symmetry. However, the reader should not lose sight of the fact that many of the simple numerical methods are successful only because they incorporate in a mechanical way the necessary group theory, together with the elementary idea that molecular orbitals should be ordered by number of nodes. Thus, any "theory" which permits a numerical calculation on benzene will of necessity give *at least* the results derived in this section. It is therefore usually worthwhile to try to separate out the group-theoretically determined part of a calculation from the whole, if only because the portion which most securely rests on simple concepts is most likely to be correct and independent of the calculational approximations used to obtain it.

PROBLEMS

11.1 Show that (neglecting overlap between nonnearest neighbors) the bond order in benzene is very nearly $\frac{2}{3}$ for any reasonable value of the overlap integral.

11.2 Verify that the molecular (symmetry) orbitals given in (11.7) have the indicated transformation properties. Assume that $C_2' = C_2^{(x)}$ in Table 8.1. Show especially that $\psi_{e_{1gx}}$ and $\psi_{e_{1gy}}$ transform as x and y under the group generators C_6 and C_2'. Then show that the appropriate products transform as given in (11.9).

11.3 Show that $J_{x\epsilon} = J_{y\xi}$ and $J_{y\epsilon} = J_{x\xi}$.
Hint: Derive the coupling coefficients for $E_1 \times E_1$ and $E_2 \times E_2$. Use these to show that the integrands $x(1)x(1)\xi(2)\xi(2) + y(1)y(1)\xi(2)\xi(2) - x(1)x(1)\epsilon(2)\epsilon(2)$ $- y(1)y(1)\epsilon(2)\epsilon(2)$ and $x(1)x(1)\xi(2)\xi(2) - y(1)y(1)\xi(2)\xi(2) + x(1)x(1)\epsilon(2)\epsilon(2) -$ $y(1)y(1)\epsilon(2)\epsilon(2)$ both transform as ϵ and are therefore both zero.

11.4 Find the bond order in cyclobutadiene.

11.2 THE HÜCKEL METHOD

The Hückel method is one way of evaluating matrix elements within the π-electron approximation. Specifically, all diagonal matrix elements $\langle \varphi_i | \mathbf{F} | \varphi_i \rangle$ are set equal to α, and all off-diagonal elements between nearest neighbors are set equal to β. The matrix elements between nonnearest neighbors are assumed to be zero. Overlap integrals are also set equal to zero, excepting, of course, the diagonal elements. Thus, the secular determinant for benzene is

$$\begin{vmatrix} \alpha - \epsilon & \beta & 0 & 0 & 0 & \beta \\ \beta & \alpha - \epsilon & \beta & 0 & 0 & 0 \\ 0 & \beta & \alpha - \epsilon & \beta & 0 & 0 \\ 0 & 0 & \beta & \alpha - \epsilon & \beta & 0 \\ 0 & 0 & 0 & \beta & \alpha - \epsilon & \beta \\ \beta & 0 & 0 & 0 & \beta & \alpha - \epsilon \end{vmatrix} \blacktriangleright (11.18)$$

Of course, the orbitals are given by (11.6) and (11.7) since they are symmetry-determined. Solving the secular equation above or explicitly evaluating the diagonal energies $\langle \psi_i | \mathbf{F} | \psi_i \rangle$ (Prob. 11.5), we find the orbital energies $\epsilon_{a_{2u}} = \alpha + 2\beta$, $\epsilon_{e_{1g}} = \alpha + \beta$, $\epsilon_{e_{2u}} = \alpha - \beta$, and $\epsilon_{b_{2g}} = \alpha - 2\beta$. Thus β is a negative quantity.

Unfortunately, this is as far as the Hückel method goes. Matrix elements such as $\langle x\epsilon || y\xi \rangle$ cannot be calculated, and in fact there is no way provided for estimating the energies of the singlet and triplet states. Using Koopmans' theorem, one may say that the ionization potential of benzene is $\alpha + \beta$, but this is not very helpful unless numerical values of the two constants are specified. Apparently, we learn nothing about benzene that we did not know before using group theory alone.

The situation is actually not quite so hopeless. First of all, for a molecule with reduced symmetry, such as butadiene, the Hückel method can be used to determine the molecular orbital coefficients. We leave it as an exercise (Prob. 11.6) to show that the molecular orbitals have energies $\alpha + 1.618\beta$, $\alpha - 0.618\beta$, $\alpha + 0.618\beta$, and $\alpha - 1.618\beta$, and that the coefficients are

$$\psi_{1b_2} = 0.525 \left[\frac{1}{\sqrt{2}} (\varphi_1 + \varphi_4) \right] + 0.849 \left[\frac{1}{\sqrt{2}} (\varphi_2 + \varphi_3) \right]$$

$$\psi_{1a_2} = 0.849 \left[\frac{1}{\sqrt{2}} (\varphi_1 - \varphi_4) \right] + 0.525 \left[\frac{1}{\sqrt{2}} (\varphi_2 - \varphi_3) \right]$$

$$\psi_{2b_2} = 0.849 \left[\frac{1}{\sqrt{2}} (\varphi_1 + \varphi_4) \right] - 0.525 \left[\frac{1}{\sqrt{2}} (\varphi_2 + \varphi_3) \right]$$

$$\psi_{2a_2} = 0.525 \left[\frac{1}{\sqrt{2}} (\varphi_1 - \varphi_4) \right] - 0.849 \left[\frac{1}{\sqrt{2}} (\varphi_2 - \varphi_3) \right]$$

(11.19)

Clearly, it is easier to solve this problem by diagonalizing the two 2×2 secular equations between symmetry orbitals then to attempt the solution of the 4×4. Explicitly, the two secular determinants for butadiene are

$$\begin{vmatrix} \alpha - \epsilon & \beta \\ \beta & \alpha + \beta - \epsilon \end{vmatrix} = 0$$

$$\begin{vmatrix} \alpha - \epsilon & \beta \\ \beta & \alpha - \beta - \epsilon \end{vmatrix} = 0$$

(11.20)

where we have given the determinant for the orbitals of b_2 symmetry first. Note that the Hückel method does not distinguish between the *cis* and *trans* forms of the molecule.

The Hückel method is also used to evaluate the total (π-electron) energy of various states. It does this in a way which is theoretically curious; we write

$$E_\pi = \sum_i N(i) \, \epsilon_i$$

▶(11.21)

Thus, the π-electron energy for benzene is $6\alpha + 8\beta$, for butadiene it is $4\alpha + 4.472\beta$, and for ethylene it is $2\alpha + 2\beta$. The collection of excited states in benzene arising from the transition $\psi_{e_{1g}} \to \psi_{e_{2u}}$ should thus have an average energy of $6\alpha + 6\beta$, so the mean excitation energy is -2β. These results may be used in two ways. First of all, it is well established experimentally that the heat content H of a molecule is to a very high degree of accuracy a simple sum of "bond energies." This rule breaks down for aromatic hydrocarbons since the bonds are intermediate between double and single. Presumably, a single bond has energy 2α, whereas a double bond has energy $2\alpha + 2\beta$. Therefore, the energy of the benzene molecule is stabilized with respect to the hypothetical structure having three isolated single bonds and three isolated double bonds by 2β. Similarly, butadiene has 0.472β extra "resonance" stabilization. This may be compared to the "resonance energy" of benzene or butadiene, that is, the experimental enthalpy minus the expected enthalpy for the hypothetical molecules. The correlation found is excellent (Murrell et al., 1965, p. 284) and gives $\beta = -0.69$ ev.

The value of β may also be determined by comparing calculated and predicted excitation energies. Once again, the correlation is superb, this time giving $\beta = -2.71$ ev. The disagreement between the two values (and among values of determined by other means), although severe, should not blind us to the fact that the Hückel theory can predict the resonance energy or excitation energy of a given aromatic hydrocarbon to an extraordinary degree of accuracy, given the proper value of β. It is most surprising, in view of the number and drastic nature of the approximations made, that the theory works at all.

In Sec. 11.4, we will investigate the reasons behind the successes of the Hückel methods. For the present, we mention two results from that section. First, Hückel theory really only works well for a special class of aromatic compounds known as alternant hydrocarbons. Secondly, for that class, Hückel theory always predicts that the gross charge densities Q_p are all unity. There are good theoretical reasons for believing that the charge densities in such compounds really *are* uniform, and one can further show that any such compound will have no long-range potentials. Consequently, the diagonal matrix elements of the Fock operator should all be roughly equal, and the off-diagonal elements should be independent of the specific nature of the two atoms involved. This, at least in part, explains why the Hückel approximations work, although the assumption $S_{ij} = \delta_{ij}$ is still inexplicable.

What is theoretically most objectionable about the Hückel method is its reliance on formulas such as (11.21). This equation is not merely an approximation; it is, in fact, clearly wrong [compare (10.11)]. It is not possible to assume that the total energy is a sum of orbital energies unless the Fock operator can be written as a sum of *independent* operators, such as

$$\mathbf{F} = \sum_{i=1}^{n} \mathbf{F}_i \tag{11.22}$$

where

$$\mathbf{F}_i \psi_i = \epsilon_i \psi_i \tag{11.23}$$

In such a situation, (11.21) holds; the difficulty arises in reconciling (11.22) and (11.1). The $\mathbf{F}_i$ *cannot* be independent since the Fock equations are coupled. A possible justification is that the electron repulsions are somehow averaged out in (11.22), but this approach is not really consistent with the Hartree-Fock method.

Another related objection to the Hückel method is that the average of the excited-state energies is *never* $\Delta\epsilon$ in the Hartree-Fock scheme. In fact, the average excited-state energy for benzene is $\Delta\epsilon - \frac{1}{2}(J_{x\xi} + J_{x\epsilon}) + \frac{1}{4}(K_{x\xi} + K_{x\epsilon})$ (Prob. 11.8). This again demonstrates the incompatability of the Hückel method with the Hartree-Fock theory. It therefore seems likely that the successes of the Hückel method are more a result of the symmetry and chemical and electronic simplicity of the chosen molecules to which it has been applied than of any real theoretical validity.

There is a class of calculational techniques known as *extended* Hückel methods which, although having nothing in common with the Hückel approximations per se, employ formulas such as (11.21). Both σ and π orbitals are considered explicitly, and overlap is not usually neglected. Such techniques have been very successful in predicting bond angles and lengths, the relative stabilities of different structures, dipole moments, and excitation energies.‡ Nevertheless, their theoretical basis remains hazy, and there is little genuine *predictive* value in them unless they are carefully calibrated on a number of closely related compounds One must always beware of pushing§ a method when the theoretical basis is uncertain.

PROBLEMS

11.5 Verify the orbital energies given for benzene.

11.6 Using the method outlined in Appendix 2, confirm the coefficients given in (11.19). Also calulate the orbital energies.

11.7 Calculate the bond orders and charge densities for butadiene. Show that the gross π-electronic charge densities Q_p are all unity.

11.8 Show that in benzene the average excited-state energy produced by the transition $\psi_{e_{1g}} \rightarrow \psi_{e_{2u}}$ is $\Delta\epsilon - \frac{1}{2}(J_{x\xi} + J_{x\epsilon}) + \frac{1}{4}(K_{x\xi} + K_{x\epsilon})$. Be careful to weigh each term by its *total* degeneracy.

11.3 THE METHOD OF PARISER, PARR, AND POPLE

In 1952, R. G. Parr proposed an approximate way of handling many-center integrals within the π-electron framework. His approximation is closely related to the Hückel assumption $S_{pq} = \delta_{pq}$, but it goes a step farther and eliminates all integrals, including specifically the two-electron integrals, which arise from the overlap of two *different* atoms. It is therefore referred to as the zero-differential overlap (ZDO) approximation;

---

‡ See especially R. Hoffmann, 1963, 1964; M. Zerner and M. Gouterman, 1966; D. G. Carroll and S. P. McGlynn, 1968.

§ According to a colleague, who for professional reasons must remain anonymous, "There is the Hückel method, the extended Hückel method, and the overextended Hückel method." Too many calculations in the first two categories (although not those listed just above) fall also in the third.

any integral containing an element of the form $\varphi_p(1)\varphi_q(1)\,dv$ is set equal to zero unless $\varphi_p = \varphi_q$.

The ZDO approximation was then applied to a series of aromatic hydrocarbons by R. Pariser and R. G. Parr (1953). Several *additional* approximations were involved, all of which are essential to their method. Shortly thereafter, J. A. Pople (1953) put the theory on a somewhat more formal basis and in particular showed that the approximations of Pariser and Parr could be incorporated into the Roothaan SCF formalism. Later modifications and applications have been made by innumerable other workers. In its fundamental form, however, the theory is due to Pariser, Parr, and Pople, and it is customary to refer to their contribution as the PPP method.

Consider the Fock operator of (11.1). We need to find expressions for matrix elements over the basis set in the general form $\langle \varphi_p | \mathbf{F} | \varphi_q \rangle$. The two-electron portion is straightforward. Expanding as in (6.100), we find

$$\langle \varphi_p | \sum_{i=1}^{n} \mathbf{G}_i | \varphi_q \rangle = \sum_{i=1}^{n} \sum_{r=1}^{m} \sum_{s=1}^{m} c_{ir}^* c_{is}(2\langle pr \| qs \rangle - \langle pr \| sq \rangle) \qquad (11.24)$$

where, for example,

$$\langle pr \| qs \rangle \equiv \langle \varphi_p(1)\varphi_r(2) | \frac{1}{r_{12}} | \varphi_q(1)\varphi_s(2) \rangle \qquad (11.25)$$

According to the ZDO approximation, $\langle pr \| qs \rangle$ is zero unless $p = q$ and $r = s$. We therefore obtain

$$\langle \varphi_p | \sum_{i=1}^{n} \mathbf{G}_i | \varphi_q \rangle = \delta_{pq} \sum_{r=1}^{m} 2\langle pr \| pr \rangle \sum_{i=1}^{n} |c_{ir}|^2 - \sum_{i=1}^{n} c_{iq}^* c_{ip}\langle pq \| pq \rangle \qquad (11.26)$$

This relationship may be further simplified by using the bond-order charge-density relationships of the last chapter. For a closed-shell system with $S_{pq} = 0$,

$$B_{pq} = 2 \sum_{i=1}^{n} c_{ip}^* c_{iq}$$
$$Q_p = P_{pp} = B_{pp} \qquad (11.27)$$

Substituting, we derive the two basic expressions

$$\langle \varphi_p | \sum_{i=1}^{n} \mathbf{G}_i | \varphi_p \rangle = \sum_{r=1}^{m} Q_r\langle pr \| pr \rangle - \tfrac{1}{2}Q_p\langle pp \| pp \rangle$$

$$\blacktriangleright (11.28)$$

$$\langle \varphi_p | \sum_{i=1}^{n} \mathbf{G}_i | \varphi_q \rangle = -\tfrac{1}{2}B_{pq}\langle pq \| pq \rangle \qquad (p \neq q)$$

If we were to handle the one-electron integrals in the same way, they would all vanish except the one-center integrals. This approximation does not seem to work very well, and instead we take only the matrix elements between nonnearest neighbors as zero. The integrals between neighbors are approximated as in the Hückel method

$$\langle \varphi_p | -\frac{\nabla^2}{2} + \sum_{k=1}^{K} \mathbf{V}_k | \varphi_q \rangle = \beta_{pq} \qquad \blacktriangleright (11.29)$$

Values of β are chosen so as to get the best possible agreement between theoretically predicted and experimentally observed quantities, and it is in this sense that PPP theory is "semiempirical." It would appear that β is at least roughly proportional to the calculated overlap integrals between two atoms (although one rarely calculates overlaps in the PPP method since all S_{pq} are neglected), and the parameter is found to be "transferable" in the sense that the best possible β for benzene will also work well on other hydrocarbon molecules with similar carbon-carbon bond distances.

The one-center, one-electron integrals may also be parametrized as in the Hückel method, but for most purposes it is preferable to use a semiempirical relation due to M. Goeppert-Meyer and A. L. Sklar (1938). Writing out the one-electron operator in full, we obtain

$$-\frac{\nabla^2}{2} + \sum_{k=1}^{K} \mathbf{V}_k = \left(-\frac{\nabla^2}{2} + \mathbf{V}_p' \right) + \sum_{\substack{r=1 \\ (r \neq p)}}^{m} \mathbf{V}_r' + \sum_l \mathbf{V}_l'' \qquad (11.30)$$

Here $\mathbf{V}_r'$ are the core-attraction operators of atoms which are charged when a π electron is removed from the atom in question (e.g., the carbon atoms in benzene), whereas the $\mathbf{V}_l''$ are the so-called "penetration operators," arising from atoms which are neutral even in the absence of π electrons, for example, the hydrogen atoms in benzene or the methyl carbon in toluene. Let us assume that φ_p is an eigenfunction of the operator in parentheses.‡ (This can be used to define φ_p.) Then, since

$$\left(-\frac{\nabla^2}{2} + \mathbf{V}_p' \right) |\varphi_p\rangle = \epsilon_p |\varphi_p\rangle \qquad (11.31)$$

we may write

$$\langle \varphi_p | -\frac{\nabla^2}{2} + \sum_{k=1}^{K} \mathbf{V}_k | \varphi_p \rangle = \epsilon_p + \sum_{\substack{r=1 \\ (r \neq p)}}^{m} \langle \varphi_q | \mathbf{V}_r' | \varphi_q \rangle + \sum_l \langle \varphi_q | \mathbf{V}_l'' | \varphi_q \rangle$$

$$(11.32)$$

‡ This assumption is valid only when $Z_p = 1$. More generally, $(-\nabla^2/2 + \mathbf{V}_p')|\varphi_p\rangle + (Z_p - 1)\langle pp\|pp\rangle = \epsilon_p|\varphi_p\rangle$, where Z_p is the core charge on atom p.

According to Koopmans' theorem, $\epsilon_p = -I_p$, the valence-state ionization potential (see Sec. 6.8) of the atom. Furthermore,

$$\langle \varphi_p | \mathbf{V}_r' | \varphi_p \rangle = -Z_r \langle pr \| pr \rangle + \langle \varphi_p | \mathbf{V}_r'' | \varphi_p \rangle \tag{11.33}$$

where $\mathbf{V}_r''$ is the one-electron operator due to the neutral atomic attraction and Z_r is the charge on atom r when all π electrons have been removed. Thus,

$$\langle \varphi_p | - \frac{\nabla^2}{2} + \sum_{k=1}^{K} \mathbf{V}_k | \varphi_p \rangle = -I_p - \sum_{\substack{r=1 \\ (r \neq p)}}^{m} Z_r \langle pr \| pr \rangle + \sum_k \langle \varphi_p | \mathbf{V}_k'' | \varphi_p \rangle$$

$$\blacktriangleright (11.34)$$

The final sum extends over *all* atoms (neutral) in the system. Combining the expressions for the matrix elements of the one- and two-electron operators, (11.28), (11.29), and (11.34), we get

$$\langle \varphi_p | \mathbf{F} | \varphi_p \rangle = -I_p + \sum_{\substack{r=1 \\ (r \neq p)}}^{m} (Q_r - Z_r) \langle pr \| pr \rangle$$

$$+ (\tfrac{1}{2} Q_p - Z_p + 1) \langle pp \| pp \rangle + \sum_k \langle \varphi_p | \mathbf{V}_k'' | \varphi_p \rangle \quad \blacktriangleright (11.35)$$

$$\langle \varphi_p | \mathbf{F} | \varphi_q \rangle = \beta_{pq} - \tfrac{1}{2} B_{pq} \langle pq \| pq \rangle$$

Note that β_{pq} is ordinarily taken as zero for φ_p and φ_q nonnearest neighbors. Equation (11.35) is valid even when $Z_p \neq 1$.

There are many modifications possible within the PPP method. For example, Pople has suggested taking $\langle pr \| pr \rangle = 1/R_{pr}$ (in atomic units). This at least makes the calculation of integrals trivial, but is not at all accurate. Another approximation, suggested by N. Mataga and K. Nishimoto (1957), is

$$\gamma_{pr} = \frac{1}{R_{pr} + 2/(\gamma_{pp} + \gamma_{rr})} \qquad \blacktriangleright (11.36)$$

where we have used the customary abbreviation $\gamma_{pr} = \langle pr \| pr \rangle$. This formula gives generally good agreement between theoretically predicted and experimentally observed quantities, although it gives poor agreement with the rigorously calculated $\langle pr \| pr \rangle$. Actually, use of the accurately calculated integrals does not give as good agreement with experimental data as does the Nishimoto-Mataga formula. Note that theoretical accuracy is not what is desired here, but rather success in applications. It has been shown (Parr, 1964, p. 68) that at least in benzene, coulombic and exchange integrals J_{ij} and K_{ij} are given with extraordinary accuracy by the ZDO approximation when the proper values of the $\langle pr \| pr \rangle$ are used, but this is partially beside the point; when the Nishimoto-Mataga values are used, the resultant J_{ij} and K_{ij} integrals can be used to predict band positions with greater success than when the purely theoretical integrals are used.

Another approximation widely used is to neglect all neutral penetration integrals $\langle \varphi_p | V_k'' | \varphi_p \rangle$. This is certainly not justifiable, since the integrals are often as large as an electron volt. (However, $\langle \varphi_p | V_r'' | \varphi_p \rangle$ drops off very rapidly with increasing distance between p and r.) Nevertheless, at least for aromatic hydrocarbons, the contribution of the neutral penetration integrals to each diagonal matrix element $\langle \varphi_p | F | \varphi_p \rangle$ is nearly the same, and so they provide a uniform shift in the orbital energies. In heteronuclear systems, containing nitrogen, oxygen, sulfur, and so on, the neglect is more serious since the penetration terms contribute differently to different atoms. In spite of this objection, neutral penetration terms are rarely *not* neglected since they are difficult to calculate.

The final, and in many ways most crucial, approximation in PPP theory concerns the one-center coulombic integrals $\langle pp \| pp \rangle$, which enter (11.35) not only through the term in Q_p, but also through the two-center coulombic integrals via (11.36).

Following Pariser (1953), it is now customary to take "empirical" values for $\langle pp \| pp \rangle$, according to the relation

$$\langle pp \| pp \rangle = I_p - A_p \tag{11.37}$$

where I_p is the ionization potential of a π electron in the appropriate valence state and A_p is the corresponding electron affinity. This expression can be "derived" by assuming that Koopmans' theorem applies to both ionization potentials and electron affinities, an assumption which, as noted in both Chaps. 6 and 10, is incorrect. Nevertheless, (11.37) is widely used. Its effect is to greatly reduce the value of $\langle pp \| pp \rangle$ employed; using Slater orbitals, we calculate $\langle pp \| pp \rangle$ for carbon as 16.93 ev, compared to the empirical $I - A = 10.53$ ev. Unquestionably, the use of this formula enhances the usefulness of PPP theory since the predictions of spectra are improved—but only at the expense of theoretical accuracy. Orloff and Sinanoglu (1965) have discussed the relationship between the success of (11.37) and the importance of correlation energy, but their discussion clearly shows that the use of (11.37) cannot be rigorously justified within a molecular orbital framework. It is interesting to note in this connection that in a valence-bond framework, empirical corrections along the lines of (11.37) can be made in a way which is very nearly rigorous, or rather, in a way in which the exact nature of the approximations is perfectly clear. This leads to the method of *atoms in molecules*, discussed very briefly in Sec. 7.5. It is therefore unfortunate that the valence-bond method is difficult to apply in a general way to large molecular systems.

PPP theory is usually referred to as semiempirical since it incorporates many experimental results and aims at predictive success rather than theoretical accuracy. The reason the method works so well remains somewhat of a puzzle. The two-electron part of the Fock operator is certainly treated in an acceptable way since, as noted, the J's and K's calculated using ZDO agree well with those calculated rigorously [provided that (11.36) and (11.37) are *not* used]. The handling of the one-electron operators is more difficult to explain, especially the approximation $S_{pq} = \delta_{pq}$. Rather involved explanations have been given by Ruedenberg (1961) and by Fischer-Hjalmars (1965); they show that if the matrix elements in the *exact* expression are expanded in terms of powers of the overlap integral, the first nonvanishing coefficients are those that multiply S^2, which is, of course, genuinely small. Note that it is therefore not valid to include overlap when the other approximations [such as (11.29)] are made. This approach gives poorer results.

Another surprise is that the parameters β are so pleasantly constant from one molecule to the next. In fact, the value $\beta = -2.4$ ev is roughly suitable for almost any molecular system, even those containing heteroatoms. On a crude basis (Prob. 11.11),

$$\beta_{pq} \approx \beta'_{pq} - \tfrac{1}{2} S_{pq}(\beta'_{pp} + \beta'_{qq})$$

$$\beta'_{pq} = \langle \varphi_p | -\frac{\nabla^2}{2} + \sum_{k=1}^{K} \mathbf{V}_k | \varphi_q \rangle \tag{11.38}$$

where β'_{pq} is the *calculated* value including overlap. Orbitals φ_p and φ_q are thus orthogonalized so their overlap is zero, and β_{pq} is then the integral between the new orbitals. It is not hard to show that β_{pq} should be insensitive to nuclear potentials $\mathbf{V}_k$ at a large distance from both p and q, although β'_{pq} itself is quite sensitive. At such distances, the approximation of Mulliken (1949),

$$\langle \varphi_p | \mathbf{V}_k | \varphi_q \rangle = \frac{S_{pq}}{2} \left(\langle \varphi_p | \mathbf{V}_k | \varphi_p \rangle + \langle \varphi_q | \mathbf{V}_k | \varphi_q \rangle \right) \qquad \blacktriangleright (11.39)$$

can be shown to hold with high accuracy. Thus, although it is understandable that β_{pq} should be roughly insensitive to the larger molecular environment, it is surprising that it is so totally constant.

The fact that PPP theory is so successful is no doubt due in large part to its flexibility and to the relatively low standards applied as criteria for success. In molecules which contain heteroatoms, PPP theory is not nearly as accurate as, say, ligand field theory, and we do not usually expect agreement between experimental and predicted band positions within 10 percent. Furthermore, there are so many published variations to the theory that a judicious selection can nearly guarantee some sort of success. Nevertheless, in alternant hydrocarbons, and in some other systems, the theory has met with excellent success, as we now demonstrate with an extended example. We choose the benzene molecule since the orbitals are determined by symmetry; thus there is no need of an iterative SCF approach. Taking the internuclear separation as 1.39 Å and choosing $\gamma_{11} = 10.53$ ev, we find $\gamma_{12} = 5.22$, $\gamma_{13} = 3.81$, and $\gamma_{14} = 3.47$ using the Nishimoto-Mataga formula. Furthermore, neglecting the neutral penetration integrals and choosing $I_p = 11.22$, it is possible to prove that

$$
\begin{aligned}
F_{11} &= -5.96 \text{ ev} \\
F_{12} &= -4.14 \text{ ev} \\
F_{14} &= +0.579 \text{ ev} \\
F_{13} &= 0.0
\end{aligned}
\tag{11.40}
$$

where $F_{pq} = \langle \varphi_p | \mathbf{F} | \varphi_q \rangle$. Using the orbitals of (11.7), we can show that

$$
\begin{aligned}
\epsilon_{e_{1g}} &= F_{11} + F_{12} - F_{13} - F_{14} = -10.67 \text{ ev} \\
\epsilon_{e_{2u}} &= F_{11} - F_{12} - F_{13} + F_{14} = -1.24 \text{ ev}
\end{aligned}
\tag{11.41}
$$

Therefore,

$$
\Delta\epsilon = 2(F_{12} - F_{14}) = 9.44 \text{ ev}
\tag{11.42}
$$

We require, in addition, the integrals

$$
\begin{aligned}
J_{x\xi} &= \tfrac{1}{4}(\gamma_{11} + \gamma_{12} + \gamma_{13} + \gamma_{14}) = 5.76 \text{ ev} \\
J_{x\epsilon} &= \tfrac{1}{12}(\gamma_{11} + 5\gamma_{12} + 5\gamma_{13} + \gamma_{14}) = 4.93 \text{ ev} \\
K_{x\xi} &= \tfrac{1}{4}(\gamma_{11} - \gamma_{12} + \gamma_{13} - \gamma_{14}) + 1.41 \text{ ev} \\
K_{x\epsilon} &= \tfrac{1}{12}(\gamma_{11} + \gamma_{12} - \gamma_{13} - \gamma_{14}) = 0.71 \text{ ev} \\
\langle x\epsilon \| y\xi \rangle &= \tfrac{1}{12}(\gamma_{11} - \gamma_{12} - \gamma_{13} + \gamma_{14}) = 0.41 = \tfrac{1}{2}(J_{x\xi} - J_{x\epsilon}) \\
\langle x\epsilon \| \xi y \rangle &= \tfrac{1}{12}(\gamma_{11} - 5\gamma_{12} + 5\gamma_{13} - \gamma_{14}) = 0.00 \\
\langle x\xi \| \epsilon y \rangle &= \tfrac{1}{12}(\gamma_{11} + \gamma_{12} - \gamma_{13} - \gamma_{14}) = 0.71 \text{ ev}
\end{aligned}
\tag{11.43}
$$

Table 11.1 Comparisons between calculated and experimental band locations in benzene

State	PPP predicted, ev (see text)	Observed,* ev	CNDO/2,† ev (see Sec. 11.5)	NEMO,‡ ev (see Sec 11.6)
$^3B_{1u}$	3.26	3.66	8.29	5.60
		⟍ 1.03 ⟋		
$^3E_{1u}$	4.09	4.69	9.18	6.84
		⟍ 1.07 ⟋		
$^3B_{2u}$	4.92	5.76	10.07	8.08
$^1B_{2u}$	4.92	4.89 $f = 0.001$	10.07	8.32
$^1B_{1u}$	6.09	6.14 $f = 0.04$	10.26	9.63
$^1E_{1u}$	6.91	7.75 $f = 0.69$	14.14	12.28

* See G. Herzberg, 1966, p. 555; D. R. Kearns, 1962; F. L. Pilar, 1968, p. 673.
† P. A. Clark and J. L. Ragle, 1967.
‡ J. L. Lippert and P. O'D. Offenhartz, unpublished results.

Using the equations of Sec. 11.1, we obtain the results quoted in Table 11.1. Comparing these with experiment, we see that for at least the singlets, the agreement is spectacular. The triplets are predicted in the right order, although about 0.6 ev below the observed levels. Furthermore, they obey a 1:1 "ratio rule" within 2 percent, surely one of the most astonishing features of the spectrum. However, the excellent agreement is marred somewhat by doubt as to the *assignment* of the experimental bands. It is very difficult to experimentally determine the irreducible representation to which an excited state belongs. Secondly, extensive a priori calculations, including complete configuration interaction (all multiply excited states) within the π-electron framework, do not give nearly as good agreement with experiment. Thus, core excitations and the like probably are important in any configuration interaction, and we are left with a feeling of surprise that the simple theory can even predict the correct ordering of the bands. Nevertheless, it is probable that the bands *are* correctly assigned and that PPP theory does work unexpectedly well.

As noted in Sec. 11.1, the $^1E_{1u}$ state is the only one arising from the transition $\psi_{e_{1g}} \rightarrow \psi_{e_{2u}}$ which is allowed. The oscillator strength is given by

$$f = 1.085 \times 10^{11} E |\mathbf{D}|^2 \qquad \blacktriangleright (11.44)$$

where E is the band energy in (centimeters)$^{-1}$. The dipole matrix element $\mathbf{D}$ (in centimeters) is easily calculated in PPP theory. We write

$$\mathbf{D} = \langle \Psi_0 | \mathbf{\mu} | \Psi(^1E_{1u}) \rangle = \frac{1}{\sqrt{2}} \{ \langle \Psi_0 | \mathbf{\mu} | \Psi_2' \rangle + \langle \Psi_0 | \mathbf{\mu} | \Psi_3' \rangle \} \qquad \blacktriangleright (11.45)$$

Since the dipole moment $\mathbf{\mu}$ is a one-electron operator,

$$\mathbf{D} = \langle \psi_x | \mathbf{\mu} | \psi_\epsilon \rangle + \langle \psi_y | \mathbf{\mu} | \psi_\xi \rangle \qquad (11.46)$$

Consistently with the ZDO approximation, we now assume that a matrix element over basis orbitals of the form $\langle \varphi_p | \mathbf{\mu} | \varphi_q \rangle$ is zero unless $p = q$. Furthermore, when $p = q$, we take

$$\langle \varphi_p | \mathbf{\mu} | \varphi_p \rangle = \mathbf{R}_p Q_p = Q_p (x\mathbf{i} + y\mathbf{j} + z\mathbf{k}) \qquad (11.47)$$

where x, y, and z are the coordinates of nucleus p and $\mathbf{i}$, $\mathbf{j}$, and $\mathbf{k}$ are unit vectors in the x, y, and z directions. In benzene, $Q_p = 1$ for all p. Therefore,

$$\mathbf{D} = \frac{1}{\sqrt{12}} (-\mathbf{R}_2 - \mathbf{R}_3 + \mathbf{R}_5 + \mathbf{R}_6) \qquad (11.48)$$

Using Fig. 11.2, we have

$$\begin{aligned}
R_2 &= \tfrac{1}{2}\sqrt{3}R\mathbf{i} + \tfrac{1}{2}R\mathbf{j} \\
R_3 &= \tfrac{1}{2}\sqrt{3}R\mathbf{i} - \tfrac{1}{2}R\mathbf{j} \\
R_5 &= -\tfrac{1}{2}\sqrt{3}R\mathbf{i} - \tfrac{1}{2}R\mathbf{j} \\
R_6 &= -\tfrac{1}{2}\sqrt{3}R\mathbf{i} + \tfrac{1}{2}R\mathbf{j}
\end{aligned} \qquad (11.49)$$

where $R = 1.39\text{Å}$ is the internuclear separation. Thus

$$\mathbf{D} = -R\mathbf{i} \tag{11.50}$$

which shows that the ${}^1E_{1ux}$ component $(1/\sqrt{2})(\Psi_2' + \Psi_3')$ is indeed x-polarized. Substituting in (11.44), we find $f = 1.3$, which is much larger than observed. The intensity is probably reduced through borrowing by other bands via configuration interaction, again demonstrating just why quantum chemists regard the agreement in Table 11.1 as astonishing.

There is thus the following dilemma: molecular orbital theory succeeds beyond expectations when we allow a certain degree of empiricism into the theory. However, the use of such empiricism cannot be fully justified theoretically. Furthermore, when we carry out theoretically sound but exhaustingly detailed calculations, they do not have nearly the predictive success of the semiempirical methods. This dilemma is as old as quantum chemistry and in part is what makes the entire field so fascinating.

In the next section, we examine in greater detail the application of PPP theory to aromatic alternant hydrocarbons. We will show that this class of compounds, which incidentally includes benzene, is special in that all carbon atoms are predicted to have $Q_p = 1$. By itself, this does not guarantee the success of PPP theory, but it does imply that a calculation on such a molecule using a grossly incorrect theory can give results as good as a far more refined method.

PROBLEMS

11.9 Verify all of the steps in the calculation of the state energies of benzene. Begin by finding the bond orders B_{12}, B_{13}, and B_{14}, and verify the results of (11.40). Also confirm the results of (11.43).

11.10 Obtain general formulas for the ground-state dipole and transition dipole moments in terms of matrix elements over basis orbitals. Simplify the formula, using the ZDO approximation.

11.11 Consider a two-orbital molecule such as ethylene. Approximately orthogonalize the two orbitals by constructing new orbitals $\chi_1 \sim \varphi_1 - (S/2)\varphi_2$ and $\chi_2 \sim \varphi_2 - (S/2)\varphi_1$. (The overlap integral between the new orbitals is $S^3/4$, presumably a small number.) Show that these orbitals lead to (11.38), to the extent that S^2 can be neglected compared to unity.

11.4 ALTERNANT HYDROCARBONS

An alternant hydrocarbon is by definition *starrable*; that is, one can place a star on a given carbon atom, omit a star on the neighboring carbon atoms, place a star on the next atoms, and so on, either starring or not starring

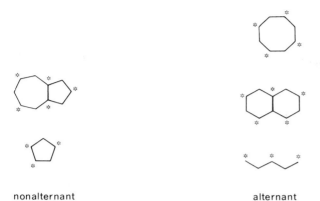

nonalternant   alternant

Fig. 11.4 Nonalternant and alternant hydrocarbons. Alternant hydrocarbons are called *odd* or *even* according to the number of carbon atoms.

every carbon atom in the molecule alternatively. Equivalently, the carbon atoms can be divided into two distinct sets, such that the atoms of one set are directly bonded only to members of the other set, and vice versa. Clearly, a hydrocarbon with an odd-membered ring cannot be alternant; this is illustrated in Fig. 11.4. Alternant hydrocarbons are further divided into odd and even alternants, depending on the total number of carbon atoms in the system. An odd-alternant hydrocarbon will in general have a doublet ground state.

Because the starred atoms are only bonded to unstarred atoms and vice versa, the Hückel equations for an alternant hydrocarbon have an especially simple form. Writing (10.49) with $S_{pq} = \delta_{pq}$, we find

$$(F_{pp} - \epsilon_i)c_{ip} + \sum_{q \neq p} F_{pq}c_{iq} = 0 \qquad \blacktriangleright (11.51)$$

In Hückel theory, F_{pq} is zero unless p and q are neighbors, in which case it equals β. Thus

$$(\alpha_p - \epsilon_i)c_{ip} + \sum_{\substack{q \\ \text{(Nearest} \\ \text{neighbors)}}} \beta c_{iq} = 0 \qquad (11.52)$$

Since, in the Hückel method, $\alpha_p = \alpha$ for all carbon atoms and β is the same for all nearest neighbor pairs, the quantity $(\alpha_p - \epsilon_i)/\beta = x_i$ is a function only of the orbital energy ϵ_i. Therefore,

$$x_i c_{ip} + \sum_{\substack{q \\ \text{(Nearest} \\ \text{neighbors)}}} c_{iq} = 0 \qquad \blacktriangleright (11.53)$$

Note that if x_i, c_{ip}, and the c_{iq} form a solution to the Hückel equations, $-x_i$, c_{ip}, and $-c_{iq}$ also form a solution; thus

$$-x_i c_{ip} + \sum_{q} (-c_{iq}) = 0 \qquad \blacktriangleright (11.54)$$

(Nearest neighbors)

Therefore, if $\epsilon_i = \alpha_p - x_i \beta$ is a solution to the equations, then

$$\epsilon_{-i} = \alpha_p + x_i \beta$$

is also a solution. The orbital energies are thus *paired* about the mean value α_p; compare the energies derived for benzene and butadiene in Probs. 11.5 and 11.6. Furthermore, if ψ_i is the bonding orbital

$$\psi_i = \sum_{r} c_{ir} \varphi_r + \sum_{r*} c_{ir*} \varphi_{r*} \qquad \blacktriangleright (11.55)$$

(Unstarred)   (Starred)

then the *paired* antibonding orbital is

$$\psi_{-i} = \sum_{r} c_{ir} \varphi_r - \sum_{r*} c_{ir*} \varphi_{r*} \qquad \blacktriangleright (11.56)$$

This follows directly from the fact that if atom p is unstarred, the nearest neighbor atoms q are necessarily starred. The notation ψ_i and ψ_{-i} is to be preferred to the standard ordering of the orbitals, since it brings out the explicit pairing property. The bonding orbitals are ordered by their energies (Fig. 11.5).

In an *odd*-alternant hydrocarbon, it is not possible to pair all the molecular orbitals since there are an odd number of them. For one orbital, which we call ψ_0, Eqs. (11.55) and (11.56) cannot be independent. Therefore one of the sums (over either starred or unstarred atoms) must vanish. From (11.53), this implies $x_0 = 0$ or, equivalently, $\epsilon_0 = \alpha_p$. The orbital is therefore nonbonding. It is further possible to prove that there are $(n + 1)/2$ nonzero coefficients in the nonbonding orbital, that is, that the orbital extends over the larger of the starred and unstarred sets.‡ By custom, in an odd-alternant hydrocarbon, the starred set is the one from which the nonbonding orbital is formed.

For an even-alternant hydrocarbon, it is possible to prove that the bond-order charge-density matrix elements B_{pq} are equal to δ_{pq}, provided that *both* p and q are starred or unstarred. Since the matrix of coefficients $\mathbf{C}$ with elements C_{pi} (note the change in order of the subscripts) is orthogonal (real unitary),

$$\sum_{i=n}^{-n} C_{pi} C_{qi} = \delta_{pq} \qquad \blacktriangleright (11.57)$$

‡ H. C. Longuet-Higgins, 1950.

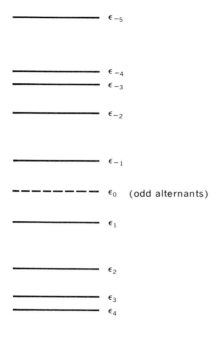

ϵ_{-5}

ϵ_{-4}

ϵ_{-3}

ϵ_{-2}

ϵ_{-1}

ϵ_0 (odd alternants)

ϵ_1

ϵ_2

ϵ_3

ϵ_4

ϵ_5

Fig. 11.5   Pairing of orbital energies in an alternant hydrocarbon (schematic).

However, $C_{pi}C_{qi} = C_{p-i}C_{q-i}$ for p and q both starred or unstarred. Therefore,

$$2 \sum_{i=1}^{n} C_{pi}C_{qi} \equiv B_{pq} = \delta_{pq} \qquad \blacktriangleright (11.58)$$

The uniform charge-density theorem [usually called the Coulson-Rushbrooke theorem (1940)] therefore remains valid, and all carbon atoms have unit π charge ($B_{pp} = 1$) in odd-alternant hydrocarbons.

Of equal interest is the implication for spin densities. Suppose ψ_0 is occupied by a single up-spin electron. Summing separately over α- and β-spin electrons, we have

$$B_{pq}^{\alpha} = \sum_{i=1}^{n} C_{pi}C_{qi} + C_{p0}C_{q0} = \tfrac{1}{2}(\delta_{pq} + C_{p0}C_{q0})$$

$$B_{pq}^{\beta} = \sum_{i=1}^{n} C_{pi}C_{qi} = \tfrac{1}{2}(\delta_{pq} - C_{p0}C_{q0})$$

$$(11.59)$$

Therefore,

$$\mathcal{S}_{pp} = B_{pp}^{\alpha} - B_{pp}^{\beta} = C_{p0}^2 = c_{0p}^2 \qquad \blacktriangleright (11.60)$$

The spin density is thus identical to its density in the nonbonding molecular orbital, and unstarred atoms have zero spin density.

The theorem for even-alternant hydrocarbons remains true in the PPP theory, neglecting the neutral penetration integrals or at least assuming they contribute equally to all carbon atoms. We begin by assuming the truth of the theorem. Then, from (11.35),

$$\langle \varphi_p | \mathbf{F} | \varphi_p \rangle = -I_p + \tfrac{1}{2} \langle pp \| pp \rangle$$
$$\langle \varphi_p | \mathbf{F} | \varphi_p \rangle = \beta_{pq} - \tfrac{1}{2} B_{pq} \langle pq \| pq \rangle \qquad (p \text{ and } q \text{ nearest neighbors})$$
$$\langle \varphi_p | \mathbf{F} | \varphi_q \rangle = 0 \qquad (p \text{ and } q \text{ both starred or both unstarred})$$
$$\langle \varphi_p | \mathbf{F} | \varphi_q \rangle = -\tfrac{1}{2} B_{pq} \langle pq \| pq \rangle \qquad (p \text{ and } q \text{ not nearest neighbors;}$$
$$\text{either } p \text{ or } q \text{ starred}) \qquad \blacktriangleright (11.61)$$

Thus, if the theorem is true, all diagonal elements are identical, and all off-diagonal elements between starred atoms or between unstarred atoms are zero. The secular equations may therefore be written in the form

$$(F_{pp} - \epsilon_i)c_{ip} + \sum_{r*} F_{pr*}c_{ir*} = 0 \qquad (11.62)$$

where the sum extends only over starred atoms if p is unstarred, and vice versa. Once again, if $(F_{pp} - \epsilon_i)$, c_{ip}, and the c_{ir*} form a solution to the PPP Fock equations, then $-(F_{pp} - \epsilon_i)$, c_{ip}, and $-c_{ir*}$ also form a solution, with $\epsilon_{-i} = 2F_{pp} - \epsilon_i$. The pairing property as expressed in (11.55) and (11.56) therefore holds, and this is sufficient to guarantee the truth of (11.58). The theorem is therefore *consistent* with the PPP method, and a formal proof can be given using inductive logic.‡

For radical odd-alternant hydrocarbons, a corresponding proof cannot be given because of the complexities of Roothaan's equations for an open shell. However, if we assume that the odd electron is in an orbital of different symmetry from all other filled orbitals in the system, Roothaan's equations are much simplified, and it is easily shown that the pairing property and the uniformity of charge densities still holds.

The Coulson-Rushbrooke theorem obviously remains true for certain excited states of alternant hydrocarbons, those produced by the excitation $\psi_i \rightarrow \psi_{-i}$. It is clear that since the orbital ψ_{-i} differs from ψ_i only by the *sign* of certain coefficients, the charge densities, which depend on the square of the coefficients, are not changed at all. Of even more interest is the fact that the transition $\psi_i \rightarrow \psi_{-i+j}$ is necessarily degenerate with the transition from ψ_{i-j} to ψ_{-i}, where j is *any* integer. Thus, only the transitions $\psi_i \rightarrow \psi_{-i}$ are nondegenerate. As it turns out, the double degen-

‡ J. A. Pople, 1953.

eracy of all but such states guarantees that *all* excited states have uniform charge densities. The Coulson-Rushbrooke theorem is therefore true even when the *correlation* of the π electrons is taken into account, a most remarkable result.‡

If the charge densities are actually uniform in alternant hydrocarbons, there can be no long-range potentials within such a molecule. This is seen in the vanishing of the first sum in (11.35). It is therefore possible that PPP theory does not correctly take such interactions into account since success with alternant hydrocarbons is guaranteed no matter what the nature of the treatment of long-range interactions. Since most demonstrations of the usefulness of PPP theory are based on success with alternant hydrocarbons, this observation has somewhat disturbing consequences for nonalternant molecules, or those containing heteroatoms. Fortunately, however, notable successes have been obtained with such difficult molecules, but generally speaking, the choice of parameters is more critical, and a search of the literature is necessary to determine which set of choices has been most successful. It is also advisable to carry out calculations on a number of closely related compounds to test the consistency of the predictions.

The reader should also note that the Coulson-Rushbrooke theorem guarantees that all alternant hydrocarbons will have near-zero dipole moments. This prediction, which has been confirmed experimentally, also implies that PPP predictions of ground-state dipole moments in other classes of molecules may be inaccurate. Furthermore, the neglect of the σ-electron dipole moment may be important, and so one should not expect that the PPP method will do more than predict trends.

PROBLEMS

11.12 Using Eq. (10.48), show that the matrix C is unitary when S is diagonal. Is C unitary when S contains off-diagonal nonzero elements?

11.13 Prove that the Coulson-Rushbrooke theorem on the uniform charge densities of alternant hydrocarbons remains true even when electron correlation is taken into account.

11.5 COMPLETE NEGLECT OF DIFFERENTIAL OVERLAP

Pople, Santry, and Segal (1965) have attempted to extend the PPP equations to include σ orbitals. This is not a simple matter; once the restriction of one orbital per atom is removed, the number of possible independent parameters goes up rapidly with the number of atoms in the system, and even the one-center integrals are not entirely trivial. Pople

‡ For an elegant proof, see A. D. McLachlan, 1959.

et al. have tried to avoid this escalation of parameters by making a large number of simplifying assumptions; it may be that in doing so they have underparametrized the theory. They begin by assuming the zero differential overlap approximation, applying it uniformly to all integrals except the two-center, one-electron matrix elements. The only remaining two-electron integrals are the γ_{pp} and γ_{pq}, so the Fock equations are enormously simplified from their original, rigorous form. As we shall see, Pople et al. do not stop at this point, but go on to make other, perhaps equally drastic, assumptions.

Let φ_p^A and φ_q^B be atomic orbitals centered on atoms A and B, respectively. We will occasionally abbreviate these as p_A, q_B, and so on. Writing the one-electron operator as $\mathbf{H}$, we see that the assumption of zero differential overlap leads to matrix elements of the form

$$F_{pp} = H_{pp} + \tfrac{1}{2}B_{pp}\gamma_{pp} + \sum_{\substack{q=1 \\ q \neq p}}^{m} B_{qq}\gamma_{pq}$$

$$F_{pq} = H_{pq} - \tfrac{1}{2}B_{pq}\gamma_{pq}$$

(11.63)

The sum over q extends over *all* m basis functions, regardless of the atom on which they are centered. B_{pq} has the usual definition,

$$B_{pq} = \sum_i N(i)c_{ip}c_{iq}$$

(11.64)

However, B_{pq} is not a "bond order" unless p and q are orbitals on different (neighboring) centers.

The sums in (11.63) are now simplified by making a number of convenient approximations. First, we assume that the electron-repulsion integrals γ_{pq} are independent of the exact nature of the orbitals p and q and depend only on the atoms on which p and q are centered. Thus, $\gamma_{p_A q_B} = \gamma_{r_A s_B} = \gamma_{AB}$. (This approximation is in some molecular systems *necessitated* by the assumption of zero differential overlap if the results are to be independent of the choice of coordinate system.) Also, $\gamma_{p_A q_A} = \gamma_{p_A p_A} = \gamma_{AA}$. We may now write

$$F_{pp} = H_{pp} - \tfrac{1}{2}B_{pp}\gamma_{pp} + \sum_B B_{BB}\gamma_{AB}$$

(11.65)

where

$$B_{AA} = \sum_{p_A} B_{p_A p_A}$$

▶(11.66)

The second of Eqs. (11.63) is essentially unchanged.

As in the PPP method, the one-electron matrix elements H_{pp} and H_{pq} are parametrized. Following (11.31) and (11.32), we have

$$H_{pp} = \langle \varphi_p^A | - \tfrac{1}{2}\nabla^2 - \mathbf{V}_A | \varphi_p^A \rangle - \sum_{B \neq A} \langle \varphi_p^A | \mathbf{V}_B | \varphi_p^A \rangle$$

$$= \alpha_p - \sum_{B \neq A} V_{AB}$$

▶(11.67)

For consistency, we have assumed that $\langle \varphi_p^A | \mathbf{V}_B | \varphi_p^A \rangle$ is independent of the specific orbital form of φ_p^A. For the off-diagonal elements, we must consider two separate cases. If φ_p and φ_q are on the *same* center, then the matrix element

$$\langle \varphi_p^A | - \tfrac{1}{2}\nabla^2 - \mathbf{V}_A | \varphi_q^A \rangle = \alpha_{p_A q_A} \tag{11.68}$$

is ordinarily zero by symmetry (this is because φ_p^A and φ_q^A belong to different irreducible representations of the full rotation group, whereas the operator is spherically symmetric). Therefore,

$$H_{p_A q_A} = \alpha_{p_A q_A} - \sum_{B \neq A} \langle \varphi_p^A | \mathbf{V}_B | \varphi_q^A \rangle = 0 \tag{11.69}$$

since all matrix elements in the sum are to be neglected in the zero differential overlap approximation. This leaves the one-electron matrix elements between φ_p^A and φ_p^B. This element is parametrized as in the PPP scheme; thus

$$H_{p_A q_B} = \beta_{p_A q_B}$$

▶(11.70)

Pople, Santry, and Segal go a step farther and let $\beta_{p_A q_B}$ be proportional to the corresponding overlap integral

$$H_{p_A q_B} = \beta_{p_A q_B} = \beta_{AB}^0 S_{p_A q_B}$$

▶(11.71)

The constant β_{AB}^0 depends only on the nature of the atoms A and B and not on the internuclear distance or the nature of the orbitals φ_p and φ_q.

Collecting together all approximations, we see that the equations become

$$F_{pp} = \alpha_p - \tfrac{1}{2}B_{pp}\gamma_{AA} + \sum_B B_{BB}\gamma_{AB} - \sum_{B \neq A} V_{AB}$$

$$F_{p_A q_A} = -\tfrac{1}{2}B_{pq}\gamma_{AA}$$

$$F_{p_A q_B} = \beta_{AB}^0 S_{p_A q_B} - \tfrac{1}{2}B_{pq}\gamma_{AB}$$

▶(11.72)

Note that the third equation reduces to the second provided that $S_{pq} = 0$ when $A = B$. The application of these equations is straightforward once it is decided what values are to be used for the parameters α_p, γ_{AB}, V_{AB}, and β_{AB}^0. In addition, the value of S_{pq} will depend on the choice of basis functions.

Two methods have been described for choosing parameters. They differ in the way both V_{AB} and α_p are treated. In the original method (1965), which the authors term CNDO/1 (the letters stand for "complete neglect of differential overlap"), V_{AB} and γ_{AB} were both calculated exactly using 2s Slater orbitals; integrals obtained in this way are close to the *average* values for 2s, $2p_x$, $2p_y$, and $2p_z$ orbitals. In the second method‡, CNDO/2, all "neutral penetration integrals" are neglected, which is equivalent to setting $V_{AB} = Z_B \gamma_{AB}$ in (11.72). (Z_B is the nuclear charge on atom B.) The second approach works considerably better than the first, so we will not give CNDO/1 any further consideration. It should be noted, however, that the method used to obtain α_p in CNDO/1 may be better than that later used in CNDO/2. In the original method, the authors set

$$\alpha_p = -I_p - (Z_A - 1)\gamma_{AA} \qquad \blacktriangleright (11.73)$$

In CNDO/2, I_p is replaced by $\frac{1}{2}(I_p + A_p)$, so that

$$\alpha_p = -\tfrac{1}{2}(I_p + A_p) - (Z - \tfrac{1}{2})\gamma_{AA} \qquad \blacktriangleright (11.74)$$

The reasoning, clearly somewhat faulty in the case of CNDO/2, is based on the assumption that Koopmans' theorem should apply to both the atom and the negative ion as follows:

$$\begin{aligned} -I_p &= \alpha_p + (Z_A - 1)\gamma_{AA} \\ -A_p &= \alpha_p + Z_A \gamma_{AA} \end{aligned} \qquad (11.75)$$

Equation (11.74) is then obtained as the *average* of the two formulas above.

Both methods make the extra approximation

$$\beta_{AB}^0 = \tfrac{1}{2}(\beta_A^0 + \beta_B^0) \qquad \blacktriangleright (11.76)$$

The β^0 are chosen to give the best possible agreement between CNDO calculations and accurate SCF calculations on diatomic molecules. Values of β^0, I_p, and $\frac{1}{2}(I_p + A_p)$ are listed in Table 11.2. Together with the use of Slater orbitals for all atoms (but $\zeta = 1.2$ for hydrogen), these constants completely define the CNDO/2 method. The explicit formulas for the matrix elements are

$$F_{pp} = -\tfrac{1}{2}(I_p + A_p) - \tfrac{1}{2}(B_{pp} - 1)\gamma_{AA} + \sum_B (B_{BB} - Z_B)\gamma_{AB}$$

$$F_{p_A q_B} = \tfrac{1}{2}(\beta_A^0 + \beta_B^0)S_{pq} - \tfrac{1}{2}B_{pq}\gamma_{AB} \qquad \blacktriangleright (11.77)$$

Of course, the overlap integrals are neglected except in the calculation of β_{AB}^0; the self-consistent Fock matrix must be determined iteratively.

The major criticism to be leveled at the CNDO method is that it is only moderately successful in predicting experimental properties. This

‡ J. A. Pople and G. A. Segal, 1966.

Table 11.2 Parameters for CNDO calculations, in electron volts*

		H	Li	Be	B	C	N	O	F
I_p	$\varphi_p = 2s$	13.06	5.39	9.32	14.05	19.44	25.58	32.38	40.20
	$\varphi_p = 2p$		3.54	5.96	8.30	10.67	13.19	15.85	18.66
$\frac{1}{2}(I_p + A_p)$	$\varphi_p = 2s$	7.176	3.106	5.946	9.594	14.051	19.316	25.390	32.272
	$\varphi_p = 2p$		1.258	2.563	4.001	5.572	7.275	9.111	11.080
$-\beta_A^0$		9	9	13	17	21	25	31	39

		Na	Mg	Al	Si	P	S	Cl
$\frac{1}{2}(I_p + A_p)$	$\varphi_p = 3s$	2.804	5.125	7.771	10.033	14.033	17.650	21.591
	$\varphi_p = 3p$	1.302	2.052	2.995	4.133	5.464	6.989	8.708
	$\varphi_p = 3d$	0.150	0.162	0.224	0.337	0.500	0.713	0.977
$-\beta_A^0$		7.720	9.447	11.301	13.065	15.070	18.150	22.330

* Pople and Segal, 1965; Santry and Segal, 1967.

is hardly surprising; the method is so full of approximations that it would be extraordinary if it were highly accurate. It has been used primarily to rationalize bond lengths, bond angles, and dipole moments, rather than to predict spectra, but no really systematic studies of its applicability and accuracy have yet been published. CNDO/2 should therefore be regarded as a foundation stone on which better theories may someday be built.

Clark and Ragle (1967) have reported that calculations on benzene using CNDO/2 predict the $\pi \rightarrow \pi^*$ states to lie far too high (see Table 11.1). This is to be expected, since CNDO/2 is parametrized to SCF calculations on small molecules, not to experimental results. However, Clark and Ragle noted that CNDO/2 also predicts $\sigma \rightarrow \sigma^*$ bands at *lower* energies than the $\pi \rightarrow \pi^*$ bands. In particular, the calculations on benzene predict a (primarily) $\sigma \rightarrow \sigma^*$ $^1E_{1u}$ band at 4.86 ev, with $f \approx 1$, in total discord with experiment. Attempts to remedy this defect by using smaller values of β° for carbon were at best only partially successful; $\sigma \rightarrow \sigma^*$ excitations were still predicted to be of great importance, in disagreement with "intuition," although not completely ruled out by the experimental data. Benzene orbital energies calculated by Clark and Ragle are compared in Table 11.6 to orbital energies calculated using a more accurate approximate method discussed in the next section. The agreement leaves much to be desired.

Several modifications of the CNDO method have been proposed. These deal primarily with open shells‡ and the calculation of excited-state geometries.§ A modification of the method, in which *all* one-center integrals are explicitly calculated, has been proposed‖; although these integrals are important in predicting the spin densities and geometries of molecules with open shells, it does not appear that including them will remedy the defects mentioned above in regard to spectral properties.

It seems clear that for *large, polar* molecules, CNDO is the only available approximate method within the Hartree-Fock framework which presently offers hope of success. The CNDO technique is unlike the extended Hückel methods in that the matrix elements of the Fock Hamiltonian are determined by a proper iterative SCF procedure. Furthermore, the total energies are to be calculated in a rigorous way. The *nature* of the approximations in the CNDO method is clear, which is, of course, only a virtue if the approximations are also successful. Programs for CNDO calculations are available from the Quantum Chemistry Program Exchange at Indiana University. These programs are written in FORTRAN, and are easily modified. This is fortunate since it is likely that many changes will be necessary before the method will achieve real predictive utility.

‡ H. W. Kroto and D. P. Santry, 1967*a*.
§ H. W. Kroto and D. P. Santry, 1967*b*.
‖ J. A. Pople, D. L. Beveridge, and P. A. Dobosch, 1967.

11.6 THE METHOD OF NEWTON, BOER, AND LIPSCOMB

In 1949, Mulliken proposed an approximate method of calculating integrals which does not depend on the assumption of zero differential overlap. Since that time, his suggestion has been generalized into the differential formula

$$\varphi_p\varphi_q = \tfrac{1}{2}S_{pq}(\varphi_p\varphi_p + \varphi_q\varphi_q) \qquad \blacktriangleright(11.78)$$

Thus, operating on the differential form with an arbitrary **O** and integrating, we have

$$\langle\varphi_p|O|\varphi_q\rangle = \tfrac{1}{2}S_{pq}(\langle\varphi_p|O|\varphi_p\rangle + \langle\varphi_q|O|\varphi_q\rangle) \qquad \blacktriangleright(11.79)$$

For **O** a coulomb or exchange operator,

$$\langle\varphi_p\varphi_q|\frac{1}{r_{12}}|\varphi_r\varphi_s\rangle = \tfrac{1}{4}S_{pr}S_{qs}(\gamma_{pq} + \gamma_{ps} + \gamma_{rq} + \gamma_{rs}) \qquad (11.80)$$

The differential formula works with varying success with different kinds of integrals and operators; probably the best results have been obtained with the two-electron many-centered integrals over $2p_\pi$ orbitals in aromatic systems. On the other hand, as noted by Mulliken, the formula does not work at all well for the kinetic-energy operator, and therefore it cannot be applied to the Fock operator.

Wolfsberg and Helmholz (1952) have suggested applying the Mulliken approximation to the Fock operator but with a empirical parameter to correct for the kinetic-energy difficulty. Specifically, they suggest

$$F_{pq} = \tfrac{1}{2}S_{pq}K_{pq}(F_{pp} + F_{qq}) \qquad (11.81)$$

with K_{pq} equal to 1.67 for σ-σ interactions and equal to 2.00 for π-π interactions. This kind of approximation forms the basis for the extended Hückel methods. However, one difficulty is that explicit calculations show that (11.81) does not hold very accurately; secondly, the extended Hückel methods are prone to a number of fundamental difficulties (see Sec. 11.2).

Newton, Boer, and Lipscomb (1966) have reinvestigated the validity of (11.79), rewritten in the more general form

$$\langle\varphi_p|O|\varphi_q\rangle = \tfrac{1}{2}S_{pq}K_{pq}(\langle\varphi_p|O|\varphi_p\rangle + \langle\varphi_q|O|\varphi_q\rangle) \qquad \blacktriangleright(11.82)$$

Using the accurately calculated values for the three matrix elements involved, they have obtained values of K_{pq} for different orbitals φ_p and φ_q and different operators **O**. As expected, the values of K so calculated show great scatter for the kinetic-energy operator and the Fock operator. However, as is shown in Table 11.3, K values for the potential-energy operator, the nuclear-attraction operator, and the two-electron operator show little scatter, with values clustered closely around unity.

Table 11.3   Values of K_{pq} for the two-center matrix elements of methane*

Fock operator	$-\frac{1}{2}\nabla^2$	Potential energy	Nuclear-attraction energy	Two-electron interaction energy
K_{1sH}	2.04	−0.01	0.83	0.92	1.14
K_{2sH}	1.46	0.37	1.05	1.00	0.98
K_{2pH}	2.10	0.54	1.00	0.93	0.91
K_{HH}	2.95	−0.04	1.19	1.09	1.05

* Calculated using Eq. (11.82). See M. D. Newton, F. P. Boer, and W. N. Lipscomb, 1966.

Let us define a total potential operator **V** by the equation

$$\mathbf{V} = \mathbf{F} + \tfrac{1}{2}\nabla^2 \qquad\qquad \blacktriangleright (11.83)$$

Then, according to the discussion above,

$$F_{pq} = T_{pq} + \tfrac{1}{2}S_{pq}K_{pq}(V_{pp} + V_{qq}) \qquad\qquad \blacktriangleright (11.84)$$

with values of K_{pq} close to unity. (The matrix elements of the kinetic-energy operator, T_{pq}, are to be calculated rigorously and not approximated.) Equation (11.84) suggests a systematic way of handling the numerous off-diagonal Fock matrix elements, *provided that some way is found to calculate the diagonal elements.* Given the latter, the elements V_{pp} and V_{qq} can be obtained by explicitly calculating T_{pp} and T_{qq}.

Unfortunately, systematic means of obtaining the F_{pp} have not yet been discovered. However, *in molecules with uniform charge densities,* it is reasonable to suppose that there are no long-range contributions to the F_{pp} and that the major contributions come from one-center and nearest-neighbor interactions. Thus, matrix elements F_{pp} *calculated* on small molecules ought to be transferable to large systems.

There are two minor difficulties. First of all, calculations show that the one-center matrix element

$$\langle \varphi_{2s} | \mathbf{V} | \varphi_{2p} \rangle$$

is decidedly nonzero although the corresponding overlap integral vanishes by symmetry. In this particular case, the authors take

$$F_{pq} = K_{pq}^{zo} \sum_r S_{pr} S_{qr} F_{rr} \qquad\qquad \blacktriangleright (11.85)$$

where the sum extends over *all* basis orbitals in the system and the zero-overlap K values K_{pq}^{zo} are calculated to be on the order of 0.5.

The second difficulty occurs in choosing the parameters $\alpha_p = F_{pp}$ which are to be transferred from the model calculations. Ideally, such parameters ought to be the same for p_x, p_y, and p_z orbitals, so that they may be used without regard for choice of axes. Unfortunately, actual calculations show, for example, that $\alpha = 0.75$ au for the $2p_\sigma$ carbon orbital in acetylene, but $\alpha = 0.17$ au for the $2p_\pi$ orbitals. The authors have been unable to devise a way of calculating the α_p for arbitrary situations, so we must either make do with *averaged* values or take those most nearly suited to the local molecular environment in a large molecule. For example, in benzene, carbon α's may be taken from calculations on the ethylene molecule. In this way, a large list of different possible α values may be prepared, and this is given in Table 11.4. Which values are used in a given calculation on a large molecular system depends on the symmetry and structure of the molecule, but the choice may in some instances be ambiguous. This only happens if the molecule has little symmetry.

A similar variety of choices is possible for the values of K_{pq}. Model calculations indicate that for a given type of interaction, K is remarkably constant. For example, the calculated constant for the carbon $2s$–hydrogen interaction only varies between 1.04 and 1.07 for the molecules CH_4, C_2H_2, C_2H_4, C_2H_6, HCN, and H_2CO. The average value 1.05 may therefore be used in *all* computations. In this way, the list given in Table 11.5 may

Table 11.4 The parameters α (diagonal Fock matrix elements) for carbon, oxygen, nitrogen, and hydrogen

Parameter	Value, au	Description
α_{1s}	-11.284	All carbon atoms except carbonyl
α_{2s}	-1.463	All carbon atoms except carbonyl
α_{1s}	-11.352	Carbonyl carbon atoms
α_{2s}	-1.545	Carbonyl carbon atoms
α_{2p}	-0.364	Alkyl carbon atoms
α_{2p_π}	-0.146	Carbon except carbonyls, propynes
α_{2p_π}	-0.199	Carbonyl carbon
α_{2p_σ}	-0.480	Carbon except carbonyls, propynes
α_{2p_σ}	-0.580	Carbonyl carbon
α_{2p_π}	-0.169	Propyne carbon
α_{2p_σ}	-0.749	Propyne carbon
α_{1s}	-15.519	Nitrogen
α_{2s}	-1.859	Nitrogen
α_{2p_σ}	-0.334	Pyrrole nitrogen
α_{2p_σ}	-0.427	Doubly bonded nitrogen
α_{2p_π}	-0.149	Doubly bonded nitrogen
α_{1s}	-20.588	Carbonyl oxygen
α_{2s}	-2.449	Carbonyl oxygen
α_{2p_σ}	-0.503	Carbonyl oxygen
α_{2p_π}	-0.179	Carbonyl oxygen
α_H	-0.537	Hydrogen
α_H	-0.570	Aldehyde hydrogen

be obtained. Alternatively, all K's may be taken as unity for a rough approximation.

It should be noted that the method we have described contains no *empirical* parameters; that is, the constants are chosen so as to mimic SCF results, not experimental numbers. Newton, Boer, and Lipscomb therefore refer to their scheme as the nonempirical molecular orbital (NEMO) method.

To date, few applications of the NEMO method have been made, although the authors note that the technique should be useful in discussing bond lengths, bond angles, ionization potentials, dipole moments, and related ground-state properties. No attempt has been made to predict spectra. To remedy this omission, we now present some of the results of a NEMO calculation‡ on benzene. The orbital energies are listed in Table 11.6, along with the π levels of PPP theory and the upper σ and π levels in the modified CNDO scheme of Clark and Ragle (1967). The disagreement between the PPP scheme and the other two should not be taken seriously since it arises from the semiempirical nature of the former. On the other hand, the comparison between the NEMO method and the modified CNDO/2 is more revealing. For the occupied orbitals, the discrepancy between the orbital energies is not too severe, with an average disagreement of only 1.5 ev. On the other hand, the virtual orbital energies show no agreement at all, even as to ordering. Since the NEMO method is carefully parametrized, we expect it to give far more accurate results than CNDO. The question is then why CNDO gives such poor results for the virtual orbital energies. Part of the answer lies in the failure of CNDO to give lower α's for σ orbitals than for π orbitals. Equations (11.71) and (11.76) for β may also be at fault.

It is not possible to use the virtual orbital approach to excited-state

‡ J. L. Lippert, private communication.

**Table 11.5   The parameters K_{pq}
for the potential-energy matrix**

K_{1s2s} (one-center)	0.66
K_{1s2s} (two-center)	0.81
K_{1s2p}	0.82
K_{2s2s}	1.02
K_{2s2p}	1.06
$K_{2p_\sigma 2p_\sigma}$	1.05
$K_{2p_\pi 2p_\pi}$ (perpendicular)	1.10
$K_{2p_\pi 2p_\pi}$ (in the molecular plane)	0.73
K_{1sH}	0.83
K_{2sH}	1.05
K_{2pH}	0.98
K_{HH}	1.18

Table 11.6 Orbital energies for benzene calculated by four different methods

Symmetry	Orbital energies, ev			
	Gaussian*	NEMO	Modified† CNDO/2	PPP
a_{1g}	-307.5	-307.16		
e_{1u}	-307.5	-307.15		
e_{2g}	-307.2	-307.12		
b_{1u}	-307.2	-307.11		
a_{1g}	-31.8	-30.80	-31.18	
e_{1u}	-28.2	-25.75	-24.81	
a_{1g}	-20.1	-21.54	-19.64	
e_{2g}	-23.0	-21.09	-20.10	
b_{2u}	-17.8	-16.08	-13.81	
b_{1u}	-18.0	-15.78	-14.26	
$a_{2u}(\pi)$	-14.6	-15.63	-15.15	-13.66
e_{1u}	-16.9	-15.02	-12.99	
e_{2g}	-14.3	-12.15	-9.79	
$e_{1g}(\pi)$	-10.15	-8.34	-9.40	-10.67
	13.61	13.95	13.34	9.43
$e_{2u}(\pi)$	3.46	5.61	3.94	-1.24
a_{1g}		15.45	8.20	
$b_{2g}(\pi)$	9.88	18.30	7.39	1.75
e_{1u}		18.55	8.50	
b_{1u}		18.94	6.16	
e_{2g}		20.03	6.41	
b_{1u}		25.54	10.74	
e_{2g}		29.63	10.05	
b_{1g}		37.73	10.27	
e_{1u}		39.51	8.88	

* J. M. Shulman and J. W. Moskowitz, 1967.
† P. A. Clark and J. L. Ragel, 1967.

energies in the NEMO scheme without additional information about the values of the two-electron integrals used in Eqs. (11.14) and (11.17); these integrals are not provided or needed in NEMO calculations. However, using the *exact* integrals over a Slater basis given by Karplus and Shavitt (1963), it is possible to calculate the energies of the three lowest singlets and triplets. There are given in Table 11.1 (along with the results of PPP and CNDO/2 calculations). The results are strikingly similar to the calculations discussed in Chap. 6, that is, all excited states are predicted to lie too high. Note that the smallest $\Delta\epsilon$ ($e_{1g} \rightarrow e_{2u}$) is 13.95 ev in the NEMO scheme. The second smallest $\Delta\epsilon$ which will produce states of the same symmetry ($^1E_{1u}$, etc) is 30.47 ev ($e_{1u} \rightarrow a_{1g}$); the $\sigma \rightarrow \sigma^*$ transition $e_{2g} \rightarrow e_{1u}$ ($\Delta\epsilon = 30.70$ ev) can also interact with the lowest $\pi \rightarrow \pi^*$ states. The lowest states therefore lies far below the lowest $\sigma \rightarrow \sigma^*$ states which can interact with them, and, contrary to the predictions of Clark and Ragle, we expect the σ and π states not to be greatly mixed.

The NEMO method is a promising start toward the goal of obtaining accurate approximate SCF wavefunctions. Before the method can become widely useful, however, it will be necessary to modify it in a number of ways. First, some method must be found for calculating α. This method must apply even in polar molecules. Secondly, it would be helpful if the technique could be extended to include double-ζ basis functions. At present, this can be done without trouble only if the ratio of the coefficients in the double-ζ function is kept fixed throughout the calculation. This only amounts to a change in the shape of the (fixed) basis set, keeping the size of the basis constant. It would be highly beneficial to remove this restriction; approximate calculations with large basis sets could lead to more accurate predictions of ground-state properties.

A recent calculation by Shulman and Moskowitz (1967) goes a long way toward a double-ζ basis for benzene. Evaluating all integrals exactly, the authors employed 162 gaussian functions, contracted to 60 different atomic orbitals (see Sec. 7.5 for definitions). This is equivalent to a basis set of from 35 to 50 Slater-type orbitals—a large basis by any standards and certainly larger than the minimum Slater basis which would contain 36 Slater functions.

The results for the benzene orbital energies are quoted in Table 11.6. Generally, the agreement between the gaussian calculation and the NEMO and modified CNDO/2 calculations are encouraging, at least for the occupied orbitals. Especially gratifying is the constancy of the difference between orbital energies of the e_{2u} antibonding (empty) orbital and the e_{1g} orbital.

Shulman and Moskowitz have also calculated the total energy of the benzene positive ion, using open-shell techniques. It is of interest to note that they obtain in this way an ionization potential of 9.74 ev for neutral benzene, compared to the experimental 9.25 ev. Note that Koopmans' theorem gives equally good agreement with experiment, that is, 10.15 ev.

In terms of the size of the basis set, the calculation of Shulman and Moskowitz is probably the largest to date. Such calculations may well become fairly routine in the future, especially if promised improvements (by R. M. Stevens) in the speed of calculation of integrals over Slater type orbitals are forthcoming. Thus, nonempirical or semiempirical calculations on benzene-sized molecules will soon be obsolete. This should not discourage those interested in approximate methods. By choosing sufficiently large molecules, it is always possible to keep one step ahead of the ab initio calculations.

PROBLEMS

11.14 Equation (11.79) holds exactly for $\mathbf{O}$, the potential of a uniform external field. Show therefore that (11.84) is not properly invariant to the choice of energy origin unless $K_{pq} = 1$.

11.15 Suppose that two orbitals φ_1 and φ_2, which are eigenfunctions of the separate Fock operators $\mathbf{F}_1 = -\nabla^2/2 + \mathbf{V}_1$ and $\mathbf{F}_2 = -\nabla^2/2 + \mathbf{V}_2$, are allowed to overlap slightly. Suppose further that the Fock operator for the pair of orbitals is now $F = -\nabla^2/2 + V_1 + V_2$. Show that $\langle \varphi_1|F|\varphi_2 \rangle = S_{12}(\epsilon_1 + \epsilon_2) - \langle \varphi_1| - \nabla^2/2 \,|\varphi_2\rangle$, where the ϵ's are the orbital energies of the separated orbitals. For what orbitals of C_2H_2 might this expression be a better approximation than Eq. (11.84)?

11.7 CONCLUSIONS: SOME PERSONAL OPINIONS

The theory of atomic and molecular orbitals is quite remarkable. As a rigorous framework for the calculation of atomic and molecular properties, it is certainly somewhat disappointing—the calculations are tedious and the predictions not especially good. On the other hand, as a framework for approximations, and as a guide to intuition, the theory can be most helpful. Not only can the theory be used to rationalize experimental results and correlate different experimental phenomena, but it can also make use of one set of results to calculate another. Ligand field theory is the most elegant example; the experimental $10Dq$ (obtained from one band of the optical spectrum) can be used to calculate the magnetic susceptibility, the electron spin resonance spectrum, and the positions of the *other* bands in the optical spectrum. This is surprising in view of the difficulty in calculating $10Dq$, and it is especially surprising that a single empirical value of $10Dq$ fits nearly all the various data.

Similar observations can be made concerning the electronic structure of free atoms and concerning the spectra of alternant hydrocarbons. In each case, *calculations* predict overly large separations among excited states, yet the order and *relative* energies of the bands are in good agreement with the theory in most cases. Furthermore, as is well known, adding a *limited amount* of configuration interaction to the orbital calculation often destroys the agreement with the ordering and the relative separations.

This suggests, very tentatively, that orbital theory somehow has extra validity beyond the Hartree-Fock limit and that semiempirical work can in some way be justified theoretically. The suggestion is supported on experimental grounds only, but these grounds are relatively broad.

There are many areas in orbital theory where intensive research is required. Certainly more minimum basis set calculations on small molecules are needed, and, equally, calculations on very small molecules using double-ζ functions would be desirable. Methods like the NEMO technique must be developed to parametrize the Hartree-Fock calculations, and ways of accurately duplicating double-ζ calculations in an approximate scheme must be found. Thus we may hope that *accurate, approximate* Hartree-Fock calculations on medium-sized molecules [FeF_6^{3-}, $Fe(CN)_6^{3-}$, pyridine, naphthalene, and so forth] will become routinely available. Finally, approximate ways of handling configuration interaction must be found; perhaps here zero-overlap methods will be useful.

On the purely theoretical side, more work needs to be done on the basic theoretical framework of orbital theory. In particular, a thorough explanation of the successes of Hückel-type theories is to be desired. The physical meaning of virtual orbitals needs clarification. The extent to which configuration interaction can be handled semiempirically must be determined, especially in view of the success of formula of the type $\gamma_{pp} = I_p - A_p$, used in PPP theory. In this connection it would be extremely helpful to have a more reasonable proof of Koopmans' theorem.

In the long run, one may hope that quantum mechanics, via some sort of orbital theory, will be capable of accurate predictions of the properties of medium-sized molecules without an oppressive amount of work and without the obscurity of merely solving Schrödinger's equation directly without the intervention of models, pictures, and physically reasonable approximations. Whether this expectation is justified remains to be seen. Undoubtably, valence-bond theory and its approximate counterparts, the atoms-in-molecules methods, should also be reconsidered in the light of success of the NEMO and CNDO approximations. Perhaps an approximate valence-bond scheme could prove more useful than any orbital method. The obstacles are formidable, however, and we say this mainly to hedge our bet against what the future will bring.

SUGGESTIONS FOR FURTHER READING

The best source of information regarding new techniques for the approximate calculation of molecular properties is the current literature. Of course, the reader must decide for himself the applicability and accuracy of any new method; as we have seen in this chapter, a method suited to the calculation of ground state properties may fail to yield useful predictions of spectra.

For a more detailed discussion of older approximate methods, two excellent texts are available.

1. Parr, Robert G.: "Quantum Theory of Molecular Electronic Structure," W. A. Benjamin, Inc., New York, 1964.

   This text contains a good discussion of π-electron methods, together with a brief introduction to the electron correlation problem. Many fundamental papers in general quantum chemistry and in π-electron theory are reprinted at the back of the book. These provide a valuable introduction to the literature.

2. Pilar, F. L.: "Elementary Quantum Chemistry," McGraw-Hill Book Company, New York, 1968.

   This is a more recent text. The last four chapters are devoted to a discussion of calculations of molecular electronic structure and represent by far the most comprehensive study of such calculations to date. Pilar's book also provides an excellent treatment of quantum chemistry in general, on a somewhat more advanced level than the present text. It is an excellent reference and, together with Parr's monograph, belongs on the shelf of every quantum chemist.

1 APPENDIX

ATOMIC UNITS: PHYSICAL CONSTANTS AND CONVERSION FACTORS‡

Constant	Symbol	Value, au	Value, conventional units
Electronic charge	e	1	1.60210×10^{-19} coulomb
			4.80298×10^{-10} esu
Electronic mass	m_e	1	9.1091×10^{-28} gram
Proton mass	m_p	1836.10	1.67252×10^{-24} gram
Bohr radius	a_0	1	0.529167×10^{-8} cm
Planck constant	$\hbar$	1	1.05450×10^{-27} erg-sec
	h	2π	6.6256×10^{-27} erg-sec
Rydberg constant	R_∞	$\frac{1}{2}$	1.0973731×10^{5} cm^{-1}
Boltzmann constant	k	3.1668×10^{-8}	1.38054×10^{-16} erg-deg^{-1}
Speed of light	c	137.039	2.997925×10^{10} cm-sec^{-1}

One electron volt is 8,065.7 cm^{-1}, 23.05 kcal-mole^{-1}, 1.6021×10^{-12} erg-molecule^{-1} or 0.036750 au.

One atomic unit of energy ($2R_\infty$) is 219,475 cm^{-1}, 627.1 kcal-mole^{-1}, 4.3594×10^{-11} erg-molecule^{-1} or 27.211 electron volts.

‡ From *Physics Today* **17**, (2):48 (1964). See also Condon and Shortley, 1935, p. 432, for a more detailed discussion of the atomic system of units.

2 APPENDIX

DIAGONALIZATION OF A REAL SYMMETRIC MATRIX ‡

The matrix equation

$$\mathbf{FC} = \mathbf{SC\varepsilon} \qquad\qquad\qquad \blacktriangleright (A.1)$$

occurs frequently in quantum theory. In form it is often disguised as a determinantal equation

$$|F_{pq} - S_{pq}\epsilon_i| = 0 \qquad\qquad\qquad (A.2)$$

but this equation follows immediately from (A.1) since the determinant of a product of matrices is the product of the individual determinants; thus $|\mathbf{FC}| = |\mathbf{F}| \cdot |\mathbf{C}|$. The equation may also be disguised by changes in nomenclature and, for example, Eq. (5.10) in the form

$$\sum_i (H_{ik} - ES_{ik})p_i = 0 \qquad \text{for any } k \qquad\qquad (A.3)$$

is equivalent to (A.1).

‡ See Löwdin, 1950. This section also owes much to lectures by Professor J. deHeer.

The matrix **C** describes the transformation from basis functions to orthonormal eigenvectors; for example,

$$\psi_i = \sum_p C_{pi}\varphi_p \tag{A.4}$$

Because of the orthogonality of the eigenvectors,

$$\tilde{\mathbf{C}}\mathbf{S}\mathbf{C} = \mathbf{I} \qquad\qquad \blacktriangleright (A.5)$$

The reader should verify this expression by considering the components of the product $\tilde{\mathbf{C}}\mathbf{S}\mathbf{C}$. Equation (A.5) acts as a subsidiary condition on (A.1).

We first treat the special case $\mathbf{S} = \mathbf{I}$. Then

$$\mathbf{F}\mathbf{C} = \mathbf{C}\boldsymbol{\varepsilon} \qquad\qquad \blacktriangleright (A.6)$$

or

$$\mathbf{C}^{-1}\mathbf{F}\mathbf{C} = \boldsymbol{\varepsilon} \qquad\qquad \blacktriangleright (A.7)$$

The matrix **C** is said to diagonalize **F**. Note that **C** is now a unitary matrix

$$\tilde{\mathbf{C}}\mathbf{C} = \mathbf{C}^{-1}\mathbf{C} = \mathbf{I} \tag{A.8}$$

A solution to (A.7) is generally possible only when **F** is Hermitian; we restrict ourselves to the case when it is in addition real and therefore symmetric.

In (A.6) we seek **C** such that $\boldsymbol{\varepsilon}$ is diagonal. Many methods for accomplishing this procedure have been proposed, but we outline here only one technique, the Jacobi method, since it is both elementary and widely used. Other, probably better, methods, have been proposed.‡ A FORTRAN program for automatically carrying out the Jacobi procedure is given in Table A.1.

Suppose that **F** is a 2×2 real matrix. Then it is not hard to show that the matrix

$$\mathbf{C} = \begin{pmatrix} \cos\theta & \sin\theta \\ -\sin\theta & \cos\theta \end{pmatrix} \tag{A.9}$$

with $\tan 2\theta = 2F_{12}/(F_{11} - F_{22})$ will diagonalize **F**. (The angle θ must be restricted to the first and fourth quadrants.) The reader should do this as an exercise, explicitly working out the expression for all components of $\mathbf{C}^{-1}\mathbf{F}\mathbf{C}$. Note that $\cos\theta = \sin\theta = \sqrt{1/2}$ when $F_{11} = F_{22}$.

For the $n \times n$ matrix, the program works by selecting the largest off-diagonal element. It then diagonalizes the 2×2 matrix consisting of this element and its two corresponding diagonal elements. This is equiv-

‡ See the catalog of the Quantum Chemistry Program Exchange, available from Indiana University.

alent to transforming the matrix $\mathbf{F}$ by the matrix $\mathbf{C}_1$, such that

$$\mathbf{F}' = \mathbf{C}_1^{-1}\mathbf{F}\mathbf{C}_1 \tag{A.10}$$

where, for example,

$$\mathbf{C}_1 = \begin{pmatrix} 1 & 0 & 0 & 0 & 0 & 0 & 0 \\ 0 & \cos\theta & 0 & 0 & \sin\theta & 0 & 0 \\ 0 & 0 & 1 & 0 & 0 & 0 & 0 \\ 0 & 0 & 0 & 1 & 0 & 0 & 0 \\ 0 & -\sin\theta & 0 & 0 & \cos\theta & 0 & 0 \\ 0 & 0 & 0 & 0 & 0 & 1 & 0 \\ 0 & 0 & 0 & 0 & 0 & 0 & 1 \end{pmatrix} \tag{A.11}$$

The program then searches once again for a new maximum off-diagonal element in $\mathbf{F}'$ and again diagonalizes the 2×2 matrix to which this element corresponds; thus

$$\mathbf{F}'' = \mathbf{C}_2^{-1}\mathbf{F}'\mathbf{C}_2 = \mathbf{C}_2^{-1}\mathbf{C}_1^{-1}\mathbf{F}\mathbf{C}_1\mathbf{C}_2 \tag{A.12}$$

The process converges; the off-diagonal element of $\mathbf{F}''$ corresponding to the original largest element in $\mathbf{F}$ is still very small. Thus, when all off-diagonal elements are below a certain test value (RAP), we have

$$\boldsymbol{\varepsilon} = \mathbf{C}_{\mathrm{NR}}^{-1}\mathbf{C}_{\mathrm{NR}-1}^{-1} \cdots \mathbf{C}_2^{-1}\mathbf{C}_1^{-1}\mathbf{F}\mathbf{C}_1\mathbf{C}_2 \cdots \mathbf{C}_{\mathrm{NR}} \tag{A.13}$$

NR is the number of "rotations" necessary to diagonalize $\mathbf{F}$. Clearly,

$$\mathbf{C} = \mathbf{C}_1 \cdot \mathbf{C}_2 \cdots \mathbf{C}_{\mathrm{NR}} \tag{A.14}$$

When the overlap matrix $\mathbf{S}$ is not the unit matrix, our problem is somewhat more complicated. It is not difficult to write down a *formal* solution, however, provided we define a matrix $\mathbf{S}^{1/2}$ by the equation

$$\mathbf{S}^{1/2}\mathbf{S}^{1/2} = \mathbf{S} \qquad\qquad \blacktriangleright(A.15)$$

We also define the inverse of this matrix,

$$\mathbf{S}^{-1/2} = (\mathbf{S}^{1/2})^{-1} \qquad\qquad \blacktriangleright(A.16)$$

We assume that both $\mathbf{S}^{1/2}$ and $\mathbf{S}^{-1/2}$ are Hermitian (and real), so that

$$\begin{aligned} \tilde{\mathbf{S}}^{1/2} &= \mathbf{S}^{1/2} \\ \tilde{\mathbf{S}}^{-1/2} &= \mathbf{S}^{-1/2} \end{aligned} \qquad\qquad \blacktriangleright(A.17)$$

Left-multiplying Eq. (A.1) by $\tilde{\mathbf{C}}$ and using (A.5), we obtain

$$\tilde{\mathbf{C}}\mathbf{F}\mathbf{C} = \boldsymbol{\varepsilon} \tag{A.18}$$

Note, however, that $\mathbf{C}$ is not a unitary matrix, and so (A.18) is *not* the same as (A.7). Let us define a matrix $\mathbf{C}'$ such that

$$\mathbf{C}' = \mathbf{S}^{1/2}\mathbf{C} \tag{A.19}$$

Then $\mathbf{C}'$ is unitary,

$$\tilde{\mathbf{C}}'\mathbf{C}' = \widetilde{\mathbf{S}^{1/2}\mathbf{C}}\mathbf{S}^{1/2}\mathbf{C} = \tilde{\mathbf{C}}\tilde{\mathbf{S}}^{1/2}\mathbf{S}^{1/2}\mathbf{C} = \tilde{\mathbf{C}}\mathbf{S}\mathbf{C} = \mathbf{I} \qquad (A.20)$$

We also define a matrix $\mathbf{F}'$,

$$\mathbf{F}' = \mathbf{S}^{-1/2}\mathbf{F}\mathbf{S}^{-1/2} \qquad (A.21)$$

Then

$$\begin{aligned} \mathbf{C}'^{-1}\mathbf{F}'\mathbf{C}' &= \tilde{\mathbf{C}}\mathbf{S}^{1/2}\mathbf{S}^{-1/2}\mathbf{F}\mathbf{S}^{-1/2}\mathbf{S}^{1/2}\mathbf{C} \\ &= \tilde{\mathbf{C}}\mathbf{F}\mathbf{C} = \boldsymbol{\varepsilon} \end{aligned} \qquad \blacktriangleright (A.22)$$

This is now a true eigenvalue equation since $\mathbf{F}'$ is Hermitian (it is the product of Hermitian matrices) and $\mathbf{C}'$ is unitary.

It is necessary only to find a suitable matrix $\mathbf{S}^{1/2}$. The solution is not unique, but a particularly useful solution can be obtained by considering the unitary matrix $\mathbf{U}$ which diagonalizes the Hermitian matrix $\mathbf{S}$. We write

$$\tilde{\mathbf{U}}\mathbf{S}\mathbf{U} = \mathbf{d} \qquad \blacktriangleright (A.23)$$

and also define a diagonal matrix $\mathbf{d}^{1/2}$ with diagonal elements $(\mathbf{d}^{1/2})_i = (\mathbf{d})_i^{1/2}$. Then

$$\mathbf{d}^{1/2}\mathbf{d}^{1/2} = \mathbf{d} = \tilde{\mathbf{U}}\mathbf{S}\mathbf{U} = \tilde{\mathbf{U}}\mathbf{S}^{1/2}\mathbf{S}^{1/2}\mathbf{U} = \tilde{\mathbf{U}}\mathbf{S}^{1/2}\mathbf{U}\tilde{\mathbf{U}}\mathbf{S}^{1/2}\mathbf{U} \qquad (A.24)$$

so

$$\mathbf{d}^{1/2} = \tilde{\mathbf{U}}\mathbf{S}^{1/2}\mathbf{U} \qquad (A.25)$$

or

$$\mathbf{S}^{-1/2} = \mathbf{U}\mathbf{d}^{-1/2}\tilde{\mathbf{U}} \qquad \blacktriangleright (A.26)$$

where $\mathbf{d}^{-1/2}$ is the diagonal matrix with elements $1/\sqrt{d_i}$. This procedure therefore defines a unique matrix $\mathbf{S}^{-1/2}$.

Putting (A.21) and (A.26) together, we now obtain

$$\mathbf{F}' = \mathbf{U}\mathbf{d}^{-1/2}\tilde{\mathbf{U}}\mathbf{F}\mathbf{U}\mathbf{d}^{-1/2}\tilde{\mathbf{U}} \qquad (A.27)$$

The matrix $\mathbf{F}'$ is then diagonalized, which determines $\mathbf{C}'$ and hence $\mathbf{C}$ and $\boldsymbol{\varepsilon}$.

A FORTRAN program for determining $\mathbf{C}$ and $\boldsymbol{\varepsilon}$ is given in Table A.2. It uses a minimum of temporary storage space during the calculation, taking the matrices S and H ($= \mathbf{F}$) as input and returning the coefficient matrix $\mathbf{C}$ in H and the (ordered) eigenvalues ϵ_i in SF. Only the three extra one-dimensional arrays $X(N)$, $SR(N)$, and $IQ(N)$ are required, so the total storage space for variables is only $\approx 2N^2 + 4N$ for an N-dimensional diagonalization. The program uses the routine HDIAG (Table A.1) to diagonalize $\mathbf{S}$ and $\mathbf{F}'$. The COMMENT cards of the program should prove self-explanatory, provided one notes that $\mathbf{S}^{-1/2}$ is called delta-to-the-minus-one-half or delta-t-t-m-o-h, and $\mathbf{F}$ is everywhere called H.

The overlap matrix **S** is positive definite; i.e., it has positive eigenvalues and its determinant is positive. This is because the basis functions are all linearly independent. The subroutine EIGEN tests for this linear independence after diagonalizing S; in this way basis function errors may be detected. The author has found this a useful way of testing for various kinds of input and programming errors in molecular calculations.

Table A.1 FORTRAN program for the diagonalization of a real symmetric matrix. Written by F. J. Corbato and M. Merwin at Massachusetts Institute of Technology

```
      SUBROUTINE HDIAG(H,N,IEGEN,U,NR,RAP)
C     FORTRAN DIAGONALIZATICN CF A REAL SYMMETRIC MATRIX BY
C        THE JACOBI METHCD.
C     AUGLST 1, 1962
C     CALLING SEQUENCE FCR DIAGONALIZATICN
C           CALL  HDIAG( H, N, IEGEN, U, NR)
C           WHERE H IS THE ARRAY TO BE DIAGONALIZED.
C     N IS THE ORDER OF THE MATRIX, H.
C      IVECTCR=0 MEANS NO EIGENVECTCRS WANTED
C
C
C     U IS THE UNITARY MATRIX USED FCR FCRMATION OF THE EIGENVECTCRS.
C
C     NR IS THE NUMBER CF RCTATIONS.
C
C
C
C     THE SUBROUTINE OPERATES CNLY CN THE FLEMENTS OF H THAT ARE TO THE
C           RIGHT CF THE MAIN DIAGONAL.  THUS, CNLY A TRIANGULAR
C           SECTICN NEECS TC BE STCRED IN THE ARRAY H.
C
      DIMENSICN H(75,75),U(75,75),X(75),IC(75)
C
C
      IF (IEGEN) 10,15,10
   10 DO 14 I=1,N
      DO 14 J=1,N
      IF(I-J)12,11,12
   11 U(I,J)=1.C
      GO TO 14
   12 U(I,J)=C.0
   14 CONTINUE
C
   15 NR = 0
      IF (N-1) 100C,100C,17
C
C     SCAN FCR LARGEST CFF DIAGCNAL ELEMENT IN EACH RCW
C     X(I) CONTAINS LARGEST ELEMENT IN ITH RCW
C     IC(I) HOLCS SECCNC SUBSCRIPT DEFINING POSITION CF ELEMENT
C
   17 NMI1=N-1
      DO 30 I=1,NMI1
      X(I) = 0.C
      IPL1=I+1
      DC 30 J=IPL1,N
      IF ( X(I) - ABS( H(I,J))) 20,20,30
   20 X(I)=ABS(H(I,J))
      IC(I)=J
   30 CONTINUE
C
C     SET INCICATOR FCR SHUT-CFF.RAP=2**-27,NR=NO. CF RCTATIONS
C     RAP=.74505RC59E-C8      FCR MAXIMUM ACCURACY BUT WE SET RAP EXTERNALLY
C     TO A LOWER VALUE FOR GREATER SPEED.
C
```

```
      HDTEST=1.0E38
C
C     FIND MAXIMUM OF X(I) S FOR PIVOT ELEMENT AND
C     TEST FOR END OF PROBLEM
C
 40   DO   70   I=1,NMI1
      IF (I-1) 60,60,45
 45   IF ( XMAX- X(I)) 60,70,70
 60   XMAX=X(I)
      IPIV=I
      JPIV=IQ(I)
 70   CONTINUE
C
C     IS MAX. X(I) EQUAL TO ZERO, IF LESS THAN HDTEST, REVISE HDTEST
      IF ( XMAX) 1000,1000,80
 80   IF (HDTEST) 90,90,85
 85   IF (XMAX - HDTEST) 90,90,148
 90   HDIMIN = ABS( H(1,1) )
      DO 110   I= 2,N
      IF (HDIMIN- ABS( H(I,I))) 110,110,100
 100  HDIMIN=ABS(H(I,I))
 110  CONTINUE
C
      HDTEST=HDIMIN*RAP
C
C     RETURN IF MAX.H(I,J)LESS THAN(2**-27)ABSF(H(K,K)-MIN)
      IF (HDTEST- XMAX) 148,1000,1000
 148  NR = NR+1
C
C     COMPUTE TANGENT, SINE AND COSINE,H(I,I),H(J,J)
 150  TANG=SIGN(2.0,(H(IPIV,IPIV)-H(JPIV,JPIV)))*H(IPIV,JPIV)/(ABS(H(IPI
     1V,IPIV)-H(JPIV,JPIV))+SQRT((H(IPIV,IPIV)-H(JPIV,JPIV))**2+4.0*H(IP
     2IV,JPIV)**2))
      COSINE=1.0/SQRT(1.0+TANG**2)
      SINE=TANG*COSINE
      HII=H(IPIV,IPIV)
      H(IPIV,IPIV)=COSINE**2*(HII+TANG*(2.0*H(IPIV,JPIV)+TANG*H(JPIV,JPI
     1V)))
      H(JPIV,JPIV)=COSINE**2*(H(JPIV,JPIV)-TANG*(2.0*H(IPIV,JPIV)-TANG*H
     1II))
      H(IPIV,JPIV)=0.0
C
C       PSEUDO RANK THE EIGENVALUES
C       ADJUST SINE AND COS FOR COMPUTATION OF H(IK) AND U(IK)
      IF ( H(IPIV,IPIV) -  H(JPIV,JPIV)) 152,153,153
 152  HTEMP = H(IPIV,IPIV)
      H(IPIV,IPIV) = H(JPIV,JPIV)
      H(JPIV,JPIV) = HTEMP
C       RECOMPUTE SINE AND COS
      HTEMP = SIGN(1.0, -SINE) * COSINE
      COSINE = ABS(SINE)
      SINE = HTEMP
 153  CONTINUE
C
C     INSPECT THE IQS BETWEEN I+1 AND N-1 TO DETERMINE
C     WHETHER A NEW MAXIMUM VALUE SHOULD BE COMPUTED SINCE
C     THE PRESENT MAXIMUM IS IN THE I OR J ROW.
C
      DO 350 I=1,NMI1
      IF(I-IPIV)210,350,200
 200  IF(I-JPIV)210,350,210
 210  IF(IQ(I)-IPIV)230,240,230
 230  IF(IQ(I)-JPIV)350,240,350
 240  K=IQ(I)
 250  HTEMP=H(I,K)
      H(I,K)=0.0
      IPL1=I+1
      X(I) =0.0
```

```
C
C      SEARCH IN DEPLETED RCW FCR NEW MAXIMUM
C
       DO 320 J=IPL1,N
       IF ( X(I)- ABS( H(I,J)) ) 300,300,320
 300   X(I) = ABS(H(I,J))
       IQ(I)=J
 320   CONTINUE
       H(I,K)=HTEMP
 350   CONTINUE
C
       X(IPIV) =0.0
       X(JPIV) =0.0
C
C      CHANGE THE OTHER ELEMENTS OF H
C
       DO 530 I=1,N
C
       IF(I-IPIV)370,530,420
 370   HTEMP = H(I,IPIV)
       H(I,IPIV) = CCSINE*HTEMP + SINE*H(I,JPIV)
       IF ( X(I) -   ABS( H(I,IPIV)) )380,390,390
 380   X(I) = ABS(H(I,IPIV))
       IQ(I) = IPIV
 390   H(I,JPIV) = -SINE*HTEMP + CCSINE*H(I,JPIV)
       IF ( X(I) -   ABS( H(I,JPIV)) ) 400,530,530
 400   X(I) = ABS(H(I,JPIV))
       IQ(I) = JPIV
       GO TO 530
C
 420   IF(I-JPIV)430,530,480
 430   HTEMP = H(IPIV,I)
       H(IPIV,I) = CCSINE*HTEMP + SINE*H(I,JPIV)
       IF ( X(IPIV) -  ABS( H(IPIV,I)) ) 440,450,450
 440   X(IPIV) = ABS(H(IPIV,I))
       IQ(IPIV) = I
 450   H(I,JPIV) = -SINE*HTEMP + CCSINE*H(I,JPIV)
       IF ( X(I) -   ABS( H(I,JPIV)) ) 400,530,530
C
 480   HTEMP = H(IPIV,I)
       H(IPIV,I) = CCSINE*HTEMP + SINE*H(JPIV,I)
       IF ( X(IPIV) -  ABS( H(IPIV,I)) ) 490,500,500
 490   X(IPIV) = ABS(H(IPIV,I))
       IQ(IPIV) = I
 500   H(JPIV,I) = -SINE*HTEMP + CCSINE*H(JPIV,I)
       IF ( X(JPIV) -  ABS( H(JPIV,I)) ) 510,530,530
 510   X(JPIV) = ABS(H(JPIV,I))
       IQ(JPIV) = I
 530   CONTINUE
C
C      TEST FCR CCMPUTATION CF EIGENVECTCRS
C
       IF (IEGEN) 540,40,540
 540   DO 550 I=1,N
       HTEMP=U(I,IPIV)
       U(I,IPIV)=CCSINE*HTEMP+SINE*U(I,JPIV)
 550   U(I,JPIV)=-SINE*HTEMP+CCSINE*U(I,JPIV)
       GO TO 40
1000   RETURN
       END
```

**Table A.2   FORTRAN program for the diagonalization
of the Hamiltonian matrix including overlap**

```
      SUBROUTINE EIGEN(S,SF,H,IS,THRESH)
      DIMENSION S(75,75),H(75,75),SF(1),SR(1)
C
C
C  SPECIAL EIGEN WITH MINIMAL TEMPORARY STORAGE USAGE
C
C
C     STORAGE SPACE SHORTENED BY PETER O'D. OFFENHARTZ AT THE UNIVERSITY OF
C     COLORADO, JULY, 1967.   PROGRAM USES STANDARD JACOBI WITH LOWDIN
C     ORTHOGONALIZATION AND FOLLOWS PLAN OF SEVERAL EXISTING PREVIOUS PROGRAMS.
C
C
      IF(IS-1) 100,200,500
  500 CONTINUE
      DO 401   I=1,IS
  401 SR(I)=H(I,I)
      DO 402   I=2,IS
      L=I-1
      DO 402   J=1,L
  402 S(I,J)=H(J,I)
C
C  OVERLAP AND HAMILTONIAN MATRICES NOW CONTAINED IN S.  DIAGONAL TERMS IN
C     HAMILTONIAN CONTAINED IN SR.
C
      CALL HDIAG(S,IS,1,H,NR,THRESH)
      DO 505   I=1,IS
      IF(S(I,I)) 502,502,505
  502 WRITE(6,503) I,I,S(I,I)
  503 FORMAT(50X,26HIMPOSSIBLE S EIGENVALUE S(,I3,1H,,I3,4H) = ,F8.3)
      WRITE(6,504)
  504 FORMAT(48H MISTAKE IN OVERLAP CALC., ENERGY MATRIX FOLLOWS)
      GO TO 890
  505 SF(I)=1.0/SQRT(S(I,I))
      DO 509   I=1,IS
      DO 509   J=1,I
      S(J,I)=0.0
      DO 509   K=1,IS
  509 S(J,I)=S(J,I)+H(I,K)*SF(K)*H(J,K)
C
C  UPPER HALF OF S NOW STORES DELTA-TO-THE-MINUS-ONE-HALF.   LOWER HALF STILL
C     STORES HAMILTONIAN  WITH DIAGONAL ELEMENTS IN SR.
C
      DO 514   J=1,IS
      DO 5131 I=1,IS
      SF(I)=0.0
      DO 5131 K=1,IS
      MI=MINO(I,K)
      MXK=MAXO(I,K)
      MJ=MINO(J,K)
      MXKP=MAXO(J,K)
      IF(I-K) 5129,5130,5129
 5129 CONTINUE
      SF(I)=SF(I)+S(MXK,MI)*S(MJ,MXKP)
      GO TO 5131
 5130 SF(I)=SF(I)+SR(I)*S(MJ,MXKP)
 5131 CONTINUE
      DO 514   I=1,J
      H(I,J)=0.0
```

```
      DO 514 K=1,IS
      MI=MINO(I,K)
      MXKP=MAXO(I,K)
  514 H(I,J)=H(I,J) +S(MI,MXKP)*SF(K)
C
C   AT THIS PCINT H-PRIME IS IN THE UPPER HALF OF H, DELTA-TO-THE-MINUS-ONE-HALF
C    IS IN THE UPPER HALF CF S.
C
      DO 407  I=1,IS
  407 SR(I)=S(I,I)
      DO 408  I=2,IS
      L=I-1
      DO 408  J=1,L
  408 H(I,J)=S(J,I)
C   DELTA-TO THE-MINUS-ONE-HALF IS NOW IN THE LOWER HALF  OF H.    DIAGONAL OF
C   DELTA-T-T-M-O-H  IS IN SR.
C
      CALL HDIAG(H,IS,1,S,NR,THRESH)
      DO 517  I=1,IS
      SF(I)=H(I,I)
  517 H(I,I)=SR(I)
C
C    SF NCW CONTAINS THE EIGENVALUES, S CONTAINS THE EIGENVECTORS CF H-PRIME,
C    H (LOWER HALF)  CCNTAINS DELTA-T-T-M-O-H.
C
      DO 5170  I=1,IS
      DO 5170  J=1,I
 5170 H(J,I)=H(I,J)
      DO 5176  I=1,IS
      DO 5174  J=1,IS
      SR(J)=0.0
      DO 5174  K=1,IS
 5174 SR(J)=SR(J)+H(K,I)*S(K,J)
      DO 5176  J=1,IS
 5176 H(J,I)=SR(J)
      NM=IS-1
      DO 740  J=1,NM
      JP=J+1
      DO 740  K=1,J
      IF(SF(K)-SF(JP)) 743,741,741
  741 HH=SF(K)
      SF(K)=SF(JP)
      SF(JP)=HH
      DO 742  I=1,IS
      SR(I)=H(K,I)
      H(K,I)=H(JP,I)
  742 H(JP,I)=SR(I)
  743 CONTINUE
  740 CONTINUE
      DO 810  I=1,IS
  810 WRITE(6,820) I,SF(I)
  820 FORMAT(1H0,51X,2HE(I3,3H) =,F12.5)
  890 RETURN
  100 WRITE(6,840)
  840 FORMAT(46H MOLECULE NEEDS RCTATION - CCEF.=0 BY SYMMETRY)
  200 SF(1)=H(1,1)
      H(1,1)=1.0
      GO TO 890
      END
```

BIBLIOGRAPHY

Anderson, J. M.: "Introduction to Quantum Chemistry," W. A. Benjamin, Inc., New York, 1969.

Bader, R. F. W., W. H. Henneker, and P. E. Cade: *J. Chem. Phys.*, **46**:3341 (1967).

Ballhausen, C. J.: "Introduction to Ligand Field Theory," McGraw-Hill Book Company, New York, 1962.

Ballhausen, C. J., and H. B. Gray: "Molecular Orbital Theory," W. A. Benjamin, Inc., New York, 1965.

Bates, D. R.: "Quantum Theory," vol. I, Academic Press Inc., New York, 1961.

Becker, R. A.: "Introduction to Theoretical Mechanics," McGraw-Hill Book Company, New York, 1954.

Berthier, G.: in "Molecular Orbitals in Chemistry, Physics, and Biology," Academic Press Inc., New York, 1964.

Born, M., and J. R. Oppenheimer: *Ann. der Physik*, **84**:457 (1927).

Browne, J. C., and F. A. Matson: *Phys. Rev.*, **135**:A1227 (1964).

Cade, P. E., and W. Huo: *J. Chem. Phys.*, **47**:614 (1967).

Carrington, A., and A. D. McLachlan: "Introduction to Magnetic Resonance," Harper & Row Publishers, Incorporated, New York, 1967.

Carroll, D. G., and S. P. McGlynn: *Inorg. Chem.*, **7**:1285 (1968).

Clark, P. A., and J. L. Ragle: *J. Chem. Phys.*, **46**:4235 (1967).

Clementi, E.: "Tables of Atomic Functions," International Business Machine Corporation, San Jose, Calif., 1965.

Clementi, E., and D. L. Raimondi: *J. Chem. Phys.*, **38**:996 (1963).

Cline, B. L.: "The Questioners," Thomas Y. Crowell Company, New York, 1965.

Cohen, M., and A. Delgarno: *Proc. Phys. Soc. (London)*, **77**:748 (1961).

Condon, E. U., and G. H. Shortley: "The Theory of Atomic Spectra," Cambridge University Press, London, 1935.

Cotton, F. A.: "Chemical Applications of Group Theory," Interscience Publishers, Inc., New York, 1963.

Coulson, C. A., and G. S. Rushbrooke: *Proc. Camb. Phil. Soc.*, **36**:193 (1940).

Coulson, C. A., and E. T. Stewart: Wave Mechanics and the Alkene Bond, in S. Patai, (ed.), "The Chemistry of Alkenes," Interscience Publishers, New York, 1964.

Dicke, R. H., and J. P. Wittke: "Introduction to Quantum Mechanics," Addison-Wesley Publishing Company, Inc., Reading, Mass., 1960.

Dirac, P. A. M.: "The Principles of Quantum Mechanics," 4th ed., Oxford University Press, London, 1958.

Eyring, H., J. Walter, and G. E. Kimball: "Quantum Chemistry," John Wiley & Sons, Inc., New York, 1944.

Fischer-Hjalmers, I.: *J. Chem. Phys.*, **42**:1962 (1965).

Freeman, A. J., and R. E. Watson: *Phys. Rev.*, **118**:1168 (1960).

Gamow, G.: "Thirty Years that Shook Physics, The Story of Quantum Theory," Doubleday & Co., Inc., Garden City, N.Y., 1966.

George, P., and D. S. McClure: *Prog. Inorg. Chem.*, **1**:381 (1959).

Gimarc, B. M.: *J. Am. Chem. Soc.*, **91**:6000 (1969).

Goddard, W. A.: *Phys. Rev.*, **157**:73 (1967).

Goeppert-Meyer, M., and A. L. Sklar: *J. Chem. Phys.*, **6**:645 (1938).

Goldstein, H.: "Classical Mechanics," Addison-Wesley Publishing Company, Inc., Reading, Mass., 1959.

Griffith, J. S.: "The Irreducible Tensor Method for Molecular Symmetry Groups," Prentice-Hall, Inc., Englewood Cliffs, N.J., 1962.

Griffith, J. S.: "The Theory of Transition Metal Ions," Cambridge University Press, London, 1961.

Hameka, H. F.: "Advanced Quantum Chemistry," Addison-Wesley Publishing Company, Inc., Reading, Mass., 1965.

Hameka, H. F.: "Introduction to Quantum Theory," Harper & Row Publishers, Incorporated, New York, 1967.

Herzberg, G.: "Molecular Spectra and Molecular Structure. I. Spectra of Diatomic Molecules," D. Van Nostrand Company, Inc., Princeton, N.J., 1950.

Herzberg, G.: "Infrared and Raman Spectra of Polyatomic Molecules," D. Van Nostrand Company, Inc., Princeton, N.J., 1945.

Herzberg, G.: "Electronic Spectra of Polyatomic Molecules," D. Van Nostrand Company, Inc., Princeton, N.J., 1966.

Hoffmann, B.: "The Strange Story of the Quantum," Dover Publications, Inc., New York, 1955.

Hoffmann, R.: *J. Chem. Phys.*, **39**:1397 (1963); *J. Chem. Phys.*, **40**:2245, 2474, 2480 (1964).

Hoffmann, R., and R. B. Woodward: *Accounts Chem. Res.*, **1**:17 (1968).

Holmes, O. G., and D. S. McClure: *J. Chem. Phys.*, **26**:1686 (1957).

Hubbard J., D. E. Rimmer, and F. R. A. Hopgood: *Proc. Roy. Soc. (London)*, **88**:13 (1966).

Huo, W.: *J. Chem. Phys.*, **43**:624 (1965).

Huo, W.: *J. Chem. Phys.*, **45**:1554 (1966).

Jahn, H. A., and E. Teller: *Proc. Roy. Soc. (London)*, **A161**:200 (1937).

James, H. M., and A. S. Coolidge: *J. Chem. Phys.*, **1**:825 (1933).

Jammer, M.: "The Conceptual Development of Quantum Mechanics," McGraw-Hill Book Company, New York, 1966.

Jørgenson, C. K.: *Acta. Chem. Scand.*, **8**:1495 (1954).

Jørgenson, C. K.: "Absorption Spectra and Chemical Bonding in Complexes," Addison-Wesley Publishing Company, Inc., Reading, Mass., 1962.

Jørgenson, C. K.: *Disc. Farad. Soc.*, **26**:110 (1958).

Kahalas, S. L., and R. K. Nesbet: *J. Chem. Phys.*, **39**:529 (1963).

Karplus, M., and I. Shavitt: *J. Chem. Phys.*, **38**:1256 (1963).

Kauzman, W.: "Quantum Chemistry," Academic Press Inc., New York, 1957.

Kearns, D. R.: *J. Chem. Phys.*, **36**:1608 (1962).

Kolos, W., and C. C. J. Roothaan: *Rev. Mod. Phys.*, **32**:205, 219 (1960a, 1960b).

Koopmans, T. A.: *Physica*, **1**:104 (1923).

Kroto, H. W., and D. P. Santry: *J. Chem. Phys.*, **47**:2736 (1967a); *J. Chem. Phys.*, **47**:792 (1967b).

Lipkin, H. J.: "Lie Groups for Pedestrians," 2d ed., North Holland Publishing Company, Amsterdam, 1966.

Longuet-Higgins, H. C.: *J. Chem. Phys.*, **18**:265 (1950).

Louguet-Higgins, H. C.: *Mol. Phys.*, **6**:445 (1963).

Longuet-Higgins, H. C., and E. W. Abrahamson: *J. Am. Chem. Soc.*, **87**:2045 (1965).

Löwdin, P. O.: *J. Chem. Phys.*, **18**:365 (1950)

Löwdin, P. O.: *Mol. Spec.*, **3**:46 (1959).

MacDonald, J. K. L.: *Phys. Rev.*, **43**:830 (1933).

Margenau, H., and G. M. Murphy: "The Mathematics of Physics and Chemistry," 2d ed., vol. I, D. Van Nostrand, Company, Inc., Princeton, N.J., 1956.

Mataga, N., and K. Nishimoto: *Z. Physik. Chem. (Frankfort)*, **13**:140 (1957).

McLachlan, A. D.: *Mol. Phys.*, **2**:271 (1959).

McWeeny, R.: *Proc. Roy. Soc. (London)*, **A235**:490 (1956).

McWeeny, R.: *Proc. Roy. Soc. (London)*, **A253**:242 (1959).

McWeeny, R.: *Rev. Mod. Phys.*, **32**:335 (1960).

Moffitt, W.: *Proc. Roy. Soc. (London)*, **A210**:245 (1951); **A218**:464 (1953); **A220**:530 (1953).

Moore, R.: "Niels Bohr, The Man, His Science, and the World They Changed, Alfred A. Knopf, Inc., New York, 1966.

Mulliken, R. S.: *J. Chem. Phys.*, **23**:1833, 1841 (1955).

Mulliken, R. S.: *J. Chim. Phys.*, **46**:497 (1949).

Murrell, J. N., S. F. A. Kettle, and J. M. Tedder: "Valence Theory," John Wiley & Sons, Inc., New York, 1965.

Murrell, J. N., and K. L. McEwen: *J. Chem. Phys.*, **25**:1143 (1956).

Nesbet, R. K., and R. E. Watson: *Phys. Rev.*, **110**:1073 (1958).

Newton, M. D., F. P. Boer, and W. N. Lipscomb: *J. Am. Chem. Soc.*, **88**:2353, 2357 (1966).

Nishimoto, K., and N. Mataga: *Z. Physik. Chem. (Frankfurt)*, **12**:335; **13**:140 (1957).

Offenhartz, P. O'D.: *J. Chem. Ed.*, **44**:604 (1967).

Offenhartz, P. O'D.: *J. Chem. Phys.*, **47**:2951 (1967a).

Offenhartz, P. O'D.: *J. Am. Soc.*, **91**:5699 (1969).

Oohata, K., H. Taketa, and S. Huzinaga: *J. Phys. Soc. Japan*, **21**:2313 (1966).

Orgel, L. E.: "An Introduction to Transition Metal Chemistry," Methuen & Co., Ltd., London, 1960.

Orloff, M., and O. Sinanoglu: *J. Chem. Phys.*, **43**:49 (1965).

Pariser, R.: *J. Chem. Phys.*, **21**:568 (1953).

Pariser, R., and R. G. Parr: *J. Chem. Phys.*, **21**:466, 767 (1953).

Parr, R. G.: *J. Chem. Phys.*, **20**:1499 (1952).

Parr, R. G.: "Quantum Theory of Molecular Electronic Structure," W. A. Benjamin, Inc., New York, 1964.

Pauling, L., and E. B. Wilson: "Introduction to Quantum Mechanics," McGraw-Hill Book Company, New York, 1935.

Petersen, A.: "Quantum Physics and the Philosophical Tradition," M. I. T. Press, Cambridge, Mass., 1968.

Pilar, F. L.: "Elementary Quantum Chemistry," McGraw-Hill Book Company, New York, 1968.

Pople, J. A.: *Trans. Farad. Soc.*, **49**:1375 (1953).

Pople, J. A., D. L. Beveridge, and P. A. Dobosch: *J. Chem. Phys.*, **47**:2026 (1967).

Pople, J. A., D. P. Santry, and G. A. Segal: *J. Chem. Phys.*, **43**:s129, s146 (1965).

Pople, J. A., and G. A. Segal: *J. Chem. Phys.*, **44**:3289 (1966).

Prichard, G., and H. A. Skinner: *J. Inorg. Nucl. Chem.*, **24**:937 (1962).

Racah, G.: *Phys. Rev.*, **61**:186; **62**:738 (1942).

Ransil, B.: *Rev. Mod. Phys.*, **32**:245 (1960).

Richardson, J. W., W. C. Nieupoort, R. R. Powell, and W. F. Edgell: *J. Chem. Phys.*, **36**:1057 (1962).

Richardson, J. W., D. M. Vaught, T. F. Soules, and R. R. Powell: *J. Chem. Phys.*, **50**:3633 (1969).

Roothaan, C. C. J.: *Rev. Mod. Phys.*, **23**:69 (1951).

Roothaan, C. C. J.: *Rev. Mod. Phys.*, **32**:179 (1960).

Roothaan, C. C. J., and P. S. Bagus: in B. Alder (ed.), "Methods in Computational Physics," Academic Press Inc., New York, 1963.

Ruedenberg, K.: *J. Chem. Phys.*, **34**:1861 (1961).

Santry, D. P., and G. A. Segal: *J. Chem. Phys.*, **47**:158 (1967).

Shewell, J. R.: *Am. J. Phys.*, **27**:16 (1959).

Shulman, J. M., and J. W. Moskowitz: *J. Chem. Phys.*, **47**:3491 (1967).

Slater, J. C.: "Quantum Theory of Atomic Structure," vols. I and II, McGraw-Hill Book Company, New York, 1960*a* and 1960*b*.

Stevenson, R.: "Multiplet Structure of Atoms and Molecules," W. B. Saunders Company, Philadelphia, 1965.

Sugano, S., and Y. Tanabe: *J. Phys. Soc. Japan*, **20**:1155 (1965).

Tinkham, M.: "Group Theory and Quantum Mechanics," McGraw-Hill Book Company, New York, 1964.

Vollmer, J. J., and K. L. Servis: *J. Chem. Ed.*, **45**:214 (1968).

Wahl, A. C., P. E. Cade, and C. C. J. Roothaan: *J. Chem. Phys.*, **41**:2578 (1964).

Walsh, A. D.: *J. Chem. Soc.*, **1953**:2260.

Watson, R. E.: *Phys. Rev.*, **119**:170 (1960*a*).

Watson, R. E., and A. J. Freeman: *Phys. Rev.*, **134**:1526 (1964).

Watson, R. E.: Iron Series Hartree-Fock Calculations, *Technical Report #12, Solid-State and Molecular Theory Group, Massachusetts Institute of Technology, Cambridge, Mass.*, 1959; see also *Phys. Rev.*, **118**:1036 (1960).

Weinbaum, S.: *J. Chem. Phys.*, **1**:593 (1933).

Wilson, E. B.: *J. Chem. Phys.*, **43**:s172 (1965).

Wilson, E. B., Jr., J. C. Decius, and P. C. Cross: "Molecular Vibrations," McGraw-Hill Book Company, New York, 1955.

Wolfsberg, M., and L. Helmholtz: *J. Chem. Phys.*, **20**:837 (1952).

Wolniewicz, L., and W. Kolos: *J. Chem. Phys.*, **41**:3663 (1964).

Wolniewicz, L., and W. Kolos: *J. Chem. Phys.*, **43**:2429 (1965).

Woodward, R. B., and R. Hoffmann: *J. Am. Chem. Soc.*, **87**:395, 2046, 2511, 4388 (1965).

Zerner, M., and M. Gouterman: *Inorg. Chem.*, **5**:1707 (1966).

Zerner, M., and M. Gouterman: *Theoret. Chim. Acta* **4**:44 (1966).

INDEXES

INDEX OF SYMBOLS

Symbol usage in the text is intended to be conventional. The result is that the same symbol is often used to represent several different quantities. The list which follows may help reduce the possibilities of confusion. Page numbers refer to the point of first usage. Subscripts on symbols are often dropped when not needed; thus, S is often used in place of S_{pq} or S_{ij}.

ROMAN AND RELATED SYMBOLS

A	Racah parameter, 134	b^k	angular factor, 130
A_3	fictitious triangular molecule, 207	C	Racah parameter, 134
		$C_n{}^m$	rotation of $2\pi m/n$, 189
A_p	electron affinity of atomic orbital φ_p, 314	C_{pi}	coefficient of the atomic (basis) orbital φ_p in the
a^k	angular factor, 130		molecular orbital ψ_i,
$\mathbf{A}$	matrix, 21		278
$\mathbf{A}\dagger$	adjoint matrix, 21	$\mathbf{C}$	coefficient matrix, 277
$\tilde{\mathbf{A}}$	transpose matrix, 21	$c_j{}^{(i)}$	perturbation theory ex-
$\mathcal{Q}$	antisymmetrizer, 59		pansion coefficient, 100
B	Racah parameter, 134	c^k	angular factor, 130
B_{pq}	bond orbitals between orbitals φ_p and φ_q, 291	c_{ip}	expansion coefficient, 149

c_{ip} coefficient of the atomic (basis) orbital φ_p in the molecular orbital ψ_i, 272

D_0 observed dissociation energy, 165

D_e electronic dissociation energy, 165

$\mathbf{D}$ first-order density matrix, 293

Dq ligand field splitting parameter, 226

E energy; eigenvalue of $\mathfrak{IC}$, 40

E group unit element, 189

E_i internal energy, 79

E_t translational energy, 79

$E_n{}^\circ$ eigenvalue of $\mathbf{H}_0$, 99

e electronic charge, 9

e natural logarithm base, 3

F_0 Slater-Condon parameter, 132, 134

F_2 Slater-Condon parameter, 132, 134

F_4 Slater-Condon parameter, 134

F^k Slater-Condon parameter, 130

F_{pq} matrix element of the Fock operator between basis functions, 151

f fractional occupancy of an open shell, 280

f oscillator strength, 317

G_{ij} matrix element of $\mathbf{G}$, 122, 128

G^k Slater-Condon parameter, 130

g Lande splitting factor, 145

$\mathbf{G}$ two-electron operator, 122, 128

H classical Hamiltonian, 32

h order of a group, 204

h_k number of elements in the class k, 205

H_{ii} matrix element of $\mathbf{H}$, 122, 128, 149

$\mathbf{H}$ one-electron operator, 122, 128, 149

$\mathbf{H}'$ perturbation Hamiltonian, 99

$\mathbf{H}_m$ magnetic field operator, 144

$\mathfrak{IC}_{ij}$ matrix elements of $\mathfrak{IC}$, 122, 128

$\mathfrak{IC}$ quantum mechanical Hamiltonian, 32

I moment of inertia, 76

I absorption intensity, 81

I_p ionization potential of atomic orbital φ_p, 314

$\mathbf{i}$ unit vector, x direction, 39

J rotational quantum number, 80

J_{ij} coulomb integral, 124, 149

$\mathbf{j}^2$ total angular momentum operator, 57

$\mathbf{j}_z$ total angular momentum operator, z component, 57

$\mathbf{j}^\pm$ total angular momentum shift operator, 57

$\mathbf{J}^2$ total angular momentum operator, n particle system, 57

K number of nuclei in a molecular system, 272

K_{ij} exchange integral, 125, 149

K_{pq} NEMO parameter, 329

k Hooke's law force constant, 69

k Boltzmann's constant, 3

$\mathbf{K}_c$ closed-shell exchange operator, 280

$\mathbf{K}_o$ open-shell exchange operator, 280

L Lagrangian, 9

R radial wavefunction, 85

R group element (operator), 199

R_i distance between nucleus and ith electron in atomic units, 106

R_e equilibrium internuclear separation, 166, 183

R_x special vector, 206

r_e equilibrium internuclear distance, 77

r_{ij} distance between electrons i and j in atomic units, 106

$\mathbf{r}$ position vector, 44

S spin quantum number, n particle system, 56

S_{ij} overlap integral, 22

S_n permutation group, 191

S_x special vector, 206

s spin quantum number, 56

$\mathbf{s}^2$ spin operator, 56

$\mathbf{s}_z$ spin operator, z component, 56

$\mathbf{s}^\pm$ spin shift operator, 56

$\mathbf{S}^2$ spin operator, n particle system, 61

$\mathbf{S}_z$ spin operator, z component, n particle system, 61

$\mathbf{S}^\pm$ spin shift operators, n particle system, 61

S_{pq} element of the spin density matrix, 293

T kinetic energy, 6

T temperature, 3

T tetrahedral group, 217

T_{pq} matrix element of the kinetic-energy operator, 330

U core potential, 127

u_i spin orbital, 62

V voltage, 6

V potential energy, 9

V_{pq} matrix element of the potential-energy operator, 330

v vibrational quantum number, 69

$\mathbf{V}$ ligand field operator, 227

$\mathbf{V}'$ ligand field splitting operator, 227

$\mathbf{V}_k$ potential operator of the kth center, 312

$\mathbf{V}'_k$ core-potential operator of the kth center, 312

$\mathbf{V}''_k$ neutral potential operator of the kth center, 312

Y_{lm} spherical harmonic functions, 46, 50

Z nuclear charge, 85

Z_k core charge of the kth center, 312

GREEK SYMBOLS

α spin coordinate, 56

α Hückel diagonal parameter, 307

α general one-center matrix element, 325, 330

α_i basis function; basis of an irreducible representation, 210

β spin coordinate, 56

β Bohr magneton, 144

β	nephalauxetic effect parameter, 256	μ	total magnetic moment, 260
β	Hückel integral between neighbors, 307	$\boldsymbol{\mu}$	dipole moment operator, 81
β	PPP core integral, 312	ν_0	vibrational frequency, 70
β_i	basis function; basis of an irreducible representation, 210	ξ_i	spin coordinate, 60
Γ	representation, 195	ξ	spin-orbit coupling operator, 141
$\boldsymbol{\Gamma}(R)$	matrix representing R, 195	φ_i, ϕ_i	spatial orbital, 60, 62
γ_i	basis function; basis of an irreducible representation, 211	Φ	arbitrary wave function, 95
γ_{pq}	coulomb integral between atomic orbitals φ_p and φ_q, 313	$\hat{\boldsymbol{\phi}}$	row vector of basis orbitals, 277
ϵ	basis of the E representation of the octahedral group, 231	$|\Psi\rangle$	ket or wavefunction for a state, 20
ϵ_i	orbital energy, 151	Ψ	ket or wavefunction for a state, 20
ζ	orbital exponent, 91	$\Psi_n{}^0$	eigenfunction of $\mathbf{H}_0$, 99
ζ_{nl}	spin-orbit energy parameter, 142	ψ	molecular orbital, 267
θ	basis of the E representation of the octahedral group, 231	$\boldsymbol{\psi}$	row vector of molecular orbitals, 275
λ	spin-orbit energy parameter, 141	χ_p	basis function, 149
μ	reduced mass, 69, 78	χ	molar susceptibility, 258
		χ	symmetry orbital, 266
		$\chi^{(i)}(R)$	character of the operator R in the representation $\Gamma^{(i)}(R)$, 203
		ω_e	vibrational energy, 181

NAME INDEX

SUBJECT INDEX

Absorption spectra:
  of atoms, 133, 146, 252
  of benzene, 305, 316–317, 328
  charge-transfer, 257
  of diatomic molecules, 184
  rotation-vibration, 83
  of transition-metal complexes,
      225, 252–258
  (*See also* Spectroscopy)
Adjoint matrix, 21
Alternant hydrocarbon, 309, 315
Ammonia molecule, 222
Angular momentum:
  addition of, 51–55
  of atoms, 115–120
  in Bohr atom, 9
  in diatomic molecules, 168–171,
      179
  eigenfunctions for, 46, 49–51
  generalized, 46–49
  and Lie groups, 188
  for many particles, 44, 51–55
  for one particle, 44–46
  orbital, 57
  ordinary, 43–46
  spin, 55–57
  total, 57
  in transition-metal complexes,
      260–264
Annihilation operator, 75

Antibonding orbital, 170, 175, 177,
    334
Antisymmetric function, 58–59
Assignments of spectroscopic
    bands, 252–258, 317
Associated Legendre polynomial,
    50, 80, 108
Atomic orbitals (*see* Orbital)
Atomic spectra (*see* Absorption
    spectra)
Atomic units, 15, 91, 337
Atoms in molecules, method of,
    186, 314
Aufbau principle, 113, 177

Basis:
  for atomic orbitals, 149
  double-zeta, 153, 183, 334-335
  gaussian, 182, 334
  for an irreducible representation,
      200–203, 206, 208–216, 231
  minimum, 183, 185, 290
  for a reducible representation,
      207, 266
  Slater, 148, 182
  symmetry-determined, 171, 266,
      300
Benzene molecule, 183, 189, 208,
    298, 306, 316–318, 333–334